Springer Handbook of Auditory Research

Series Editors: Richard R. Fay and Arthur N. Popper

Springer Handbook of Auditory Research

Volume 1: The Mammalian Auditory Pathway: Neuroanatomy
Edited by Douglas B. Webster, Arthur N. Popper, and Richard R. Fay

Volume 2: The Mammalian Auditory Pathway: Neurophysiology
Edited by Arthur N. Popper and Richard R. Fay

Volume 3: Human Psychophysics
Edited by William Yost, Arthur N. Popper, and Richard R. Fay

Volume 4: Comparative Hearing: Mammals
Edited by Richard R. Fay and Arthur N. Popper

Volume 5: Hearing by Bats
Edited by Arthur N. Popper and Richard R. Fay

Volume 6: Auditory Computation
Edited by Harold L. Hawkins, Teresa A. McMullen, Arthur N. Popper, and Richard R. Fay

Volume 7: Clinical Aspects of Hearing
Edited by Thomas R. Van De Water, Arthur N. Popper, and Richard R. Fay

Volume 8: The Cochlea
Edited by Peter Dallos, Arthur N. Popper, and Richard R. Fay

Volume 9: Development of the Auditory System
Edited by Edwin W Rubel, Arthur N. Popper, and Richard R. Fay

Volume 10: Comparative Hearing: Insects
Edited by Ronald Hoy, Arthur N. Popper, and Richard R. Fay

Volume 11: Comparative Hearing: Fish and Amphibians
Edited by Richard R. Fay and Arthur N. Popper

Volume 12: Hearing by Whales and Dolphins
Edited by Whitlow W.L. *Au, Arthur N. Popper, and Richard R. Fay*

Volume 13: Comparative Hearing: Birds and Reptiles
Edited by Robert Dooling, Arthur N. Popper, and Richard R. Fay

Volume 14: Genetics and Auditory Disorders
Edited by Bronya J.B. Keats, Arthur N. Popper, and Richard R. Fay

Volume 15: Integrative Functions in the Mammalian Auditory Pathway
Edited by Donata Oertel, Richard R. Fay, and Arthur N. Popper

Volume 16: Acoustic Communication
Edited by Andrea Simmons, Arthur N. Popper, and Richard R. Fay

Volume 17: Compression: From Cochlea to Cochlear Implants
Edited by Sid P. Bacon, Richard R. Fay, and Arthur N. Popper

Volume 18: Speech Processing in the Auditory System
Edited by Steven Greenberg, William Ainsworth, Arthur N. Popper, and Richard R. Fay

Volume 19: The Vestibular System
Edited by Stephen M. Highstein, Richard R. Fay, and Arthur N. Popper

Volume 20: Cochlear Implants: Auditory Prostheses and Electric Hearing
Edited by Fan-Gang Zeng, Arthur N. Popper, and Richard R. Fay

Volume 21: Electroreception
Edited by Theodore H. Bullock, Carl D. Hopkins, Arthur N. Popper, and Richard R. Fay

Continued after index

Arthur N. Popper
Richard R. Fay
Editors

Sound Source Localization

With 72 illustrations

Springer

Arthur N. Popper
Department of Biology
University of Maryland
College Park, MD 20742
USA
apopper@umd.edu

Richard R. Fay
Parmly Hearing Institute and Department of Psychology
Loyola University of Chicago
Chicago, IL 60626
USA
rfay@wpo.it.luc.edu

Series Editors:
Richard R. Fay
Parmly Hearing Institute and Department of Psychology
Loyola University of Chicago
Chicago, IL 60626
USA

Arthur N. Popper
Department of Biology
University of Maryland
College Park, MD 20742
USA

Cover illustration: Illustration of a high-frequency narrow-band noise that is delayed to the right ear with respect to the left. From Trahiotis et al., Fig. 7.1.

ISBN-13: 978-1-4419-2022-5 Printed on acid-free paper.
e-ISBN-13: 978-0-387-28863-5

Printed in the United States of America. (MP)

9 8 7 6 5 4 3 2 1

springeronline.com

We are pleased to dedicate this book to Dr. Willem A. van Bergeijk, a pioneer in the comparative study of sound source localization. Dr. van Bergijk developed new models of localization and also provided important insights into potential mechanisms of hearing and sound localization by fishes. In addition, he made important contributions to studies of auditory mechanisms in amphibians and provided seminal thinking about the evolution of hearing.

Series Preface

The *Springer Handbook of Auditory Research* presents a series of comprehensive and synthetic reviews of the fundamental topics in modern auditory research. The volumes are aimed at all individuals with interests in hearing research including advanced graduate students, postdoctoral researchers, and clinical investigators. The volumes are intended to introduce new investigators to important aspects of hearing science and to help established investigators to better understand the fundamental theories and data in fields of hearing that they may not normally follow closely.

Each volume presents a particular topic comprehensively, and each serves as a synthetic overview and guide to the literature. As such, the chapters present neither exhaustive data reviews nor original research that has not yet appeared in peer-reviewed journals. The volumes focus on topics that have developed a solid data and conceptual foundation rather than on those for which a literature is only beginning to develop. New research areas will be covered on a timely basis in the series as they begin to mature.

Each volume in the series consists of a few substantial chapters on a particular topic. In some cases, the topics will be ones of traditional interest for which there is a substantial body of data and theory, such as auditory neuroanatomy (Vol. 1) and neurophysiology (Vol. 2). Other volumes in the series deal with topics that have begun to mature more recently, such as development, plasticity, and computational models of neural processing. In many cases, the series editors are joined by a co-editor having special expertise in the topic of the volume.

RICHARD R. FAY, Chicago, Illinois
ARTHUR N. POPPER, College Park, Maryland

Volume Preface

Sound source localization is arguably one of the most important functions of the auditory system in any hearing animal. During the course of their evolution, both invertebrates and vertebrates have developed a number of different strategies to enable them to determine the position of a sound source around them. In almost all cases, this strategy has required two ears that detect sound and a central processing system that extracts direction from the tiniest differences in the signals detected at the two ears.

In the many volumes in the *Springer Handbook of Auditory Research* series, various authors have dealt with sound localization, but the information has always been in one chapter or part of a chapter in any given volume. Because there is such a large body of knowledge about localization, it became apparent that the topic was worth considering in a single volume that explores localization not only comparatively but also from the perspective of models and understanding general computational mechanisms involved in localization. Moreover, the current volume updates a number of chapters from earlier volumes (e.g., Colburn in Vol. 6—*Auditory Computation* and Brown in Vol. 5—*Comparative Hearing: Mammals*).

In Chapter 1, Fay and Popper provide a detailed overview of the book. The diversity of hearing and localization mechanisms in insects is considered in detail by Robert in Chapter 2, in which he demonstrates that localization is likely to have arisen at multiple independent times in different insects. Because sound in water is almost five times faster than in air, the binaural cues used by terrestrial vertebrates and invertebrates are not generally available to fishes. At the same time, fishes have available to them a set of cues (particle motion) not readily available in air. In Chapter 3, Fay discusses current ideas about fish sound localization, and this serves as an excellent comparative perspective toward not only insects, but also terrestrial vertebrates. Localization by nonmammalian terrestrial vertebrates and the broad range of mechanisms used by these animals are discussed in Chapter 4 by Christensen-Dalsgaard. Of course, the best known localization mechanisms are found in mammals, and these are treated in several chapters. In Chapter 5, Brown and May provide a comparative approach to knowledge of mammalian hearing. In this chapter, the authors consider the extensive psychophysical data on sound localization and all aspects of directional hearing.

Although there is no doubt considerable diversity in peripheral localization mechanisms among terrestrial vertebrates, and especially in amniotes, there is much more stability between species in central nervous system processing of such signals. Such mechanisms and their development are described in detail in Chapter 6 by Kubke and Carr.

The last two chapters continue with discussions of mammalian localization but emphasize computational mechanisms. Although these chapters are primarily aimed at mammalian systems (and especially humans), it is possible that a better understanding of computation associated with localization in mammals may ultimately be extrapolated to other vertebrates as well. In Chapter 7, Trahiotis, Bernstein, Stern, and Buell extend their consideration to computational models of binaural processing and interaural correlation. This is extended even further in Chapter 8 by Colburn and Kulkarni, who treat computational models for many aspects of binaural hearing and sound localization.

As indicated previously, sound localization has been discussed in a number of other volumes of this series. Thus, the reader interested in a deeper understanding of localization would find it useful to seek out chapters in companion volumes. For example, in Volume 3 (*Human Psychophysics*), Wightman and Kisler consider human localization from the perspective of psychophysics as did Brown in Volume 4 (*Comparative Hearing: Mammals*). The chapter in this volume by Roberts is nicely complemented by a chapter on insect localization by Michelsen in Volume 10 (*Comparative Hearing: Insects*) whereas the chapters by Christensen-Dalsgaard and Kubke and Carr extend an earlier chapter on bird and reptile localization by Klump in Volume 13 (*Comparative Hearing: Birds and Mammals*). Physiological mechanisms of sound localization are not covered extensively in this book but are much of the focus of Volume 16 (*Integrative Functions in the Mammalian Auditory Pathway*).

Arthur N. Popper, College Park, Maryland
Richard R. Fay, Chicago, Illinois

Contents

Contributors

LESLIE R. BERNSTEIN
Departments of Neuroscience and Surgery, University of Connecticut Health Center, Farmington, CT 06030, USA

CHARLES H. BROWN
Department of Psychology, University of South Alabama, Mobile, AL 36688, USA

THOMAS N. BUELL
Departments of Neuroscience and Surgery, University of Connecticut Health Center, Farmington, CT 06030, USA

CATHERINE E. CARR
Department of Biology, University of Maryland, College Park, MD 20742, USA

JAKOB CHRISTENSEN-DALSGAARD
Center for Sound Communication, Institute of Biology, University of Southern Denmark, DK-5230 Odense M, Denmark

H. STEVEN COLBURN
Hearing Research Center and Department of Biomedical Engineering, Boston University, Boston, MA 02215, USA

RICHARD R. FAY
Parmly Hearing Institute and Department of Psychology, Loyola University of Chicago, Chicago, IL 60626, USA

M. FABIANA KUBKE
Department of Anatomy with Radiology, Faculty of Medical and Health Sciences, University of Auckland, Auckland 1020, New Zealand

ABHIJIT KULKARNI
Hearing Research Center and Department of Biomedical Engineering, Boston University, Boston, MA 02215, USA

BRADFORD J. MAY
Department of Otolaryngology, The Johns Hopkins School of Medicine, Baltimore, MD 21205, USA

ARTHUR N. POPPER
Department of Biology, University of Maryland, College Park, MD 20742, USA

DANIEL ROBERT
School of Biological Sciences, University of Bristol, Bristol, BS8 1UG, United Kingdom

RICHARD M. STERN
Department of Electrical and Computer Engineering and Department of Biomedical Engineering Engineering, Carnegie Mellon University, Pittsburgh, PA 15213, USA

CONSTANTINE TRAHIOTIS
Departments of Neuroscience and Surgery, University of Connecticut Health Center, Farmington, CT 06030, USA

1

Introduction to Sound Source Localization

Richard R. Fay and Arthur N. Popper

The sense of hearing has evolved and been maintained so that organisms can make use of sound in their environment not only for communication, but also to glean information about the general acoustic milieu that enhances general survival (e.g., Fay and Popper 2000). Information in the environment enables animals to learn about sources that are in many different directions, and particularly signals that are outside of the detection range of other senses. Still, while sound is inherently important for overall survival, its value would be very limited if the receiving organism did not know the position of the source. In effect, to make maximum use of a sound from a predator, the receiver must know not only that the predator is present, but also where the predator is in order to escape most effectively.

As a consequence, one of the fundamental and most important features of sound source detection is the ability of an animal to estimate source location as a first step in behaving appropriately in response to the sound. The need for sound source localization thus has become a fundamental feature of hearing in most hearing organisms, and one could argue that it is inconceivable that a sense of hearing could have evolved at all without the ability to locate and segregate sources.

At the same time, the strategies for computation of sound location must be different in different species, and must depend on the nature of the information arriving from the ears to the brain, and this in turn must depend on interactions between the physics of sound and the characteristics of the receivers.

This volume treats sound source localization from this comparative and evolutionary perspective. The auditory receivers of vertebrates and invertebrates vary widely (e.g., Hoy et al. 1998; Manley et al. 2004), from the tympanic, pressure receiver ears of insects, birds, and mammals, to the otolithic ears of fishes that are inherently directional, to the pressure gradient receiver ears of amphibians (and some reptiles and birds) that function like inherently directional pressure receivers. In spite of these differences in the ways that different ears respond to sound, there apparently are only a few acoustic cues for source location: interaural time differences (ITD), interaural intensity differences (IID), the shape of the sound spectrum reaching the ears, and the axis of acoustic particle motion. The latter cue can be used in conjunction with inherent receiver directionality to create neural representations that are functionally equivalent to those provided by the interaural acoustic cues (IID and ITD) (Fay, Chapter 3). Several chapters of this volume make clear the diversity of mechanisms that have developed to exploit these acoustic cues in sound source localization.

Our understanding of hearing mechanisms in general has been guided by our extensive knowledge of human hearing (at a behavioral or functional level), and as such a considerable amount of work on localization in general has focused on the cues and mechanisms that operate in the human auditory system. Thus, we understand the importance of the interaural cues in determining azimuth and, more recently, the role of spectral shape in determining elevation. Colburn and Kulkarni (Chapter 8) authoritatively summarize the sorts of models that have developed in an attempt to account for sound source localization as it occurs in human hearing. The fundamental question that is asked by Colburn and Kulkarni is, "How does the localizer interpret the received signals to determine the location of the sound source?" The authors evaluate several theoretical schemes that locate the sound source while at the same time extracting information about the acoustic environment, making use of a priori and multimodal information about the original acoustic signal and environment, and estimating other properties of the signal. Models are roughly categorized as those that find the maxima in a process called "steering the beam" (e.g., cross-correlation models) and those that find the minima in the array output by "steering the null" (e.g., equalization and cancellation model). Modeling is generally most successful when there is only one source of sound, no reverberation, no conflicting cues, and no unusual spectra (for elevation judgments), and when the model is restricted to judgments of lateral position.

Trahiotis, Bernstein, Stern, and Buell (Chapter 7) evaluate binaural hearing, broadly defined, from the point of view of interaural correlation as originally suggested by Lloyd A. Jeffress (1948). After defining the indices of interaural correlation and their application to binaural perception, the concept of the three-dimensional cross-correlation function is introduced and it is shown how it is possible to understand binaural perception in terms of various types of pattern processing operations on this function. The authors then present neurophysiological evidence for cross-correlation mechanisms in binaural hearing, a 50-year-old model (Jeffress 1948) that has maintained its value both as a useful guide to the quantitative understanding of binaural hearing in humans and as a neurobiological explanation for the computation of sensory maps in the auditory system (see Kubke and Carr, Chapter 6). Trahiotis at al. use insights that arise from functional/behavioral investigations on binaural hearing mechanisms to support the principles of "sloppy workmanship" and the "principle of diversity" (Huggins and Licklider 1951) applied to the mechanistic implementations of binaural processing. The principle of sloppy workmanship refers to the danger of postulating a neural structure that is precisely arranged in detail; it is important to recognize that the postulated mechanism need function only in a statistical sense. The principle of diversity states that there are many ways to skin a cat and that the nervous system may use all of them. The implication for theory, here, is that conceptions that appear to be alternatives may supplement one another.

These principles are illustrated by Kubke and Carr (Chapter 6), who examine the development of auditory centers responsible for sound localization in birds and mammals. In both taxa, the developmental processes that shape the basic plan of the auditory circuit are complemented by plastic modifications that fine tune the neuronal connections to adapt to the experience of each individual animal. The resulting neuronally computed auditory space map associates particular binaural cues with specific sound source locations. But since binaural information will be determined by head size and shape of each individual, the auditory system must be able to adapt the basic connectivity plan to each animal. Thus, accurate associations between binaural cues and space assignments can develop only after the requirements of each individual are determined. The process therefore requires experience-dependent plasticity. In general, the neural circuits responsible for sound source localization can be recalibrated throughout life.

Comparative psychoacoustics is the link between the neural mechanisms responsible for localization and models of sound source localization applied to human hearing. Brown and May (Chapter 5) review the literature on behavioral studies of sound source localization in mammals with a focus on the cues operating for judgments in azimuth, elevation, and distance (proximity). It is clear from comparative work that the mechanisms for coding these three dimensions of directional hearing are entirely different, and may have had quite distinct evolutionary histories. The interaural cues of intensity and time are the primary ones for determination of azimuth for most mammalian species, including humans. The cues for elevation are less well understood, but seem to be related to the spectrum of the received sound, as filtered by the head-related transfer function (HRTF). It is remarkable, really, that the cues for estimating azimuth (ILD and ITD), and those for estimating elevation (the HRTF) are so fundamentally different. Processing of the HRTF depends on the tonotopic axis of the cochlea and the acuity of frequency analysis (the cue is essentially mapped onto the cochlea), while processing ITDs is independent of the acuity of frequency analysis. One could argue that processing ILDs is at least partially a matter of frequency analysis because the cue could exist in one frequency band and not in another, and its existence depends on the precision of frequency analysis. But judgments of elevation are a direct consequence of processing the spectrum, while the ILD cue must be processed in the level domain to be useful. The cues for distance or proximity are subtle and there is no direct connection between a given cue and the distance it specifies. Thus, the mechanisms for estimating azimuth, elevation, and distance are different, and it is possible that each may have had a different evolutionary history and that acuity along these dimensions may vary independently from species to species.

It is remarkable that vertebrates have so much in common when it comes to sound source localization. Even sound localization by fishes, with their otolithic ears and nonhomologous brainstem nuclei, can be viewed as a variation on a theme shared by many species (see Fay Chapter 3). Sound source localization among fishes is incompletely understood, but what we do understand seems familiar. First, localization in azimuth appears to be matter of binaural processing. Interaural acoustic cues are small or nonexistent owing to the high speed of sound underwater. In addition, fishes have close-set ears and (in some cases) indirect stimulation of the two ears via an unpaired swim bladder. Yet, the peripheral auditory system of fishes seems to reconstitute the functionally equivalent interaural cues through ears that are inherently directional. For example, while ILDs probably do not exist for fishes, the neurally coded output of the ears represents interaural response differences by virtue of the ears inherent directionality. Since response latency is a function of response magnitude in some primary afferents, there is also the equivalent of ITDs in the response of the ears. Just as in terrestrial vertebrates, the coding for azimuth and elevation seems to be independent of one another. In fishes, the ear is not tonotopically organized, but rather is organized directly with respect to sound source elevation (hair cell orientation). Tetrapod ears, having tonotopic organization, estimate elevation through an analysis of the sound spectrum as filtered by the HRTF. Fish ears analyze elevation directly through an across-fiber peripheral code that reflects hair cell orientation.

The inherent directionality of the ears of fishes puts them into the category of pressure-gradient receivers. It has become clear that pressure gradient or pressure-difference receivers are more widespread among animals than previously thought. Pressure-difference receivers are inherently directional receivers that can be of two basic types. In tetrapods the most familiar type consists of a tympanic ear for which sound may find pathways to both sides, creating a pressure difference across the membrane. Depending on the phase and amplitude of sound reaching both sides of the membrane, interaural differences in time and magnitude can become quite large. These interaural differences do not amplify the interaural differences that normally would accompany close set ears, but rather determine interaural differences essentially arbitrarily. These ears are thought to be inherently directional simply because their response varies as an arbitrary function of azimuth or elevation. The less familiar type of pressure-gradient receiver consists of an array of particle velocity receivers, such as hair cells with particular directional orientations (in fishes), or insect antennae, filiform hairs, or terminal cerci that respond directly and in a directional manner to acoustic particle motion (Robert, Chapter 2). These are properly referred to as pressure-gradient receivers because acoustic particle motion occurs to extent that pressure gradients exist. These receptors require a different sort of central processing and computation than is required for pure pressure receivers. Rather than, for example, computing azimuth based on the differences in the time and intensity of sound reaching the two ears, the axis of acoustic particle motion is probably estimated as the "best" axis from the population of active fibers varying in most sensitive axis.

These pressure difference receivers occur widely among species, and are probably primitive among vertebrates and invertebrates alike. Thus, as shown by Robert (Chapter 2), mosquitos, caterpillars, and crickets are known to have non-tympanal auditory receptors, and many insects have pressure-gradient tympanal ears. The ears of fishes are all pressure-gradient receivers, responding directly to acoustic particle motion (see Fay, Chapter 3). Those fishes that have pure pressure receivers (only confirmed in one fish taxa, the Ostariophysi) are a distinct minority. Among anuran amphibians, pressure gradient ears are ubiquitous and are thought to represent the primitive condition. Reptilian and avian ears are characterized by interaural canals, and although experiments must be done to confirm pressure difference hearing in each species, it is possible that most species detect pressure gradients, at least at low frequencies (see Christensen-Dalsgaard, Chapter 4). Thus, mammals are the only vertebrate group to have lost sensitivity to pressure gradients and to this mode of directional hearing. The pure pressure receiver characteristic of the mammalian ears seems to have given up inherent directionality and interaction with the respiratory system for sensitivity to high frequencies and all the advantages that come with high-frequency hearing.

References

Fay RR, Popper AN (2000) Evolution of hearing in vertebrates: The inner ears and processing. Hear Res 149:1–10.

Hoy R, Popper AN, Fay RR (eds) (1998) Comparative Hearing: Insects. New York: Springer-Verlag.

Huggins WH, Licklider JCR (1951) Place mechanisms of auditory frequency analysis. J Acoust Soc Am 23:290–299.

Jeffress, LA (1948) A place mechanism of sound localization. J Comp Physiol 41:35–39.

Manley GA, Popper AN, Fay RR (eds) (2004) Evolution of the Vertebrate Auditory System. New York: Springer.

2

Directional Hearing in Insects

DANIEL ROBERT

1. Introduction

In insects, like in most other auditory animals, the presence of two bilateral auditory receivers in the sound field and their relative position on the animal's body constitute elemental initial conditions in the process of directional hearing. The problem faced by insects is intimately related to their size and the physics of sound propagation; for a vast and complex array of reasons embedded in their phylogenetic histories insects are small compared to other auditory animals, and also compared to the wavelength of most biologically relevant sounds. The ears of insects can be set so close together that the conventional cues for directional hearing become, also by human standard, barely detectable. With an interaural distance as small as the diameter of a dot on an "i," for instance, the maximum time difference is in the order of 1 μs, a time scale admittedly delicate to handle by any nervous system. Similarly, the amplitude difference in sound pressure between the two ears can be immeasurably small. The constraint of size may thus cause severe difficulties to the processing of directional sound information. Constraints, in the course of evolutionary adaptation, however, also constitute multiple necessities that are the source of a multitude of innovations.

It is becoming increasingly apparent that, no matter how anatomically simple or how minute the auditory organs of insects may be, their sense of hearing is an act of sensation requiring great accuracy (Robert and Göpfert 2002). As astutely pointed out by Hudspeth (1997), hearing may be the most sensitive of the senses in terms of levels of detectable energy. Quantitatively, mechanoreceptor cells can detect mechanical displacements in the subnanometer range, involving energy levels close to thermal noise, or some 4×10^{-21} Joules (De Vries 1948; Bialek 1987; Hudspeth 1997). Some insect ears—like those of mosquitoes—may operate at similarly low levels (Göpfert and Robert 2001). In addition, audition is also designed to monitor acoustical events often more dynamic and transient than the spiking activity of neurons (for insects, see Pollack 1998; Schiolten et al 1981). Much work has been committed to the question of what are, for insects, the adequate cues—the physical quantities—that betray the direction and/or the location of a sound source, and how do insects go about to detect them. And, crucially, can these cues be converted, and if so how, into information that coherently represents the acoustical geometry of the outside world? Witness the chapters in this volume, the question of directional hearing has a long history, the problems are admittedly complex and the vast literature to date may only herald the promising depths of future research.

Probably only sifting near the surface of a rich pool of innovations, this chapter presents the mechanisms responsible for directional hearing in insects, and attempts to advance some ideas on how to explore this pool further. This chapter intends to present the constraints imposed on insects and explain the structures and functions known to operate in the process of directional hearing in insects. At times, some subjects will not be treated with the length and depth they deserve; this is not to occlude the concepts with a barrage of data. At those moments, recommendation will be made to consult recent reviews and key original articles to gather complementary insight. Insect hearing has been the subject of several recent reviews (Yager 1999). Of particular relevance is Volume 10 in the *Springer Handbook of Auditory Research* entitled *Comparative Hearing: Insects* edited by Hoy, Popper, and Fay (Hoy et al. 1998), that presents an authoritative overview. A collection of articles published as multi-author topical journal issue (see Robert and Göpfert 2004) addresses, among varied aspects, the latest research on insect auditory anatomy (Yack 2004), neurobiology (Hennig et al 2004), and psychoacoustics (Wyttenbach and Farris 2004).

2. What about Insect Ears?

2.1 Two Basic Types But Numerous Variations

The ears of insects can be categorized into two basic types, the tympanal ears and the flagellar ears. Both types, nearly always occurring as a bilateral pair, are highly sensitive to airborne vibrations and in their own way fulfill the functions of hearing organs. Figure 2.1 provides a very brief account of the diversity of insect auditory organs. Perhaps among the better known, the tympanal ear of the locust is among the "largest" of its type found in insects, yet, its tympanum spans only about 1 to 2 mm (Fig. 2.1A).

Tympanal ears can be found virtually anywhere on the general insect body plan, on the mouthparts (hawk moths), the tibia (field and bushcrickets), the abdomen (locusts and moths), the anterior thorax (parasitoid flies), the wing base (butterflies), the ventral thorax (mantisses), and the base of the neck (beetles) (reviews: Yack and Fullard 1993; Hoy and Robert 1996; Yack 2004). The basic bauplan of a tympanal ear consists in a thin cuticular membrane backed with an air-filled cavity with a mechanosensory chordotonal organ directly or indirectly in mechanical contact with the tympanum (Robert and Hoy 1998). The morphology of the tympanum, the associated air sacs, and the mechanosensory organ display a diversity that is allegedly bewildering and perhaps

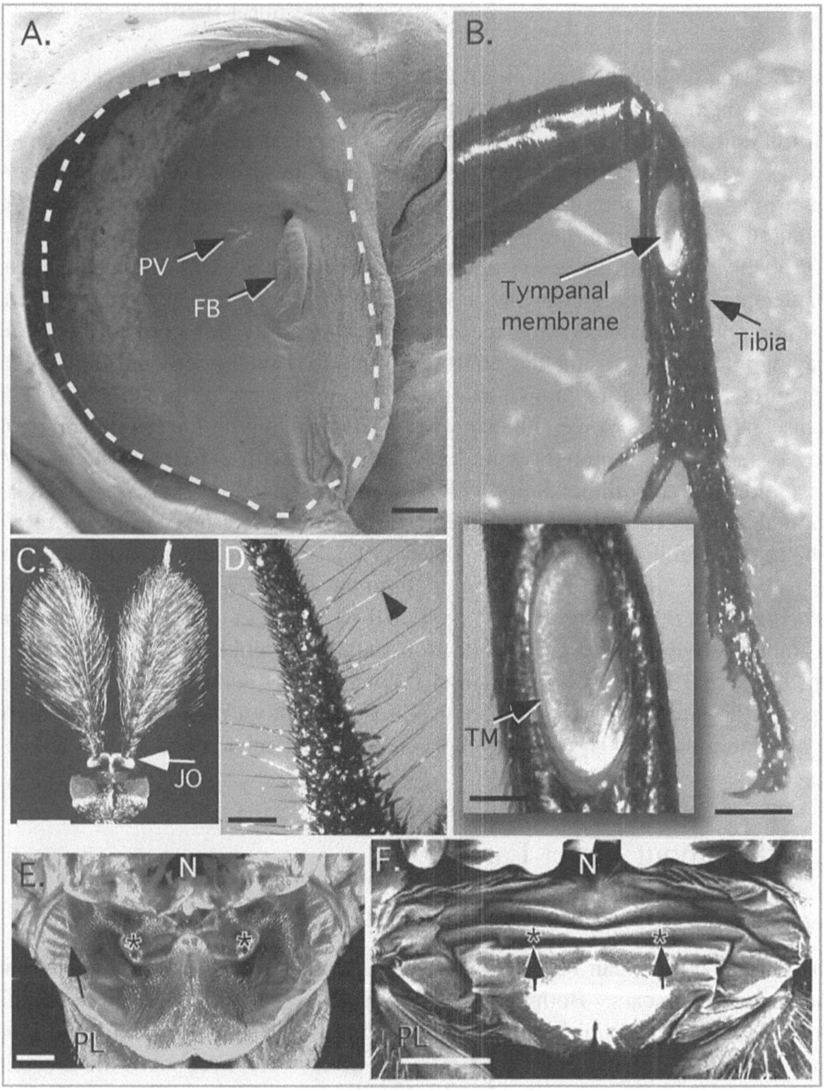

FIGURE 2.1. External auditory anatomy of a locust, a cricket, a mosquito, the cercal system of the cricket, and two parasitoid flies. (**A**) The tympanal ear of the locust. PV is the pyriform vesicle; FB the folded body to which high and low-frequency receptors respectively attach. The tympanal membrane is delineated by the *white stippled line*. Scale bar = 200 μm. (**B**) Posterior tympanum on the tibia of the first pair of legs of the cricket *Gryllus bimaculatus*. Scale bar = 1 mm (*inset*: 250 μm). (**C**) Light scanning micrograph of the antennae of the male mosquito *Toxorhynchitis brevipalpis*. The ball-like structures at the base of the antenna are the mechanoreceptive organs. Scale bar = 500 μm. (**D**) Filiform hairs on the cercus of the field cricket *G. bimaculatus*. Scale bar = 200 μm. (**E**) Tympanal ears of a tachinid fly (*Ormia*). *Arrow*, tympanal membrane; *arrowhead*, insertion point of the mechanoreceptive organ; N, neck; PL, prothoracic leg. Scale bar = 200 μm. (Light scanning micrograph by D. Huber). (**F**) Tympanal ears of a sarcophagid fly (*Emblemasoma*). *Arrows* show insertion points of mechanoreceptive organs on the tympanal fold. Scale bar = 200 μm.

unique to insects (Yack 2004). A partial and logically ultimate explanation for that variation is that tympanal ears have evolved independently perhaps as many as 19 times, and in at least 7 out of the approximately 25 orders of insects (Fullard and Yack 1993; Hoy and Robert 1996; Yager 1999). In view of the diversity of insects and their morphological adaptability, this figure may simply reflect the lack of research in the remaining "atympanate" orders. Notably, the absence of tympanal ears in the well-studied hymenoptera (ants, bees, and wasps), and odonata (dragonflies and damselflies) remains puzzling, and hence may deserve renewed attention. As tympanal hearing is only one of the two known methods for audition, it may be tempting to speculate somewhat, and suggest the possibility that notoriously atympanate insects (mostly the very speciose beetles and flies, but also little known insect orders) may be endowed with a yet unfamiliar sense of hearing, possibly based on the flagellar type, or some variation of it (Fig. 2.1).

2.2 The Directionality of Particle Velocity Receivers

The anatomical basis of nontympanal hearing has been known for quite some time and takes the form of antennae (in mosquitoes, Johnston 1855), filiform hairs borne on the body wall (caterpillars, Tautz 1977), or the terminal cerci (crickets, Gnatzy and Tautz 1980) (Fig. 2.1). This type of auditory receiver is said to be inherently directional. This is in part because it responds to the particle velocity component of the sound field, which is a physical attribute that is vectorial (as opposed to the pressure component which is a scalar quantity). For one part, directional information is thus contained by the bulk oscillations of the air particles that take place in the direction of sound propagation, in the far field. Another component of directionality relates to the anatomical arrangement of hairlike structures and antenna. Indeed, the mobility of the sound receptor may not be isotropic. An asymmetry may then confer some directionality to the system oscillating in the sound field. In effect, as shown in the caterpillar of noctuid moths, some filiform hairs display a distinct inherent directional response, and some do not (Tautz 1977). Since the particle velocity portion of acoustic energy dominates near the sound source, these organs have also been called near-field detectors. This is not to say, however, that they will detect only sound in the near field of a sound source (from one sixth to one wavelength away from it). If endowed with enough sensitivity, particle velocity receivers may well detect sound in the far field, where particles also oscillate, albeit with much smaller magnitude. Because sound fields are usually strongly divergent close to small sound sources, such as a female mosquito, bilateral particle velocity receivers (the antennae of a male mosquito) may experience vastly different vector fields depending on their distance from the sound source and the orientation of their auditory organs' axis of best sensitivity. This alone may affect the directionality extractable by two bilaterally symmetric antennal detectors. In effect, the direction of a particle velocity vectors in a sound field near the source depends on the type of source (monopole, dipole). As a result, at any point in space, the velocity field may or may not directly point to the sound source. The problems associated with the processing of such vectorial acoustic information have been little investigated in insects, but may well be of similar nature to those encountered by fish (Edds-Walton et al 1999). The capacity of flagellar auditory organs to detect sound directionally (mosquitoes, Belton 1974) has received only little attention recently. The physical basis of their directionality, in terms of their viscous interaction with a vectorial sound field, their own—sometimes variable—asymmetry and the coding of primary mechanoreceptors, remains unknown to date.

3. The Effects of Size on Directional Cues

The main acoustic cues used for directional hearing are interaural differences in intensity (IID) and in time (ITD). In addition, variations in the spectral composition of incident sounds can provide directional information about the source of sound (Middlebrooks and Green 1991; Wightman and Kistler 1997). Requiring significant diffractive effects to take place, this possibility is quite unlikely for the smaller (grasshoppers, flies), but is not excluded for larger auditory insects (locusts and bushcrickets). Another, nontrivial requirement is, of course, some sensory capacity for frequency analysis. The coding of frequency by auditory receptors and interneurons has been well documented in a variety of insects, in particular in grasshoppers and crickets (for review see Pollack 1998). Although some capacity for frequency discrimination has been demonstrated for field crickets (behavior, Wyttenbach et al 1996; neural coding, Pollack and Imaizumi 1999), directional sound detection based on spectral variation, as known from vertebrates, has received little consideration. Such a mechanism would possibly necessitate quite a fine resolution in the frequency analysis and the sequential comparison between sounds of varying frequency composition, a processing feat that has not been demonstrated in insects.

Amplitude and time domain cues, IID and ITD, are mainly determined by the spatial separation between the ears and their position relative to the sound source. In insects, the interaural distance can vary from 1 cm (locusts, bushcrickets) to only a few millimeters (crickets, grasshoppers, moths, cicadas), or a mere 500 μm (parasitoid flies). Consequently, interaural differences in the time of arrival of a sound wave (ITD) can easily vary from 30 μs to 1.5 μs. Interaural distances in insects are also often considerably smaller than the wavelength of the relevant sound, a fact that bears consequences for the other main cue (IID). Acoustical theory states that diffraction between an object of size r and a sound wave of wavelength l becomes significant when the ratio r:l exceeds 0.1 (Morse and Ingard 1968). Experiments exploring this theoretical prediction have been rarely rigorously conducted in insects; but when they have (moths, Payne et al 1966; locusts, Robert 1989; locusts and grasshoppers Michelsen and Rohrseitz 1995; flies, Robert et al. 1999), they showed diffractive effects that, as sound frequency increases, result in growing sound pressure variations and

IIDs. A systematic quantification of acoustic diffraction in a free field, using probe microphones commensurate with the task of measuring the microacoustics around the body of an insect has yet to be undertaken. Thus far, it appears that the main limiting problem has been the excessive size of probe microphones. The possible use of diffraction-related frequency cues may deserve some renewed attention in light of recent psychoacoustical evidence suggesting that the cricket may detect the direction of incident waves in the elevational plane (Wyttenbach and Hoy 1997).

Another important consequence of small body size is the reduced amount or absence of dense tissue between the auditory organs. Tympanal ears are always associated with large air sacs that generate some acoustical transparency across the body. Even for a large insect such as the locust, diffraction has limited effects (see Fig. 1 in Michelsen and Rohrseitz 1995). When ears are close together, little space is left for sound absorbing tissue to acoustically isolate the ears from each other. In locusts, some of the interindividual variation measured in the interaural transmission was attributed to differences in the amount of fat tissue between the ears (Miller 1977). In parasitoid flies, the bilateral ears even share a common air sac (Robert et al. 1996). More complete descriptions and quantifications of the biomechanics of sound propagation and the generation of cues for directional hearing in insects can be found in earlier reviews (Michelsen 1992, 1996, 1998, Robert and Hoy 1998).

4. Directional Receivers in Insects

Research in insect audition has uncovered a rich diversity of structures and functions that serve the purpose of directional sound detection. Taking advantage of the amenability of insects to a variety of experimental work—biomechanical, behavioral, anatomical or neurophysiological—the study of insect audition has fostered the discovery and the intimate understanding of alternative, original, mechanisms for directional hearing, such as pressure difference receivers and mechanically coupled receivers.

4.1 Pressure Receivers

Tympanal ears operating as pure pressure receivers are found on insects that are relatively large compared to the wavelength of the sound frequencies of behavioral relevance (either the courtship and mating songs or the high-frequency echolocation cries of bats). These ears are deemed to be pressure receivers because sound pressure is thought to act only on one face of the tympanal membrane, usually the external one (yet, the internal one in bushcrickets) (Fig. 2.2A) (see Michelsen 1998). In such situation, the insect's body is large enough to generate diffractive effects, resulting in overpressures and underpressures at the location of, respectively, the ear nearer and further from the sound source. Interaural pressure differences (or IIDs) are thus generated that constitute suf-

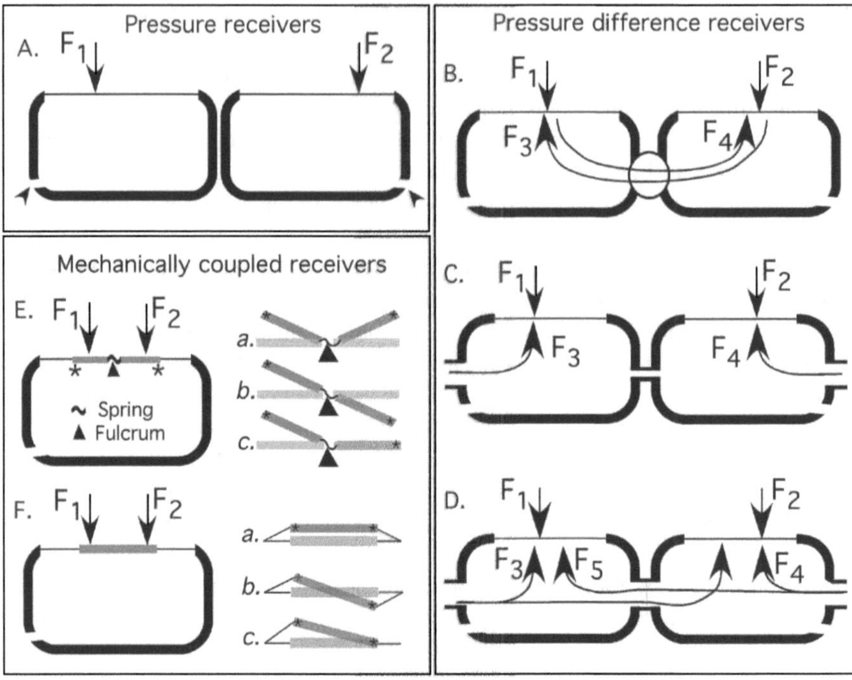

FIGURE 2.2. Directional receivers in insects. The ears are symbolized by a pair of boxes each with one thin wall representing the tympanal membrane and one small opening for static pressure equalizing (*arrowhead*). For simplicity, the action of pressure on the tympanal membrane is reduced to a point force *F*. (**A**) Pure pressure receivers. Forces act on the external side of the tympanal membranes. The two ears are acoustically independent. (**B–D**) Pressure difference receivers. For all cases, the forces act on both sides of the tympanal membranes. Internal sound pathways are shown by *long arrows*. (**B**) In locusts and grasshoppers. (**C**) In bushcrickets. (**D**) In field crickets. (**E**) Mechanically coupled pressure receivers in tachinid flies, the seesaw model. Forces act on the external side of each tympanum, generating asymmetrical deflections of the intertympanal bridge. (**F**) Mechanically coupled pressure receivers in sarcophagid flies. The deflection mode diagrams illustrate the asymmetrical mechanical behavior for both fly families. Responses are shown for three frequencies; *a*, bending at 4 kHz; *b*, rocking at 7 kHz; *c*, combination of the above at 15 kHz. *Asterisks* show the points of attachment of the mechanoreceptors to the tympanal system. See text for further explanations.

ficient cues for directional hearing. In a large noctuid moth, IIDs of some 20 to 40 dB were measured as a result of the substantial diffraction of sound with frequencies similar to those used by echolocating bats (30 to 60 kHz) (Payne et al. 1966). Such large IIDs are most practical for directional hearing, but they are not a prerequisite for it, nor do they indicate that the animal in question uses directional information to evade its aerial predators. Small IIDs can convey adequate information for localization or lateralization in insects. For some small grasshoppers, IIDs as small as 1 to 2 dB have been shown to be sufficient to elicit reliable orientation toward attractive calling song (von Helversen and Rheinlander 1988; see Hennig et al. 2004). Forcibly, as the frequency of sound decreases, diffraction-related cues vanish, a constraint that has often been invoked for the tendency of insects to design calling and mating calls with high frequencies (to the inevitable cost of transmissibility) (Michelsen 1996; Bennet-Clark 1998). It is worth noting that quantitative measurements of intra-aural (i.e., behind the tympanum) pressure variation, with reference to the external pressure variation, of insect ears have proven to be very difficult and have thus far not been undertaken in noninvasive conditions. Hence, the notion of pure pressure receiver seems to rely on the recognition of only one acoustic input into the ear (the tympanum), and the absence of evidence for an alternative input.

4.2 Pressure Difference Receivers

Pressure difference receivers are distinctly more complicated, and more variable in their anatomy and modes of operation (Fig. 2.2B–D) than pressure receivers. Pressure difference receivers are typically found in insects with body sizes only a fraction of the wavelength of interest. As illustrated schematically in Figure 2.2, pressure difference receivers can take one of three forms. By definition, their mode of operation relies on the action of sound pressure on both sides of a tympanal membrane. Requiring more than one input per ear, such a mechanism was proposed a while ago as a solution to the problem of directional hearing by small animals, and in particular, among arthropods, by grasshoppers and locusts (Autrum 1940). The principle relies on ears being each endowed with two or more acoustic inputs. These supplementary acoustic inputs are adapted to conduct pressure waves to the internal side of the tympanum (Fig. 2.2B). The internal sound pressure, owing to its travel in a tracheal tube or across air sacs, undergoes different degrees of attenuation or amplification and some phase shift as a result of alterations in propagation velocity. In such a system, the force driving the tympanal membrane is the difference between the external and internal pressures (or notionally, forces) (Fig. 2.2B, F_1 and F_3). Notably, because the pressures involved are periodic, a force still acts on the tympanal membrane when internal and external pressures happen to be equal (no attenuation or amplification through internal travel) but have a phase difference. Of course, in such a system, a combination of both phase shift and amplitude difference is likely to take place, and to affect the ipsilateral and contralateral ears differentially. In theory, well-adjusted phase shifts and amplification factors could lead to constructive and destructive interference at the tympanal membranes that may greatly enhance the contrast between the two ears.

The first type of pressure difference receiver relies on the acoustic coupling between the two ears achieved by acoustically conductive tissue, or air sacs, situated between the tympanal cavities (Fig. 2.2B). For this particular anatomy, the pressure acting on the internal side of the tympanal membrane travels from the contralateral side of the animal. This anatomical arrangement has been identified in the locust (Miller 1977). Because internal sound conduction has been shown to decrease with increasing frequency, the locust ear is deemed to operate as a pressure receiver for high frequencies and a pressure difference receiver for low frequencies. Two valuable studies have investigated in greater detail the mechanisms at work for directional hearing in large (*Schistocerca gregaria*) and small (*Chorthippus biguttulus*) grasshoppers (Michelsen and Rohrseitz 1995; Schul et al 1999). The premise of the first study was that the biomechanical mechanisms employed by large and small grasshoppers, should be scalable, owing to size and differences in the frequency range used for communication. A model was proposed that could predict directionality as a function of auditory inputs and internal interaural transmission gains and delays. Measurements failed to satisfy a simple scaling law, and it was concluded that directionality cues were poor at 5 kHz for the small grasshopper. The model would not apply as such to the smaller of the two auditory systems. This was attributed to an insufficient transmission delay in the internal interaural passage of sound. Yet, *C. biguttulus* can reliably orient to 5 kHz sound in the context of phonotactic experiments (von Helversen and Rheinlander 1988; Pollack 1998). In an effort to test the general validity of the model, Schul et al. (1999), using behavioral, acoustical, and electrophysiological methods, determined the contribution of the internal pathway responsible for the bilateral transfer of sound pressure. The acoustical measurements of Schul et al. yield transmission delays that substantially differ from those of former studies for the small species, but coincide for the large species. In particular, the delay incurred by the internal interaural sound transmission is identified—and quantified—as being essential to the generation of interaural differences, bringing to agreement phenomenology and model predictions. Therefore, it seems that the proposed two-input model featuring an interaural delay line is valid for grasshoppers and their pressure difference receiver system.

Another type of pressure difference receiver can be found in bushcrickets (katydids), which have their auditory organ (two tympana per organ, associated with one mechanoreceptive organ) on the foreleg tibia (Fig. 2.2C). Katydids can be quite large compared to the frequency of their calling songs, yet, because the tympana are borne on thin legs clear from the body wall, reduced diffractive effects can be expected. As in the large desert locusts (Robert 1989), significant diffraction-related over- and underpressures occur near the body surface. Exploiting these diffractive effects, bushcrickets possess additional acoustic inputs on their thorax, some specialized spiracles and horn-shaped atria connecting to tracheal tubes that lead to the internal side of the tympanal membranes but also to the mechanoreceptive organ (Lewis 1983). Thus, unlike grasshoppers, the pressures acting on either side of the tympanum both originate from the same side of the body (compare Fig. 2.2B, C). Notably, the tracheal tube has the shape of an exponential horn and acts like one; sound transmitted through it is amplified. The result is that the internal sound pathway dominates the force field driving the tympanal membranes (Lewis 1983). This type of pressure difference receiver has not received as much biomechanical attention, and uncertainty remains concerning the amplification process in the horn-shaped acoustic trachea, the action of the pressure on the mechanoreceptor organ, and the role of the oft-present thin tracheal connection between the tracheal horns (Fig. 2.2C).

Another, much studied example of a pressure difference receiver is the field cricket *Gryllus bimaculatus* (Michelsen et al 1994) for which body size and wavelength of relevant calling songs are, respectively, about 0.8 cm and 7 cm. Although the interaural distance is difficult to estimate in this species—the tympanal ears are situated on the highly moveable tibia (Fig. 2.1B)—it is apparent that insufficient ITDs and IIDs are available for directional detection. The anatomical arrangement found in field crickets is the most complex known to date, employing no fewer than four inputs and one internal interaural transmission channel (Fig. 2.2D). One important operational characteristic of that system is that the two more important force inputs to the anterior tympanal membrane are the external sound pressure (Fig. 2.2D, F_1), and the internal sound pressure (F_5) originating from the contralateral sound input (Fig. 2.2.D), unlike the situation in bushcrickets for which F_3 is the dominant, ipsilateral input. In field crickets, the tracheal connection between the hemilateral tracheal tubes is larger than in bushcrickets and displays a thin septum at the midline (Fig. 2.2D). This septum has been suggested to play a crucial role in enhancing the time delay in the internal interaural transmission line (Löhe and Kleindienst 1994; Michelsen and Löhe 1995). In passing, it is worth noting that in field crickets, the anatomical relationship between the tympanal membrane and the mechanosensory organ is quite complex (Yack 2004). To what degree the vibrations of the tympanal membrane translate into mechanical actuation of the sensory organ (in magnitude and phase), and what role is played by sound pressure in the tracheal system adjacent to it, remain unknown. This problem also amounts to that of the current difficulty of measuring sound pressures in small cavities.

Multiple studies have revealed the robust capacity of crickets to readily locate the source of a calling song in intact situations but also when different parts of their tracheal anatomy and identified acoustical inputs were destroyed or plugged (Schmitz et al 1983; Weber and Thorson 1989; Doherty 1991; Michelsen and Löhe 1995). Taking nothing away from the biomechanical measurements, this behavioral evidence indicates that the four input pressure difference system is sufficient, but is not, at least in certain conditions, necessary for directional hearing.

A necessary word of caution should emphasize that the drawings of Figure 2.2 are notional and by no means intend to reflect the anatomical complexity and the actual paths taken by the multiple pressure waves propagating in cavities and long tapered tubes. If granted an explanatory value, these schematic representations are meant to illustrate the operational principles of various pressure difference receivers. Variations on that theme, with intermediate forms, or entirely new forms of internal sound transmission are likely to exist in other insects. Several accounts of the biomechanics of pressure difference receivers have been published (Michelsen et al 1994; Michelsen 1992, 1996, 1998; Schul et al 1999), and a recent review covers the neural processing in directional hearing (Hennig et al. 2004)

Arguably, ears operating as pressure difference receivers, at least in their low-frequency range of sensitivity, may well be the most common type of auditory receiver in insects. Indeed, most auditory insects are in the biometric range (one centimetre and much smaller) that could require the presence—if not justify the evolution—of a pressure difference receiver system. Both conditions of limited diffraction and interaural acoustical coupling may thus be often fulfilled. With this in mind, the presence of sound transmission inside adapted acoustic tracheae in some insects raises interesting possibilities for other types of auditory receivers. Intriguingly, could internal sound transmission elicit vibrations of the tracheal wall that, in turn, could be detected by an associated chordotonal organ? Formulated more precisely, a thinning of the tracheal wall, or air sac, accompanied by the selective absence of taenidia (ridges acting as structural buttressing) and a few scolopidial mechanoreceptive units could act as a pressure or pressure difference receiver.

4.3 Mechanically Coupled Pressure Receivers

For some small insects, both interaural distance and body size are simply too small to produce IIDs and ITDs of significant magnitudes. In the little tympanate parasitoid tachinid fly *Ormia ochracea*, the ears are so close together that they link up at the midline of the animal (Fig. 2.1 E, F). For this fly, the best possible ITD has been measured to amount to 1.45 μs ($\pm$0.49, SD, $N = 10$) (Robert et al 1996). For such a small insect, body size to wavelength ratio of 1:35 at best precludes significant diffractive effects (Robert et al 1999). Yet, the fly can very accurately locate her host acoustically, a field cricket singing its 5-kHz calling song (Cade 1975) using tympanal auditory organs (Robert et al. 1992; Müller and Robert 2001). Biomechanical and physiological evidence has shown that these ears are directional, and have revealed the mechanism by which they achieve this directionality (Miles et al 1995; Robert et al 1996). The process is based on the mechanical coupling between two adjacent tympanal membranes, an unconventional mechanism that is so far known to occur only in flies (Robert and Hoy 1998; Robert and Göpfert 2002). The mechanism involves the coupling of the tympana by a flexible cuticular lever; this coupling has the effect of amplifying tiny acoustic cues into more substantial interaural differences that can be processed by the nervous system. In response to the host cricket song, a trill with a carrier frequency at 4.8 to 5 kHz, this coupled tympanal system undergoes asymmetrical mechanical oscillations. Using scanning laser Doppler vibrometry, it could be shown that the oscillations arise from the linear combination of two resonant modes of vibration. Rocking like the two arms of a floppy seesaw (see Fig. 2.2E), the coupling lever and the two tympanal membranes attached to it move out of phase and at different amplitudes at frequencies close to that of the cricket song (Miles et al 1995; Robert et al 1996). Remarkably, the mechanical ITD measured between the tympanal membranes is 50 to 60 μs and the mechanical IID is 3 to 12 dB for sounds delivered at 90° azimuth. This mechanical ITD is thus about 40 times longer than the 1.5-μs acoustical ITD. It is as if the ears of the fly were located some 20 mm from each other (instead of the real interaural distance of 520 μm). Operating as a mechanical ITD and IID amplifier, this unconventional system converts small acoustical cues into larger mechanical cues. Recent evidence reveals how these mechanical cues are used for the reliable neural coding of sound direction (Mason et al 2001; Oshinsky and Hoy 2002). These studies provide precious insight into the neural mechanisms that allow the hyperacute coding of acoustic information, a subject that is presented in Section 5 hereafter.

Directional hearing by mechanical coupling between two tympanal membranes is not unique to tachinid flies; it has also been described for a fly of another dipteran family (the sarcophagidae) illustrating a remarkable case of convergent evolution (Robert et al. 1999). As a parasitoid of cicadas, the fly (*Emblemasoma* spp.) also possesses a hearing organ on its prothorax (Fig. 2.1F). The mode of operation of this auditory organ is analogous to that of the tachinid fly *O. ochracea*, but it is not identical (Robert et al. 1999).

Phenomenologically, these two auditory systems achieve asymmetrical tympanal deflections, a prerequisite for directional hearing in these systems, but not in the same way. The tachinid and sarcophagid systems present several crucial anatomical differences that determine tympanal mechanics. In the tachinid system, intertympanal coupling is achieved by the presternum, an unpaired sclerite that spans across the midline where it is anchored to the immobile probasisternum (Fig. 2.1E). The mechanoreceptive organs attach at the end of each arm of the presternum (labeled * in Figs. 2.1E, F and 2.2E, F) Using microscanning laser Doppler vibrometry, it was shown that this sclerite acts as mechanical lever coupling the two ears. The lever consists of two beams that are joined medially by a torsional spring (marked ~) and supported by a fulcrum (a pivot point marked by a black triangle in Fig. 2.2E). Biomechanical evidence shows that such a lever system has two degrees of freedom, resulting in a rocking mode and a bending mode. At low frequencies, the presternum undergoes bending (flexion at the immobile fulcrum), whereby both arms of the lever move together (Fig. 2.2E, *a*) (Miles et al 1995; Robert et al 1996). The deflection shapes of this tympanal system have been measured; the end points of the lever (*; attachment locations of mechanoreceptor organ) experience displacements of similar amplitude at frequencies below approximately 4 kHz (Robert et al 1996). As a point of comparison, at such frequencies the tympanal system of the sarcophagid fly deflects inwards and outwards with only little bending (Fig. 2.2F, *a*). In effect, the deep folding running across the tympanal membranes and the presternum of the sarcophagid ear (Fig. 2.1F) generates a stiffness anisotropy making the entire system prone to oscillate about the animal's midline. Deflecting as a single beam unsupported medially, both tympana move together with only slightly different displacement amplitudes. The translational mode observed for low frequencies in sarcophagids (Fig. 2.2F, *a*) is thus equivalent to the bending mode described for tachinid flies (Fig. 2.2E, *a*). At intermediate frequencies (approximately 7 kHz), both tympanal systems oscillate in a rocking mode; outward displacements at one end of the presternum are accompanied by inward displacements at the other end (*b* in Figs. 2.2E, F). In this rocking mode, both tachinid tympana oscillate about the midline, but owing to the flexibility provided by the torsional spring, they do so with some phase delay and amplitude difference (Miles et al 1995; Robert et al 1996). Notably, in the sarcophagid fly the rocking mode occurs in the absence of a fulcrum anchored at the midline of the animal. For both systems, the mechanical ITDs and IIDs (differences between one side of the tympanal system and the other) increase as frequency increases. For higher frequencies (15 kHz), a combination of the two modes dominates the motion and the side contralateral to the incident sound wave experiences low displacements (*c* in Fig. 2.2E). For the sarcophagid ears the single beam formed by the tympanal fold sways about its contralateral end (*c* in Fig. 2.2F) (Robert et al 1999). Deflection shapes thus differ between the tachinid and the sarcophagid systems, yet the deflections experienced by the points of insertion of the mechanoreceptive organs are similar (compare asterisks in Fig. 2.2E, F). The single, unpaired air space backing the tympanal system of both tachinid and sarcophagid flies cannot be a priori excluded to play a role in sound transmission similar to a pressure difference system. This question has been addressed in tachinid flies where it was shown, using acoustical and direct mechanical actuation, that interaural mechanical coupling did not depend on the presence of a finite air-filled cavity (Robert et al 1998). That study concluded that the mode of operation of these ears relies on mechanical coupling only, excluding the action of a pressure difference mode.

Both auditory systems achieve asymmetrical tympanal deflections despite interaural distances of the order of 1 mm. The interaural mechanical coupling relies on a particular morphological design that provides an anisotropy in stiffness. Through functionally convergent but anatomically divergent evolutionary innovations, these two fly families have independently solved the problem of the directional detection of low-frequency sounds by tympanal membranes separated by a fraction (1:130) of the wavelength. Other tachinid flies, from other genera have been reported to use ears to detect calling songs of their hosts at higher frequencies (10 to 15 kHz) (Lakes-Harlan and Heller 1992); from morphology alone it is likely that they use mechanical coupling for directional hearing. The morphological design and mode of action of the sarcophagid tympanal membranes show, in principle, how a millimeter-size ear can be directional by virtue of one or several folds on its membranes. Again and yielding to speculation, folds and creases along a thin tracheal tube may provide in other insects the adequate substrate, if linked to mechanoreceptor neurons, for internal auditory organs with directional characteristics. For mechanically coupled pressure receivers, the exact mechanical characteristics of the tympanal membranes, such as stiffness distributions and anisotropies, tolerances for bilateral differences, and their contributions to directionality remain uninvestigated. It is also worth noting that some 43 species in 7 genera of tachinid parasitoids have been shown to possess a wide variety of modifications of their prosternal anatomies very reminiscent of mechanically coupled hearings organs (Huber and Robert unpublished results).

5. Temporal Hyperacuity in Insect Directional Hearing

In some sensory modalities, such as hearing and electroreception, the time scale of events can be far shorter than the conventional millisecond-range of neural signaling. In hearing, localization tasks near the midline often involve microsecond-scale ITDs. Defined as the capacity for submillisecond coding, temporal hyperacuity has been documented for barn owls (Knudsen and Konishi 1979; Moiseff and Konishi 1981) and electric fish (Rose and Heiligenberg 1985). Essentially, the underlying neural mechanisms have been proposed to rely on the convergence of many sensory afferents onto an interneuron acting as a coincidence detector. Interneuronal spiking would be elicited only by the coherent firing of an ensemble of afferents within a narrow window of temporal coincidence. The spiking accuracy and reliability of the primary afferents is therefore crucially important in that scheme (de Ruyter van Steveninck et al 1997). Accordingly, events following each other within microseconds are most relevant to the microscale ears of fly *Ormia ochracea*. As seen in Section 4.3, in the best-case scenario (sound source 90° to the side of the animal) the tympanal system amplifies temporal acoustic cues (ITDs) by about 40 times, yielding mechanical ITDs of some 50 to 60 μs. It was shown at the mechanical level that the system of intertympanal mechanical coupling could vary its response as a function of the angle of sound incidence. Naturally, when the sound source is near the midline of the animal mechanical ITDs and IIDs decrease to values smaller than 1 μs (Robert et al. 1996). The demonstration that the flies can use temporal sound cues at the submicrosecond scale came from a series of behavioral and neurophysiological experiments by Mason et al. (2001). Flies were tethered on their pronotum and brought to walk on an air-cushioned spherical treadmill that would record the flies' locomotory activity (Fig. 2.3A). Flies could produce walking responses oriented toward the sound source, and quite unexpectedly, they could reliably do so even though the deviation from the midline was 1 to 2° (Fig. 2.3B). When the amount of turning was measured as a function of the angle of azimuth of the sound source, a sigmoid response curve was revealed that displayed a smooth transition near the midline (Fig. 2.3C). This distribution of turning angles is expected from a system endowed with high accuracy of localization (as opposed to lateralization) along the midline (azimuth zero). Finally, the proportion of correct responses as a function of stimulus azimuthal angle was evaluated, revealing a remarkable reliability and repeatability at angles as low as 2 to 3 degrees (Fig. 2.3D). It must now be considered that interaural cues, when calculated for an angle of 2° and an interaural distance of 520 μm, amount to a mere 50 ns for acoustical ITDs, and 2 μs for the mechanical ITD (owing to a mechanical amplification factor of 40) (Mason et al. 2001). How

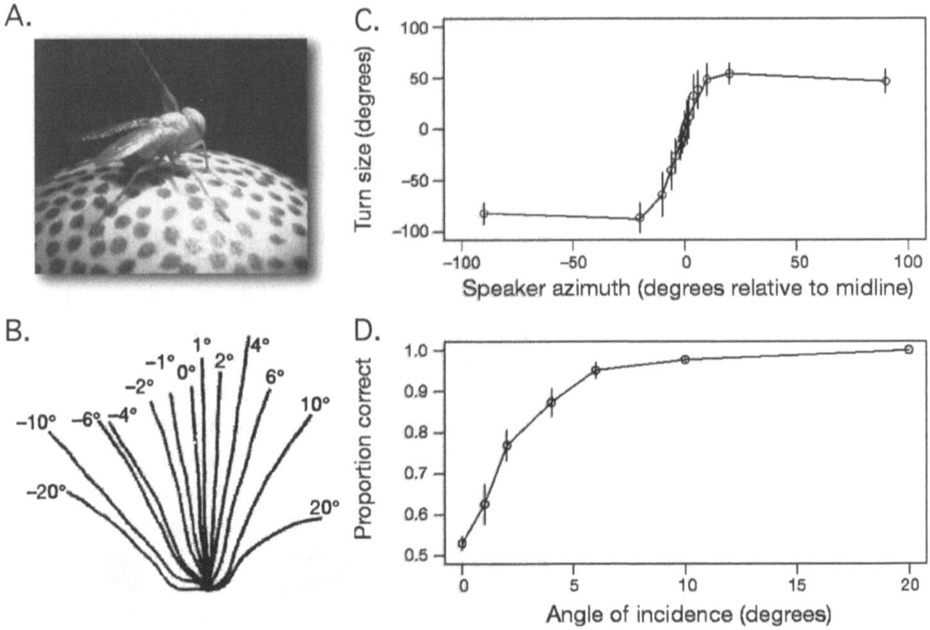

FIGURE 2.3. Phonotactic accuracy in the fly *O. ochracea* on a walking treadmill. (**A**) Videophotograph of the fly tethered on top of a Styrofoam ball supported by an air cushion. The locomotory activity of the fly is recorded by the resulting motion of the ball. (From Mason et al 2001, with permission.) (**B**) Mean paths of locomotion of one fly in response to cricket's song delivered at different angles. (**C**) (N = 7 flies, 10 trials per fly per angle, ± 1 SD). Different azimuthal angles resulted in distinct phonotactic trajectories. (**D**) Proportion of correct turns as a function of azimuthal angle of incidence. A reliable oriented response occurs for angles as little as 2 to 3° (N = 19 flies, 20 responses per fly per angle). (**B–D** modified from Mason et al. 2001; © Nature Publishing Group.)

is the observed phonotactic behavior possible in view of such small directional cues? The answer required further knowledge on the actual response characteristics of the fly's auditory primary afferent receptor neurons (Mason et al. 2001; Oshinsky and Hoy 2002).

The differences in the spike latency between left and right receptor neurons have been measured for different sound locations (Mason et al. 2001). For 90° azimuth, neural ITDs, measured as summed action potentials, amount to 150 μs and, as the angle decreases, drop by 3.5 μs per degree, predicting a neural ITD of 7μs at 2° azimuth. Hence, in view of the observed phonotactic behavior, and perhaps allowing for some degree of error in the measurements, the fly's primary afferent neurons seem capable of reliably encoding temporal events separated by a mere 10μs. The studies by Mason et al. and Oshinsky and Hoy together provide key evidence that such capacity is based on a remarkably fast spike time code. First, most of the afferent neurons that were recorded were shown to respond to acoustical stimulation with a single spike (type I afferents) (Fig. 2.4A, 90 dB sound pressure level (SPL), and have very low probability of spontaneous activity (Oshinsky and Hoy 2002). Characteristically of this category of afferents, only one spike is released, irrespective of the duration of the stimulus. The latency of that single spike, measured as the time between stimulus onset and spiking, increases as stimulus amplitude decreases, resulting in a "time/intensity tradeoff" observed in numerous sensory systems (Fig. 2.4A). Such effect is useful to generate directionality. Owing to the asymmetrical mechanical deflections of the tympanal system, the primary afferents, in addition to enhanced

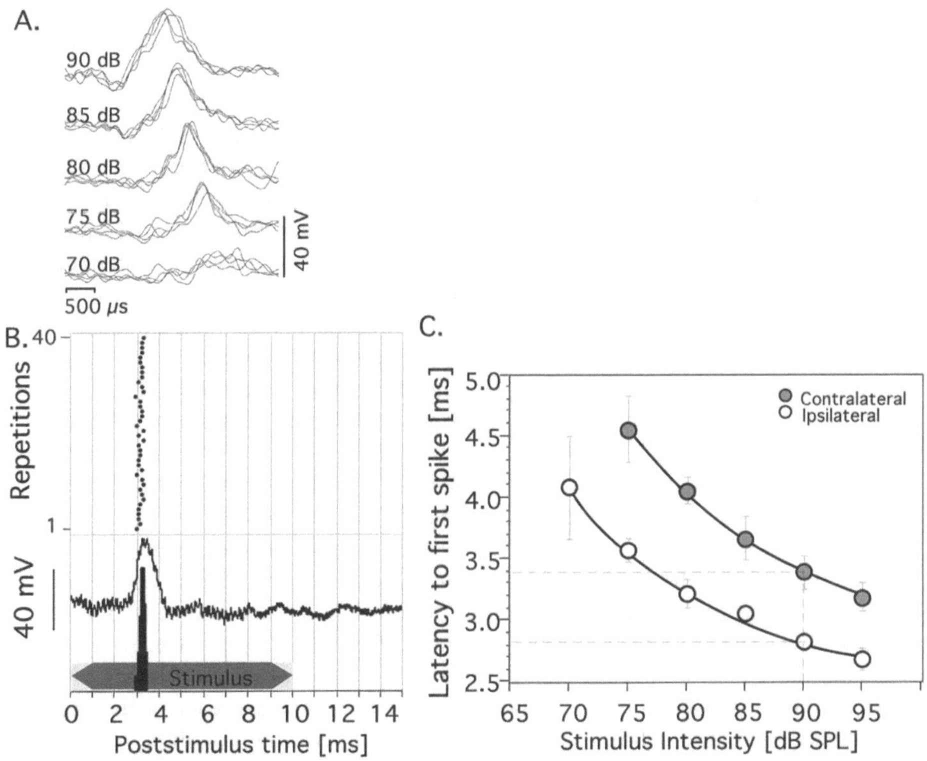

FIGURE 2.4. Temporal coding of mechanoreceptor neurons in the fly *O. ochracea*. (**A**) Multiple action potentials of a single neuron as a function of the amplitude of the sound stimulus in dB SPL. Low-amplitude stimuli result in a delayed action potential. (**B**) Response of a single receptor to a series of 40 stimulus presentations. Raster plot of the repetition, oscillogram of a single spike, and poststimulus histogram illustrate the high repeatability of the neuron's signalling. (**C**) Latency of receptor spiking as a function of stimulus amplitude for ipsilateral and contralateral stimulation. For a 90-db SPL stimulus, a receptor shows different spiking latencies, depending on whether it is ipsilateral or contralateral to the sound source. (**A–C** modified from Oshinsky and Hoy 2002, © 2002 by the Society for Neuroscience.)

mechanical ITDs, experience different interaural stimulus amplitudes. Yet, in detail, it is unknown whether the vibration amplitude of receptor neurons is linearly related to that of the tympanal membrane. A nonlinearity may signify a larger contrast of directionality for some range of amplitudes, generating a "foveal" zone of acuity at some stimulus amplitude. Importantly though, a difference in spiking delays is generated between the individual afferents situated in the ear ipsilateral or contralateral to the sound source (Fig. 2.4C). The response latency of single afferents from one ear was measured in response to a series of sound stimuli varying in amplitude (95 to 75 dB SPL) first broadcast ipsilateraly to that ear, and then, at the same sound pressure, from the side contralateral to that ear. This way, the difference in the latencies between ipsilateral and contralateral afferents could be estimated. For 90 dB SPL, the difference in afferent response latency was approximately 600 μs, a delay that seemingly becomes manageable for neural processing (Fig. 2.4C).

Finally, a key observation was that the variation in spiking time (jitter) was remarkably low compared to other invertebrate sensory systems. In effect, in response to 40 successive 5-kHz tones, spiking latency was 3164 μs with a jitter (the standard deviation of the latency distribution) of 95 μs (Oshinsky and Hoy 2002)(Fig. 2.4B). The jitter measured for seven animals ranged from 12 μs to 121 μs with an average of about 70 μs (Mason et al 2001). Thus, the uncertainty of the spike code may be about ten times larger than the temporal event it is required to code for (about 10 μs). At this stage, this is a task that a population of primary afferents could achieve, reaching temporal hyperacuity by the coherent pooling of a large number—in the fly maybe 50 to 100—of afferent neurons (Mason et al. 2001). In this respect, the influence of stimulus amplitude on spiking jitter bears some importance.

Although it would be expected, it is unclear if and to what exact degree the spiking jitter of a receptor neuron (and hence that of the afferent population) increases as stimulus amplitude decreases (Fig. 2.4A). Such dependence could also contribute to the putative capacity for hyperacute coincidence detection of first-order binaural auditory interneurons. Critical temporal coincidence may be reached earlier and with a higher probability for the auditory channel experiencing more intense mechanical vibrations. In *O. ochracea* the primary afferents project exclusively ipsilaterally, and in the three, fused thoracic neuromeres (Oshinsky and Hoy 2002). To further evaluate the enticing possibility of coincidence detection in an insect auditory system, the neuroanatomy of the first-order interneurons, and their temporally hyperacute physiological capacity to integrate afferent signaling, need to be further studied.

6. Insect Psychophysics and Auditory Space Perception

6.1 Psychoacoustics and the Third Dimension

Until recently, studies of directional hearing in insects were mostly concerned with directional cues in the azimuthal plane. The reception of sound at the level of the tympanal membranes, the extraction of directional cues from the sound field, and their neural coding in the elevational plane have not been given as much attention. This is perhaps a possible consequence of the tacit assumption that insect ears may not be up to the task. Compared to vertebrates and mammals in particular, insects dedicate fewer mechanoreceptor neurons (with the exception of mosquitoes) and seem to be endowed with a relatively limited capacity for signal analysis in the frequency domain (see review by Pollack 1998). In animals with bilaterally symmetrical ears, the primary cues in the elevational plane reside in the binaural (or even monaural) comparison of fine spectral characteristics of the incoming sound and their comparison over time (Middlebrooks and Green 1991, Wightman and Kistler 1997; Kulkarni and Colburn 1998). For insects, the task may be regarded as computationally demanding and hence challenging. But again, as stimulus variables and processing mechanisms may be entirely different in insects, the task is by no means impossible.

In passing, it is worth noting that acoustic events occurring at a longer temporal scale are also relevant to the sense of directional hearing in insects. For instance, crickets in simulated tethered flight show a distinct sensitivity to the precedence effect, a capacity for echo suppression that may enhance directional sound detection in some situations (Wyttenbach and Hoy 1993). In crickets again, it was shown that auditory receptor neurons are liable to habituation during long bouts of simulation (8 to 10 s) in an intensity-dependent manner. This habituation process can, surprisingly, reverse the sign of the interaural difference that results from conventional auditory processing (Givois and Pollack 2000).

Some elegant experiments have shown, in effect, that Polynesian field crickets (*Teleogryllus oceanicus*) can detect and discriminate between sounds delivered at different elevational angles (Wyttenbach and Hoy 1997). In that study, the minimum audible angle (MAA) was taken as a measure of spatial auditory acuity. As a standard descriptor in the field, MAA was defined as the smallest angular separation at which two sounds are perceived as coming from two distinct sources (Fay 1988). Remarkably, the experimental paradigm of choice to test discrimination in crickets was that of habituation–dishabituation. When presented with pulses of ultrasound mimicking echolocating bats, crickets initiate steering maneuvers that are part of a startle/avoidance behavior (Fig. 2.5A) (Moiseff et al 1978). This behavior is liable to habituation; the response amplitude in effect decreases with stimulus repetition (Fig. 2.5B, stimuli 1 to 5). Complying with criteria of habituation, the response decreases exponentially at a rate dependent on stimulus amplitude and repetition rate, can recover spontaneously, and with the presentation of a novel stimulus (Fig. 2.5B, stimuli T and P). Experiments required the cricket to habituate to a series of ultrasound pulses from a particular loudspeaker location, and then recorded whether the test (T) stimulus—a single ultrasonic pulse from another location, or with any other acoustical attributes—could cause dishabituation (Wyttenbach et al 1996; Wyttenbach and Hoy 1997). Importantly, dishabituation as such was measured as the response to a probe pulse (P) identical to the habituating pulses (Fig.

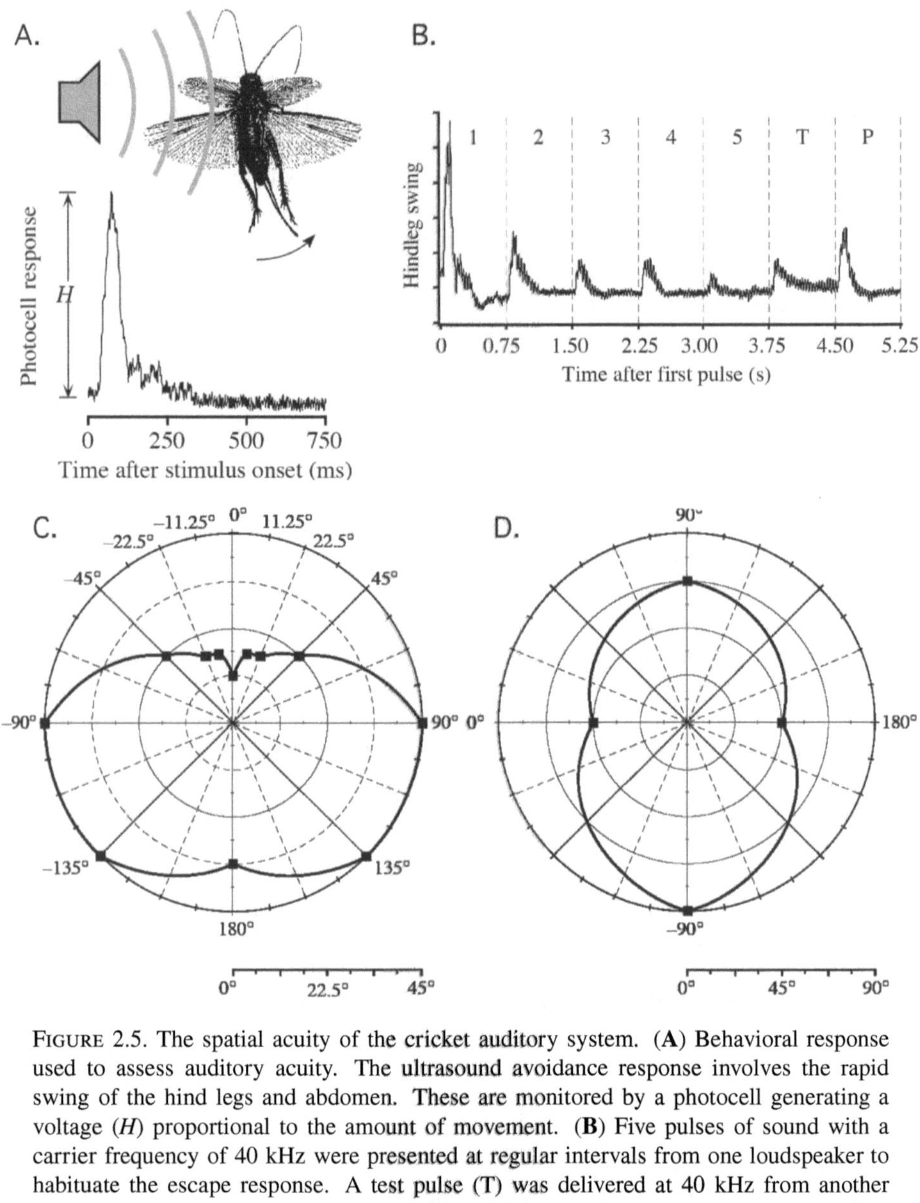

FIGURE 2.5. The spatial acuity of the cricket auditory system. (**A**) Behavioral response used to assess auditory acuity. The ultrasound avoidance response involves the rapid swing of the hind legs and abdomen. These are monitored by a photocell generating a voltage (*H*) proportional to the amount of movement. (**B**) Five pulses of sound with a carrier frequency of 40 kHz were presented at regular intervals from one loudspeaker to habituate the escape response. A test pulse (T) was delivered at 40 kHz from another loudspeaker, followed by a probe pulse (P), identical to pulse 1, from the initial loudspeaker. The minimum audible angle (MAA) for a position was defined as the smallest angular separation of the loudspeaker that would evoke dishabituation. (**C**, **D**) Polar diagram displaying MAA, shown as the distance from the center of the diagram. (**C**) Acuity in azimuth, is best (11.25°) around 0° and worst (45°) at 90° and 135°. Data on the *left* and *right* sides of this graph are mirror images. (**D**) Acuity in elevation. Acuity is best (45°) in the front and rear and worst (90°). (Modified from Wyttenbach and Hoy 1997 © Company of Biologists Ltd., with permission.)

2.5B). The use of the probe stimulus establishes whether the dishabituating pulse (T) is perceived as different from the habituating pulse, although it may not elicit any behavioral response. Hence there is a need for a probe pulse to uncover the presence of dishabituation. Using such a procedure, the minimum angular separation of the acoustic stimulus source required for dishabituation to occur was interpreted as the MAA. Quantitatively, the dishabituation index was calculated by taking the difference between the response magnitudes to the probe pulse and the last habituating pulse, and by dividing this difference by the response magnitude to the first pulse in the habituating series. In the plane of azimuth, experiments with flying tethered crickets yielded MAAs of some 11° in front of the animal and some 45° to the side (Fig. 2.5C). In the elevational plane, dishabituation took place when the sound source would be displaced by 45° in front or rear of the animal (Fig. 2.5D). This indicates that the animal can differently perceive sound stimuli broadcast from different elevations. However, changes in the position of the sound source, or switching sound sources may introduce some experimental confounding factors, such as changes in stimulus intensity, that require appropriate controls. In the present case, changes in stimulus intensity were ineffective in the range tested (Wyttenbach and Hoy 1997). In the plane of azimuth, these results concur with an earlier study, which found that a loudspeaker deviation of 10° in front of a tethered cricket did not elicit any behavioral response, while larger angles did (Pollack and Plourde 1982).

The habituation–dishabituation experimental paradigm is particularly useful because it allows a quantification of sensory acuity that choice experiments do not provide (Dooling and Brown 1990; Wyttenbach and Hoy 1997). To date, the psychoacoustical approach in all its diversity and power has been underexploited in the study of insect sensory biology (Wyttenbach and Farris 2004); it is quite likely that many insect species and other modalities are amenable to such tests.

Directional hearing in the elevation plane in field crickets makes sense in a sensory ecological context; this is also during flight at dusk and dawn that female crickets are to localize males calling from the ground. Multiple biomechanical, behavioral, and electrophysiological evidences exist that different cricket species are directionally sensitive and can perform some form of frequency analysis of incoming sound waves (Hill and Boyan 1977; Pollack and Plourde 1982; Michelsen et al. 1994). Because of the relative purity of their calling songs, field crickets may well rely on different, or unusual, stimulus variables that are perhaps related to the frequency domain and/or the multiple inputs to their auditory system, but that have thus far eluded experimental testing. Yet, the question of whether crickets, or any other insects (Pollack and Imaizumi 1999) and notably cicadas (Fonseca et al 2000), can use spectral cues for the purpose of directionality remains unanswered.

6.2 Directional Hearing and Range Detection

The detection of sound in the three-dimensional space is also related to the capacity to evaluate the distance of a sound source (Moore 1997). Like directional hearing, range detection may not be a necessity but it could constitute an adaptive advantage to the organisms endowed with such capacity. Would insects—or some of them—be capable of acoustic range detection?

Again, attention turns to a nocturnal insect that performs phonotaxis: the tachinid fly *O. ochracea*. As a parasitoid, the female must find a suitable host for her eggs of larvae. Using her prothoracic ears, the fly finds her host in the dark, homing in on the host's calling song (Cade 1975). As the phonotactic fly flies at some height above the ground (some 1 to 2 m) (D. Robert, personal observation), and the cricket sings on the ground, the task of acoustical localization may well be a problem to solve in the three-dimensional space. The fly's task seems reminiscent of the behavior of the barn owl (Knudsen and Konishi 1979). To address the question of acoustic orientation in the dark, the three-dimensional flight trajectories of *O. ochracea* were quantified as the fly was induced to home in on the source of a cricket song placed on the ground (Müller and Robert 2001). The phonotactic flight paths were recorded in three dimensions using a stereo infrared video tracking system (Fry et al 2000). This system also allowed for controlling the delivery of sound stimuli as a function of the fly's position in space (Fig. 2.6A) (Fry et al 2004). As the phonotactic behavior is performed in the dark, it was thus possible to assess the free-flight fly's reaction to alterations in acoustic stimulation taking place at predetermined and replicable times and spaces in the flight arena. In particular, the control software of the tracking system was designed to incorporate a virtual representation of the experimental arena *in silico*. In this representation, diverse volumetric objects (such as a sphere; Fig. 2.6A) could be defined and be assigned a logical function analogous to that of a conventional light barrier. Experimental conditions could thus be programmed to change automatically and online as the animal's trajectory (its X, Y, Z coordinates in the virtual representation) would mathematically intersect the description of the virtual object. Local experimental conditions could thus be controlled without physically cluttering the real flight and acoustic environment in order to test the animal's reactions to novel stimuli, or the absence of them.

Tracking experiments testing the phonotactic capacity of the fly in complete darkness show that, interestingly, flies do not take the shortest path between the starting platform and the sound source. Flies do not fly a beeline to the cricket (Fig. 2.6B). Rather, trajectories comprise three distinct phases: a brief takeoff phase; a cruising phase in which course and altitude remain quite constant; and finally a terminal, landing phase. Taking place as the fly is nearer but still above the sound source, this terminal approach is characterized by a steep spiraling descent. The accuracy of the flies' phonotactic behavior is remarkable: at the end of a flight bout of about 4 m, they landed at a mean distance of 8.2 cm

(SD ± 0.6 cm, N = 80 landings) from the center of the loudspeaker. One particular, and simple, experiment brought to light some unsuspected and intriguing characteristics of this fly's phonotactic capacity. As the fly was on its way to the loudspeaker, the acoustic stimulus was interrupted, thus removing the only navigational cue available. Surprisingly, the phonotactic behavior was not drastically affected, or disrupted, by the interruption of the acoustic stimulus (Fig. 2.6C). In effect, irrespective of her position in the flight room at the time of stimulus interruption, the fly initiates the descent maneuver (spiraling drop) at the appropriate time and location, not far above the loudspeaker. This results in a landing close to the now silent loudspeaker. Since other possible navigational cues are absent (visual and/or olfactory), these experiments suggest that, at the time of stimulus interruption, the fly had acquired sufficient acoustic information to navigate accurately to the sound source. It must be noted here that flies can localize the sound source without prior experience and also display no improvement (through learning) in their phonotactic abilities (Müller and Robert 2001).

Depending on their position in the flight arena, the free-flying flies respond in different ways to stimulus interruption. Most remarkably, stimulus interruption taking place whilst the fly is far away (e.g., 1.8 m) from the loudspeaker does not prevent the fly from landing close to it (Fig. 2.7). To achieve this, the fly maintains the same flight course and only initiates the spiraling descent at the appropriate time and place, landing relatively close to the loudspeaker (Fig. 2.6C). By contrast, if stimulus interruption takes place when the fly is close to target (0.6 m or less), she initiates her spiraling descent at a shorter time delay after stimulus interruption. Thus while the time of stimulus interruption is no predictor of the onset of the landing response, the fly's position relative to the loudspeaker is. Notably, these landing maneuvers are initiated at a time when sound cues are completely absent and thus seem to result from autonomous decisions. It could also be seen that, probably because of noise in the sensory and motors systems, the earlier the stimulus is interrupted, the less accurate the landing becomes (Fig. 2.7). It could thus be shown that the cessation of the acoustic stimulus, by itself, does not elicit the spiraling trajectory indicative of landing. From this it can be concluded that the fly must gather sufficient information about the spatial position of the sound source before stimulus cessation. Although it seems to rely on some form of motor memory instructed by sensory inputs, this behavior is notably different from idiothetic orientation, known of spiders and fossorial mammals, in that the fly has never been to the location of the loudspeaker before. The nature of the information gathered by the fly prior to stimulus interruption remains elusive to date.

Other experiments were conducted to test the fly's capacity to localize a sound source in midair (Fig. 2.8) (Müller and Robert, unpublished results). In a control situation, the fly was first attracted to a loudspeaker situated on the ground (ZS). In a second trial, the fly was asked to repeat the performance; as she did so, she entered the space of a virtual sphere (gray sphere; Fig. 2.8) that served as

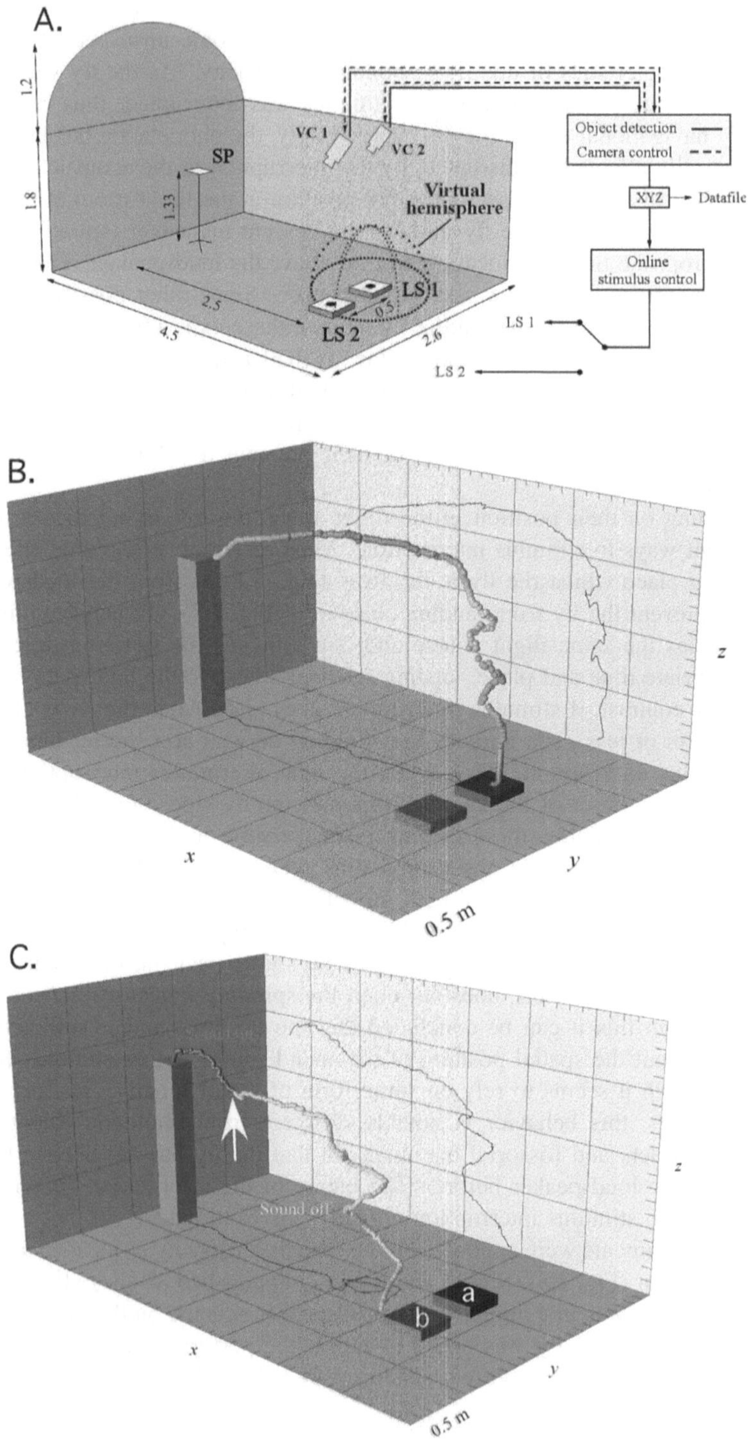
A.
1.2
1.8
SP
1.33
VC 1
VC 2
Virtual hemisphere
Object detection
Camera control
XYZ
Datafile
Online stimulus control
LS 1
LS 2
2.5
4.5
0.5
2.6
B.
x
y
z
0.5 m
C.
Sound off
a
b
x
y
z
0.5 m

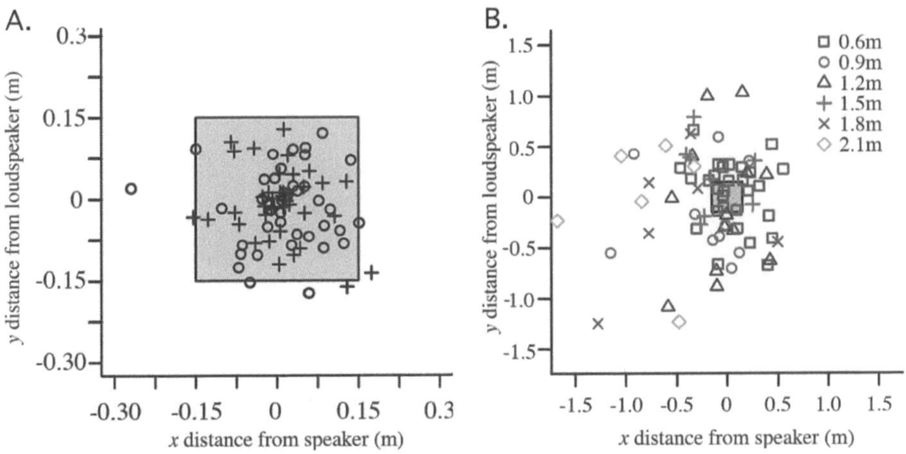

FIGURE 2.7. Landing accuracy of the fly in response of a continuous and an interrupted sound stimulus. (**A**) Continuous stimulus simulating a cricket song. Landings to the right (+) and left (o) loudspeaker are pooled. *Shaded area* is surface area of the loudspeaker box. After a flight approximately 3 m long, flies land within 8 cm of the center of the loudspeaker. (**B**) Accuracy of phonotaxis as a function of the distance of stimulus interruption. Symbols indicate the distance from the sound at which the flies were at the time of stimulus interruption. Remarkably the flies succeed at finding the sound source, even without sound cues. Accuracy decreases with increased distance. (Data modified from Müller and Robert 2001.)

FIGURE 2.6. Directional hearing as a behavior in view of three-dimensional sound localization and range finding. (**A**) Setup used for studying free-flight phonotaxis in insects. SP, Starting platform, VC1, 2: infrared computer-controlled pan-tilt video cameras. LS1, 2: Loudspeakers. The video signals of each camera are processed frame by frame as a stereo pair to determine the position of the fly on each frame, and used to instruct the tracking motions of both cameras. The flight path is computed in Cartesian coordinates (50 Hz sampling frequency) to yield the *X*, *Y*, *Z* coordinates of the animal's trajectory. This data are also used to control alterations of experimental conditions online (see Fry et al. 2004). (**B**) Phonotactic trajectory of the fly *O. ochracea* toward a loudspeaker broadcasting a cricket song. Under infrared darkness, the fly navigates to land on the active loudspeaker, depicted by (*a*) in diagram (**C**). (**C**) Experiment in which the same fly is lured toward the other loudspeaker (*b*) for a short time, until the sound stimulus is interrupted (*arrow*). In the absence of acoustic, visual or olfactory stimuli, the fly navigates to the originally active sound source.

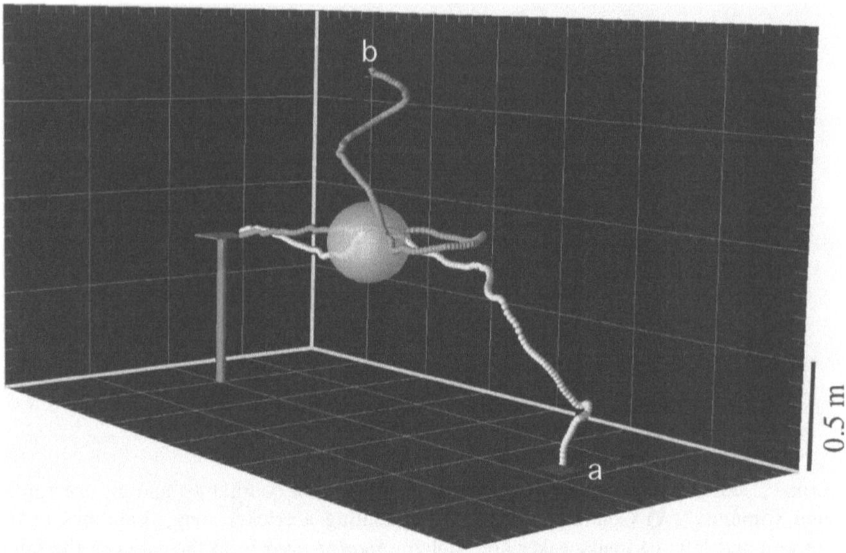

FIGURE 2.8. Orientation in the three-dimensional space in the fly *O. ochracea*. In as Fig. 2.6A and B, the fly is induced to perform a bout of phonotactically oriented flight toward a loudspeaker situated on the ground (*a*). This flight path is registered (trajectory *a*) and a virtual sphere is positioned so that it intersects that trajectory. The fly is brought back to the platform, loudspeaker *a* is activated again. The fly commences her phonotactic flight toward loudspeaker *a*, and as it enters the virtual sphere, sound is switched over to loudspeaker *b* situated directly above the sphere. Trajectory *b* indicates that the fly carries on to fly for a little while (several tens of milliseconds) before engaging in a level turn and then a spiraling ascent reaching loudspeaker *b*. Although this fly species parasitizes field crickets that sing on the ground only, the fly's auditory system is capable of finding a sound source in the three-dimensional space.

the trigger for the sound stimulus to be switched over to a target sound source (TS) straight above the sphere. The fly proved capable of pinpointing the source of sound, and of reaching it by spiraling vertically toward it.

These results seem to suggest the presence of a mechanism of acoustic detection that enables the fly to gauge both the direction (perhaps a three-dimensional vectorial representation?) and the distance of a sound source in three dimensions. Such capacity is reminiscent of the barn owl's capacity of localizing acoustic targets in both azimuth and elevation (Knudsen and Konishi 1979). Unlike the owl, the fly's ears are bilaterally symmetrical; a symmetry that was shown earlier to be a prerequisite for the mechanism used for directional hearing (Fig. 2.1E; Robert et al 1996). To date, the mechanisms supporting the proposed ability of the fly to perform three-dimensional audition and evaluate distance to target remain unknown, but certainly deserve more attention.

Recent work has shown that a conjunction of both psychophysical and sensory ecological approaches can provide precious information, and sometimes reveal the unique and sophisticated mechanisms, and unsuspected capacities, by which insects sense their world as they rapidly pass through it (Srinivasan 1998).

7. Some Outstanding Issues and Prospects

Small size may well have imposed severe constraints on the mechanics of auditory receivers, but also obviously on the number of nerve cells behind them. The principles of economy employed by insects do not necessarily signify simple or crude modes of operation or reduced processing power. Rather, principles of economy can imply the implementation of alternative, possibly cunning and efficient, mechanisms that are exquisitely adapted to the task. This point has been repeatedly illustrated in insect sensory research, be it for vision, audition, olfaction, or the lesser-known modalities of thermoreception or magnetoreception. In audition, a well-known model for accurate information processing in the time domain is the Webster–Jeffress cross-correlation model for binaural processing, requiring a complex array of neurons (Jeffress 1948). It would indeed be interesting to see whether this model finds its counterpart in insect hearing, and if so, which particular form it may take at the level of interneurons. One alternative seems to involve a neuronal network that relies on the intrinsic integration properties of coincidence detector, rather than a series of delay lines temporally tuned by the differential length of transmission lines. As the question remains open, it may be useful to consider that insects have generally come up with solutions that are economical in evolutionary terms and that are more often than not, computationally undemanding but efficient.

The capacity of insects to perform some sort of auditory scene analysis, allowing them to situate themselves in space and time within their acoustic environment deserves more attention. With this respect, adapted psychophysical experimental paradigms and techniques of behavioral monitoring in unrestrained animals may be very applicable to insect systems to address testable hypotheses on the complex issues of mechanical and neural sound processing in frequency, time, and space. These studies could even be conducted in conjunction with extracellular, and intracellular electrophysiology on primary auditory afferents as well as identified interneurons. Hypotheses may address the enticing possibility that some insects can extract three-dimensional information using two symmetrical ears only. In particular, the capacity for auditory space perception—especially considering species other than *O. ochracea*—and the type of coding involved (owl-like, or else) may be particularly interesting at both fundamental and comparative levels.

Finally, one corollary and emergent outcome of insect hearing research is the extraction of operation principles for bioinspired acoustic detection technology. In the course of their evolution, insects have acquired the capacity to do small things very well. In due course, it may become a reality to see insect-inspired microsensors equip microrobots. Indeed, fly-inspired microsensors are currently under development that seek to emulate the key characteristics of their natural analogs, such as miniaturization, accuracy, and economy of operation.

Future studies will carry on revealing the fundamental mechanisms arthropods in general, and chiefly insects, spiders, and crustaceans, employ to sense vibrations directionally, in pressure or velocity. The diversity of mechanisms used to detect sound directionally may still be much richer than presently known. Some 36 years ago, David Pye depicted insect audition (Pye 1968) in a series of verses. In a recent update to his long-lived prose, commenting on diversity, he concluded: "This list is long, the contrasts strong, and may go on for ever. And so we end with no clear trend—For Nature is so clever" (Pye 2004). Indeed, insect research contributes to enrich our knowledge of sensory systems, but also continues to impress upon us their magnificent efficiency and ever surprising diversity.

References

Autrum H (1940) Über die Lautäusserung und Schallwahrnehmung bei Arthropoden. II. Das Richtungshören von *Locusta* und Versuch einer Hörtheorie für Tympanalorgane vom Locustidentyp. Z Vergl Physiol 28:326–352.

Belton P (1974) An analysis of direction finding in male mosquitoes. In: Browne LB (ed), Analysis of Insect Behaviour. Berlin: Springer-Verlag, pp. 139–148.

Bennet-Clark HC (1998) Size and scale effects as constraints in insect sound communication. Philos Trans R Soc Lond B 353:407–419.

Bialek W (1987) Physical limits to sensation and perception. Annu Rev Biophys Chem 16:455–478.

Cade WH (1975) Acoustically orienting parasitoids: fly phonotaxis to cricket song. Science 190:1312–1313.

de Ruyter van Steveninck RR, Lewen GD, Strong S, Koberle R, Bialek W (1997) Reproducibility and variability in neural spike trains. Science 275:1805–1808.

De Vries H (1948) Brownian movement and hearing. Physica 14:48–60.

Doherty JA (1991) Sound recognition and localization in the phonotaxis behavior of the field cricket, *Gryllus bimaculatus* (Orthoptera: Gryllidae). J Comp Physiol 168:213–222.

Dooling RJ, Brown SD (1990) Speech perception by budgerigars (*Melopsittacus undulatus*): spoken vowels. Percept Psychophys 47:568–574.

Edds-Walton PL, Fay RR, Highstein SM (1999) Dendritic arbors and central projections of auditory fibers from the saccule of the toadfish (*Opsanus tau*). J Comp Neurol 411: 212–238.

Fay RR (1988) Hearing in Vertebrates. A Psychophysics Databook. Winnetka, Il: Hill-Fay Associates.

Fonseca PJ, Münch D, Hennig RM (2000) How cicadas interpret acoustic signals. Nature 405:297–298.

Fry SN, Bichsel M, Müller P, Robert D (2000) Three-dimensional tracking of flying insects using pan-tilt cameras. J Neurosci Methods 101:59–67.

Fry SN, Müller P, Baumann HJ, Straw AD, Bichsel M, Robert D. (2004). Context-dependent stimulus presentation to freely moving animals in 3D. J Neurosci Methods 135:149–157.

Fullard JH, Yack JE (1993) The evolutionary biology of insect hearing. Trends Ecol Evol 8:248–252.

Givois V, Pollack GS (2000) Sensory habituation of auditory receptor neurons: implications for sound localization. J Exp Biol 203:2529–2537.

Gnatzy W, Tautz J (1980) Ultrastructure and mechanical properties of an insect mechanoreceptor: stimulus-transmitting structures and sensory apparatus of the cercal filiform hairs of *Gryllus*. Cell Tissue Res 213:441–463.

Göpfert MC, Robert D (2001) Nanometre-range acoustic sensitivity in male and female mosquitoes. Proc R Soc Lond B 267:453–457.

Hennig M, Franz A, Stumpner A (2004) Processing of auditory information in insects. Micr Res Tech 63:351–374.

Hill KG, Boyan GS (1977) Sensitivity to frequency and direction of sound in the auditory system of crickets (Gryllidae). J Comp Physiol 121:79–97.

Hoy RR (1998) Acute as a bug's ear: an informal discussion of hearing in insects In: Hoy RR, Popper AN, Fay RR (eds), Comparative Hearing: Insects. New York: Springer-Verlag, pp. 1–17.

Hoy RR, Robert D (1996) Tympanal hearing in insects. Annu Rev Entomol 41:433–450.

Hudspeth AJ (1997) Mechanical amplification by hair cells. Curr Opin Neurobiol 7:480–468.

Jeffress LA (1948) A place theory of sound localization. J Comp Physiol Psych 41:35–39.

Johnson C (1855) Auditory apparatus of the *Culex* mosquito. Q J Microsc Sci 3:97–102.

Knudsen EI, Konishi M (1979) Mechanisms of sound localisation in the barn owl (*Tyto alba*). J Comp Physiol 133:13–21.

Kulkarni A, Colburn HS (1998) Role of spectral detail in sound-source localization. Nature 396:747–749.

Lakes-Harlan R, Heller K-G (1992) Ultrasound sensitive ears in a parasitoid fly. Naturwissenschaften 79:224–226.

Lewis B (1983) Directional cues for auditory localisation. In: Lewis B (ed), Bioacoustics, a Comparative Approach. London: Academic Press, pp. 233–257.

Löhe G, Kleindienst HU (1994) The role of the medial spetum in the acoustic trachea of the cricket *Gryllus bimaculatus*. II. Influence on directionality of the auditory system. J Comp Physiol 174:601–606.

Mason AC, Oshinsky ML, Hoy RR (2001) Hyperacute directional hearing in a microscale auditory system. Nature 410:686–690.

Michelsen A (1992) Hearing and sound communication in small animals: evolutionary adaptations to the laws of physics. In: Webster DM, Fay RR, Popper AN (eds), The Evolutionary Biology of Hearing. New York: Springer-Verlag, pp. 61–78.

Michelsen A (1996) Directional hearing in crickets and other small animals. In: Schildberger K, Elsener N (eds), Neural Basis of Behavioural Adaptations. Stuttgart and New York: Fischer, pp. 195–207.

Michelsen A (1998) Biophysics of sound localization in insects. In: Hoy RR, Popper AN, Fay RR (eds), Comparative Hearing: Insects. New York: Springer-Verlag, pp. 18–62.

Michelsen A, Löhe G (1995) Tuned directionality in cricket ears. Nature 375:639.

Michelsen A, Rohrseitz K (1995). Directional sound processing and interaural sound transmission in a small and a large grasshopper. J Exp Biol 198:1817–1827.

Michelsen A, Popov AV, Lewis B (1994) Physics of directional hearing in the cricket *Gryllus bimaculatus*. J Comp Physiol 175:153–164.
Middlebrooks JC, Green DM (1991) Sound localization by human listeners. Ann. Rev Psychol 42:135–59.
Miles RN, Robert D and Hoy RR (1995) Mechanically coupled ears for directional hearing in the parasitoid fly *O. ochracea*. J Acoust Soc Am 98:3059–3070.
Miller LA (1977) Directional hearing in the locust *Schistocerca gregaria* Forskål (Acrididae, Orthoptera). J Comp Physiol 119:85–98.
Moiseff A, Konishi M (1981) Neuronal and behavioral sensitivity to binaural time differences in the owl. J Neurosci 1:40–48.
Moiseff A, Pollack GS, Hoy RR (1978) Steering response of flying crickets to sound and ultrasound: mate attraction and predator avoidance. Proc Nat Acad Sci USA 75: 4052–4056.
Moore BCJ (1997) An Introduction to the Psychology of Hearing. London: Academic Press, pp. 213–231.
Morse PM, Ingard KU (1968) Theoretical Acoustics. New York: McGraw-Hill, pp. 412–422.
Müller P, Robert D (2001) A shot in the dark: the silent quest of a free-flying phonotactic fly. J Exp Biol 204:1039–1052.
Oshinsky ML, Hoy RR (2002) Physiology of the auditory afferents in an acoustic parasitoid fly. J Neurosci 22:7254–7263.
Payne R, Roeder KD, Wallman L (1966) Directional sensitivity of the ears of noctuid moths. J Exp Biol 44:17–31.
Pollack GS (1998) Neural processing of acoustic signals. In: Hoy RR, Popper AN, Fay RR (eds), Comparative Hearing: Insects. New York: Springer-Verlag, pp. 139–196.
Pollack GS, Imaizumi K(1999) Neural analysis of sound frequency in insects. Bioessays 21:295–303.
Pollack GS, Plourde N (1982) Directionality of acoustic orientation in flying crickets. J Comp Physiol 146:207–215.
Pye JD (1968) How insects hear. Nature 218:797.
Pye JD (2004) On the variety of auditory organs in insects. Microsc Res Tech 63:313–314.
Robert D (1989) The auditory behaviour of flying locusts. J Exp Biol 147:279–302.
Robert D, Göpfert MC (2002) Novel schemes for hearing and acoustic orientation in insects. Curr. Opin. Neurobiol 12:715–720.
Robert D, Göpfert MC (2004) Introduction to the biology of insect hearing: diversity of forms and functions. Microsc Res Tech 63:311–312.
Robert D, Hoy RR (1998) The evolutionary innovation of tympanal hearing in Diptera. In: Hoy RR, Popper AN, Fay RR (eds), Comparative Hearing: Insects. New York: Springer-Verlag. pp. 197–227.
Robert D, Amoroso J, Hoy RR (1992) The evolutionary convergence of hearing in a parasitoid fly and its cricket host. Science 258:1135–1137.
Robert D, Miles RN, Hoy RR (1996) Directional hearing by mechanical coupling in the parasitoid fly *Ormia ochracea*. J Comp Physiol 179:29–44.
Robert D, Miles RN, Hoy RR (1998) Tympanal mechanics in the parasitoid fly *Ormia ochracea*: intertympanal coupling during mechanical vibration. J Comp Physiol 183: 443–452.
Robert D, Miles RN, Hoy RR (1999) Tympanal hearing in the sarcophagid parasitoid fly *Emblemasoma* sp: the biomechanics of directional hearing. J Exp Biol 202:1865–1876.

Rose G, Heiligenberg W (1985) Temporal hyperacuity in electric fish. Nature 318:178–180.

Schiolten P, Larsen ON, Michelsen A (1981) Mechanical time resolution in some insect ears. J Comp Physiol 143:289–295.

Schmitz B, Scharstein H, Wendler G (1983) Phonotaxis in *Gryllus campestris* L. (Orthoptera, Gryllidae). II. Acoustic orientation of female crickets after occlusion of single sound entrances. J Comp Physiol 152:257–264.

Schul J, Holderied M, von Helversen D, von Helversen O (1999) Directional hearing in grashoppers: neurophysiological testing of a bioacoustic model. J Exp Biol 202:121–133.

Srinivasan MV (1998) Insects as Gibsonian animals. Ecol Psychol 10:251–270.

Tautz J (1977) Reception of medium vibration by thoracal hairs of the caterpillar of *Barathra brassicae* L. (Lepidoptera, Noctuidae). I. Mechanical properties of the receptor hairs. J Comp Physiol 118:13–31.

von Helversen D, Rheinlander J (1988) Interaural intensity and time discrimination in an unrestrained grasshopper: a tentative behavioural approach. J Comp Physiol 162: 330–340.

Weber T, Thorson J (1989) Phonotactic behavior of walking crickets. In: Huber F, Moore T, Loher W (eds), Cricket Behavior and Neurobiology. Ithaca, NY: Cornell University Press, pp. 310–339.

Wightman FL, Kistler DJ (1997) Monaural sound localization revisited. J Acoust Soc Am 101:1050–1063.

Wyttenbach RA, Farris HE (2004) Psychophysics of insect hearing. Microsc Res Tech 63:375–387.

Wyttenbach RA, Hoy RR (1993) Demonstration of the precedence effect in an insect. J Acoust Soc Am 94:777–784.

Wyttenbach RA, Hoy RR (1997) Acuity of directional hearing in the cricket. J Exp Biol 200:1999–2006.

Wyttenbach RA, May ML, Hoy RR (1996) Categorical perception of sound frequency by crickets. Science 273:1542–1544.

Yack JE (2004) The structure and function of auditory chordotonal organs in insects. Microsc Res Tech 63:315–337.

Yack JE Fullard JH (1993) What is an insect ear? Ann Entomol Soc 86:677–682.

Yager DD (1999) Structure, development and evolution of insect auditory systems. Microsc Res Tech 47:380–400.

3

Sound Source Localization by Fishes

RICHARD R. FAY

1. Introduction

General understanding of the sense of hearing in nonhuman species probably arises from human experiences, introspections, and experiments on human listeners. Whenever we hear a sound, a little attention to it usually reveals that its source exists somewhere in the space around us, and our ability to point toward the source is reasonably good in most simple environments. The locations of multiple, simultaneous sources form a sort of spatial image perceptually, an "auditory scene," that is analogous to the visual scene (Bregman 1990). We then can remember where the source was and we are able to behave appropriately. Of course, we can be fooled; sounds don't always emanate from the perceived source location. But we are correct most of the time and tend not to make severe errors in simple acoustic environments. We can even often judge the degree to which the perceived source location is likely to coincide with the actual location (e.g., when hearing a fire siren in the car with the windows up).

The intuitive understanding of sound source localization permits us to be confident in psychoacoustic measures of localization performance, such as measuring the minimum audible angle (MAA). We can listen to two sounds emanating from different locations and judge them to be "different" if their angular separation is greater than the MAA. We have little doubt that the difference we hear is in a spatial dimension and that the perceived difference is in spatial location and not in some other qualitative dimension such as timbre. The localization abilities of humans and many terrestrial vertebrates (and some invertebrates) are derived from neural computations using the differences or correlations between the neural representations arriving at the brain from the two ears. Many terrestrial species (and some invertebrates) behave appropriately with respect to actual sound sources. They approach sound sources along substantially direct pathways and are able to discriminate between different source locations with reasonable accuracy, as some of the chapters in this volume illustrate. Our everyday experiences with pets can provide convincing confirmation that what we mean by sound source localization can be applied as well to a dog, for example, as a human. We have measured MAAs in various species (e.g., Heffner and Heffner 1982) and confidently interpret the results as an aspect of auditory spatial acuity.

Of course, it is also quite clear that binaural processing is subject to several essential ambiguities that must be solved using other strategies. These include the fact that all sources on the median sagittal plane result in equal (zero) interaural differences, and that, in general, "cones of confusion" exist bilaterally on the surfaces of which interaural cues are equal for many possible source locations. These ambiguities seem to be solved adequately among humans and many terrestrial species that use head or pinnae movements, information from other senses, judgments of the plausibility (Rakerd and Hartmann 1985) of potential locations, and a processing of the head-related transfer function (HRTF) through frequency analysis (e.g., Wightman and Kistler 1993).

How confident can we be that the human notion of sound source localization can be applied to fishes? Many of the experiments and theories of sound source localization in fishes assume that, as with a pet dog or canary, this putative capacity is usefully defined for fishes as it has been defined for humans. But because of the contradictory and confusing history of observations in this field, I suggest that this assumption may not be appropriate. For example, a cod can be conditioned to respond whenever a signal sent to one loudspeaker is switched to another loudspeaker in a slightly different location (e.g., Chapman and Johnstone 1974), and find an MAA of 15° azimuth. Our interpretation has been that the cod is capable of sound source localization in the way it is understood for humans described above, but that its acuity is poor compared with our own (in the vicinity of 1° in azimuth for humans). Thus we may believe that the cod knows roughly where the two sound sources are and could be called on to behave appropriately with respect to their locations in space (e.g., to approach them or avoid them). But this sort of confirmatory experiment has rarely, if ever, been carried out in fishes. It makes great intuitive and biological sense that the ability to behave appropriately with respect to the locations of sound sources and acoustic events may be one of the most important and adaptive capabilities of auditory systems. But is this assumption useful? Should we require that this type of behavioral experiment be done with fishes, even though it apparently hasn't been required in the study of many other nonhuman animals?

This chapter evaluates the empirical and theoretical literature on sound source localization in fishes and concludes with a call for further fundamental experimentats. The earliest experimenters failed to demonstrate sound source localization in the European minnow (*Phoxinus laevis*), and they went on to explain this failure in a convincing way (von Frisch and Dijkgraaf 1935). Later experimenters assumed a localization capacity, and went on to develop rather complex theories that could explain their MAA measurements (e.g., Schuijf 1975; Schuijf and Buwalda 1975; Schellart and De Munck 1987). Recently, numerous physiological experiments have been carried out to understand better how information about a source's location is represented in the nervous system of a fish (e.g., Fay 1984; Lu et al. 1998; Edds-Walton et al. 1999; Weeg et al. 2002; Edds-

Walton and Fay 2003). Some directional information is so robustly encoded in the fish's ears, auditory nerves, and brain (e.g., Wubbles and Schellart 1997; Fay and Edds-Walton 1999) that it is difficult to believe that fishes cannot use these representations for directional hearing.

It is worth questioning whether fishes determine sound source location by acquiring and processing information that could be used at a later time to direct intentional behavior, or whether they may only be able to react to sound signals immediately in some directional manner. The literature on the Mauthner cell (M-cell) system (the reticulospinal system that contributes to very fast, sound-induced escape responses in many species) considers the question of sound source localization only from this latter point of view (e.g., Canfield and Eaton 1990). Is there more to localization than this? One possibility is that fishes have solved the computational problems of source localization at least twice, once at the reticulospinal level (e.g., Eaton et al. 1995) and again at the level of the ascending auditory system, which could contribute to more subtle or intentional behaviors. It seems that conditioning studies demonstrating that fish can discriminate between sources based only on their location (reviewed below) strongly suggest that there is more to localization than immediate, reflex behaviors.

In his analysis of the structures of acoustic near fields, Kalmijn (1997) explains sound source localization as a series of subtle responses based on immediate sensory information. Is sound source localization in fishes more than a series of immediate behaviors that combine to produce a successful approach? Our views of sound source localization abilities and localization mechanisms in fishes will be determined by data from behavioral, neuroanatomical, and physiological experiments. In this spirit, the major quantitative behavioral experiments on this topic are reviewed below. In the end, we shall see that the nature of sound source localization in fishes remains a profound mystery.

2. Modern-Era Experiments

The earliest modern-era experiments on sound source localization in fishes were carried out on the European minnow (*Phoxinus laevis*), first by Reinhardt (1935) in a laboratory tank, and then by Karl von Frisch and Sven Dijkgraaf (1935) in a shallow lake environment. Von Frisch and Dijkgraaf introduced their paper with two arguments. First, the current dominant theory of source localization in humans was that azimuth was encoded and represented by minute interaural time differences (ITD). But, they went on, this seemed "hardly imaginable" for fish because there is no external ear, the inner ears are close together, and sound travels more than four times faster in water than in air. Furthermore, they pointed out that this minnow detects sound pressure indirectly using the swim bladder, an unpaired structure that would ensure that both ears were always activated equally and simultaneously, regardless of source location. Second, they argued that the conclusion that fish should not be able to recognize the direction of a sound "displeases" the biologists. They asked, Of what use is the great auditory acuity of the fish (their minnow) if it could not recognize the direction of a sound source? Later, Pumphrey (1950) voiced a similar opinion based on thoughts on the fundamental biological significance of hearing in any organism.

Von Frisch and Dijkgraaf carried out their behavioral conditioning observations at the shore of Lake Wolfgang, Germany in water 40 cm deep. Their sound source was an automobile horn (klaxon) made waterproof and situated on the gravely bottom. The sound signal was described as having the pitch of "e1" on the musical scale and having very high intensity, although the level was not measured. They could easily hear the sound above the water, and they could feel the source's near field with the hand underwater as much as 30 cm away from the horn. Initially, two identical feeding trays were set up on the lake bottom with the horn, invisible, under one of them. The horn was activated for 2 minutes and then cut-up earthworms were dropped onto the corresponding feeding tray. These trials were repeated every 10 minutes or so. Although the authors demonstrated an effect of pairing the sound and the reinforcing worms on the fishes behavior (conditioned general activation), they did not observe the fishes to select (move toward) the feeding station over the active horn based on the sound alone, even after 55 trials over several days of training. In addition, following several changes in the number, appearance, and positioning of the feeding stations and after reducing the level of the sound stimulus, they found no good evidence that the minnows were more likely to approach the feeding station with the sound source on than the silent feeding stations. Von Frisch and Dijkgraaf concluded that these fishes cannot determine the location of a sound source and that orientation behaviors that occurred when the fish were very close to the source were probably due to sense organs other than the ears (they suggested the skin since they could feel the sound field with the hand).

3. Post-Modern-Era Experiments

The question of sound source localization in fishes was not systematically taken up again until Moulton and Dixon (1967) reported a remarkable experiment on the ear's role in the tail-flip escape response (presumed to be mediated by the M-cells) in goldfish (*Carassius auratus*). In initial experiments, Moulton and Dixon observed the tail-flip direction in fishes with respect to the location of a small underwater sound source. Consistently, the animals flipped the tail in a manner that took them away from the source. In the next experiments, they paired sound emanating from the same source with the delivery of food in the same location in a glass fish tank. After several training sessions, conditioned fish were observed to flip the tail in the opposite direction so that they tended to approach the source. Signals of 100, 150, and 1500 Hz were used and directional responses occurred with respect to the source at all frequencies. In the next experiments, conditioned and unconditioned fish were restrained in the center of a tank and the direction of the tail flip was automatically recorded. In most cases, unconditioned fish flipped the tail consistent with a movement away from the source, while previously conditioned fish flipped the tail in the opposite direction with respect to the source location. Sound sources directly in front of and behind the animals elicited approximately equal numbers of tail flips in both directions. In the final experiment, the saccular and lagenar nerves were severed uni- or bilaterally. Animals with unilateral nerve sections responded as if the source were on the side of the intact nerve. Animals with bilateral sections did not produce any normal or interpretable tail flips at all in response to the sounds. Moulton and Dixon designed these experiments so that the source-to-fish distance at 1500 Hz was greater than the extent of the near field, as defined by Harris and van Bergeijk (1962). The authors concluded that directional hearing, defined as directional orienting responses with respect to discrete sound sources, was possible in the far field and that the saccule (and/or possibly the lagena) transduced the acoustic signal eliciting the response. Furthermore, these experiments showed that this aspect of directional hearing in the horizontal plane was most likely a result of binaural processing, as is the case for most terrestrial vertebrates.

Although Moulton and Dixon assumed that the directional tail flips they observed were mediated by the Mauthner cells of the lower brainstem (Furshpan and Furukawa 1962), they did not confirm this experimentally. This issue is important because there may be neural processing mechanisms other than the M-cell circuit underlying sound source localization in fishes. The M-cells of many fishes may mediate reflex orienting responses, but are very unlikely to be responsible for any sound source localization capacities that are similar to those well known in human listeners: the information about sound source locations that can be used later for appropriate intentional behaviors. The M-cell system is a descending circuit that is not known to relay information upward in the ascending auditory pathway where it can be used to direct delayed or other nonobligatory behaviors. In other words, if fishes are capable of directional hearing of the type discussed in the introduction to this chapter, the M-cells are not likely to be involved. Thus, fishes may have at least two circuits for directional hearing: a descending one mediating reflexive or obligatory responses and an ascending one possibly mediating intentional behaviors. Moulton and Dixon's experiments demonstrated binaural directional hearing in the far field, but did not clearly distinguish between M-cell mediation or other auditory processing as its foundation. The fact that the directionality of the response could be reversed by conditioning suggests that the M-cell circuit might not be a necessary component of the response. Perhaps those experiments should be repeated and extended.

These experimental results have for decades been largely forgotten. This neglect may have arisen, in part, because this report appeared as a book chapter rather than in the peer-reviewed literature. However, the great influence that Willem van Bergeijk (1964, 1967) had on the field may have overshadowed the results of Moulton and Dixon. Taking up the initial analysis of von Frisch and

Dijkgraaf (1935), van Bergeijk argued that hearing be defined as responsiveness to a sound pressure signal and that the only reasonable mechanism for sound pressure detection (and thus hearing) was movement of the walls of a gas bubble (e.g., swim bladder) in a fluctuating pressure field. Since pressure at a point is a scalar quantity, without direction, and since the swim bladders of most fishes impinge on both ears equally, there would be little or no possibility of directional hearing for fishes. As did von Frisch and Dijkgraaf (1935) before him, van Bergeijk argued that some other sensory system must be implicated in the ability of fish to orient to sources at close range (e.g., Kleerekoper and Chagnon 1954).

Based on his "acoustico–lateralis" hypothesis and the embryological evidence known at the time, van Bergeijk suggested that this other receptor system must be the lateral line (e.g., van Bergeijk 1967). Earlier, Harris and van Bergeijk (1962) demonstrated that the lateral line neuromast organs of the killifish (*Fundulus heteroclitus*) responded in proportion to near-field particle motion generated by a nearby dipole (vibrating sphere) source. Since the lateral line system is spatially distributed over the body surface, it would be an ideal system for transducing and representing the steep gradient of near field particle motion amplitude surrounding any moving or vibrating object. In addition, since the lateral line organs appeared to van Bergeijk to derive from the same embryonic tissue (anlage) as the labyrinth, lateral line and auditory function should be thought of as parts of a single system (the so-called acoustico–lateralis system) that can compute source location within its acoustic near field. No mention was made of the M-cell circuit here, nor subsequently of the results of Moulton and Dixon (1967) showing directional responses in the far field mediated by the ears. However, in the published discussion following Moulton and Dixon's chapter (Moulton and Dixon 1967), both Dijkgraaf and van Bergeijk noted that the results pointed to the conclusion that the ear's otolithic organs could also respond to near-field particle displacements. In the same discussion, G.G. Harris noted the important point that the question in any particular case is not whether the fish was in the near or far field, but whether the receptor organs respond indirectly to sound pressure, or directly to particle motion (either displacement, velocity, or acceleration). This latter point has also been essentially ignored for several decades. It is now clear that the otolith organs of fishes are exquisitely sensitive to oscillatory motion of the head and ears (i.e., acoustic particle motion), with saccular nerve fiber sensitivities to low-frequency displacements as small as 0.1 nm, root mean square (e.g., Fay 1984; Fay and Edds-Walton 1997a). At 100 Hz, displacements of this magnitude accompany a propagating sound wave in the far field at 100 dB re: 1 μPa.

In the light of our current understanding, van Bergeijk's error (1967) was his assumption that ear-mediated hearing in fishes was a matter only of processing the sound pressure waveform using the swim bladder or other gas bubble acting as a pressure-to-displacement transformer. It is accepted today that if this were the case, sound localization by the ears in the near and far fields would indeed be impossible for fishes. However, it is also now accepted that the ears of fishes function primitively in all species as inertial accelerometers (de Vries 1950;

Dijkgraaf 1960) in both the near and far fields, and not only as receivers of sound pressure–mediated motions from a gas bladder in some specialized species. As discussed below, the otolith organs of fishes are exquisitely sensitive to acoustic particle motions at audio frequencies, and enough directional information is encoded to compute the axis of acoustic particle motion (Fay 1984; Edds-Walton et al. 1999).

4. Contemporary Experiments and Theories

4.1 Directional Unmasking

Chapman (1973) studied an interesting aspect of directional hearing in haddock (*Melanogrammus aeglefinus*), pollack (*Pollachius pollachius*), and ling (*Molva molva*) by investigating directional unmasking, a phenomenon closely related to the masking level difference (MLD) studied in human listeners (e.g., Hirsh 1948). In Chapman's experiments, fish were confined in midwater in a free-field acoustic test range about 21 m deep, and classical cardiac conditioning was used to determine tone detection thresholds in the presence of masking noise. Masked thresholds were determined as a function of the angular separation of signal and masking noise sources in azimuth. It was found that masked thresholds were highest when the signal and noise sources were within 10° of one another, but that a 7.7-dB release from masking occurred when the sources were separated by 85°. For human listeners, this release from masking has been interpreted as a central effect based on the processing of different interaural time and intensity patterns for the signal and masker.

Chapman and Johnstone (1974) reinvestigated directional unmasking for cod and haddock using the same location, apparatus, and procedures. They found that a signal/masker source separation greater than 10° led to a significant release from masking. The experiments of Chapman (1973) and Chapman and Johnstone (1974) could be viewed as an investigation of the "cocktail party effect" (Cherry 1953). For one-eared listeners, the interfering effects of noise and reverberation on speech reception are much greater than for normal hearing listeners. This is thought to be due to a loss of the binaural processing advantage used to recover signals (e.g., the voice of a talker) from noise (e.g., the combined voices of other talkers) in noisy listening situations [an important aspect of auditory scene analysis (Bregman 1990)]. A similar release from masking was demonstrated for cod and haddock at 45° signal and noise source separation by Chapman and Johnstone (1974). Hawkins and Sand (1977) later demonstrated this sort of directional unmasking in a median vertical plane for cod. These results for fishes imply, but do not require, sound source localization mechanisms. These unmasking experiments are most interesting as a demonstration of a functional parallel between fishes and terrestrial animals in directional hearing. The peripheral mechanisms of this unmasking effect appear to be quite different in fishes and humans, but the functional consequences are similar:

spatial filtering useful for signal detection in noise and possibly useful for resolving an auditory scene.

4.2 MAA and Source Distance Discrimination in Fishes

A revolution in experiments and theories on sound source localization in fishes was initiated by Schuijf and his colleagues in the 1970s (e.g., Chapman 1973; Chapman and Johnstone 1974; Schuijf 1975; Schuijf et al. 1972; Schuijf and Buwalda 1975). In one of the very first psychophysical experiments on sound source localization, Schuijf et al. (1972) studied the Ballan wrasse (*Labrus berggylta*) using appetitive conditioning carried out in a deep fjord near Bergen, Norway at a depth of about 4 m. Two sound sources were separated in azimuth by two angles: 10° and 71°. For each angle, a conditioning trial consisted of a brief change in the loudspeaker emitting a train of ever-repeating tone bursts (1500 ms in duration, at 115 Hz). A blind observer decided whether the animal responded in any way. Every positive response during a trial was rewarded with a piece of food. The statistically significant responses at both source angle differences were interpreted as an indication that the fish detected the event of the tone burst switching from one loudspeaker to the other, and this event was assumed to result in the perception of a purely spatial change. As the authors point out, however, this experiment demonstrated the detection of a spatial change, but did not demonstrate that the fish could correctly determine the locations of the sources. Any difference in perception arising from switching the activation between the two loudspeakers could have resulted in these observations, and it is simply an assumption that this difference in perception corresponded to two different perceived source locations. Thus, this type of experiment is only a weak demonstration of sound source localization, and will always be open to alternative interpretations. In other experiments, Chapman and Johnstone (1974) found that azimuthal angular separations of 20° or more were required for the fish to discriminate between source locations.

Schuijf (1975) demonstrated that cods could be conditioned to discriminate between different azimuthal source locations with an accuracy of 22°, and that two, intact pars inferior (includes both sacculus and lagena) of the ears were necessary for these behaviors. The MAA of 22° was determined using two- and four-alternative spatial choice designs in which the fish was required to move toward the location of the active sound source to be rewarded. Schuijf recognized that the cods could possibly solve this problem by recognizing the identity of each sound projector through timbre difference cues. The cod could then associate a correct response location with each projector without being able to determine the actual locations of the sources. Although he effectively ruled out the source timbre hypothesis, it remains possible that the differences in sound source location provided other, location-dependent sensory cues adequate for discrimination, but inadequate for source localization. Nevertheless, these experiments are among the best evidence we have that sound source localization, as we think of it in human experience, is a capacity that fish may have, and in addition, that azimuthal discrimination depends in some way on binaural processing, but probably not on interaural time cues. In this sense, putative localization by fishes appeared to have some important elements in common with localization as we understand it among most terrestrial vertebrates investigated. The finding of the necessity of binaural computation (as Moulton and Dixon (1967) had demonstrated earlier in goldfish) was consistent with the observations of Sand (1974) (see also Enger et al. 1973) showing that the microphonic potentials from each ear in response to oscillation along various axes on the horizontal plane resulted in a directional pattern (a near cosine) with the long axis roughly parallel to the orientations of the saccular organs in the fish's head.

In the same acoustic test range, and using similar methods to Chapman and Johnstone (1974), Hawkins and Sand (1977) measured an MAA for elevation from the horizontal plane to be about 16° at the highest signal-to-noise ratio tested at 110 Hz. The authors note that these data show an angular resolving power in elevation that is at least as good as that shown by Chapman and Johnstone (1974) in the horizontal plane. From earlier experiments on the microphonic potentials of the ear, Sand (1974) had suggested that while two intact ears are required for azimuthal localization, an elevation discrimination could be possible using only one ear. This hypothesis has not been tested, but is consistent with more recent physiological data on the peripheral encoding of directional information in *Opsanus tau*, the oyster toadfish (e.g., Fay and Edds-Walton 1997a).

Schuijf and Hawkins (1983) addressed the question of source distance determination in cod using classical cardiac conditioning. For 113-Hz tone pulses at a moderate sound pressure level (2 dB re: 1 Pa), two cod were able to discriminate between two sound sources at two distance differences, but both at 0° azimuth and elevation (4.5 m vs. 7.7 m, and 4.5 m vs. 1.3 m). This distance discrimination was interpreted to be based on the distance-dependent phase angle between sound pressure and acoustic particle acceleration within the near-field of a sound source. It is also possible that the discrimination is based on processing the amplitude ratios between these two acoustic components rather than phase differences. The authors calculated that these ratio differences were less than 4 dB for their sources and that this difference was near or below the level discrimination threshold for cod, determined independently (Chapman and Johnstone 1974). It is significant that this amplitude ratio is essentially equal to the cod's level discrimination threshold at 110 Hz. In addition, a simultaneous comparison of amplitudes could be more acute than the successive discrimination measured by Chapman and Johnstone (1974). Thus, there is some reason to believe that this distance discrimination could be based on the processing of simultaneous amplitude ratios between pressure and particle motion. These observations are consistent with the hypothesis that these fish have a sophisticated and truly three-dimensional directional hearing sense, but are not critical experiments in the sense of directly demonstrating that the fish could correctly locate the test sound sources.

4.3 Experiment and Theory on the "Phase Model" for Directional Hearing

The emerging understanding of directional hearing in fishes was that otolith organs are stimulated by the motions of the fish's head in a sound field as they take up the acoustic particle motions of the surrounding water medium (de Vries 1950; Dijkgraaf 1960). Such motion of the fish would occur in both the acoustic near and far fields provided the acoustic particle motions were of sufficient magnitude. In this case of "direct" ear stimulation, the axis of motion across the hair cell cilia could be resolved, in principle, by the pattern of response magnitude over a population of hair cells and otolith organs with widely dispersed polarization axes. Hair cells were known to be structurally and physiologically polarized since the work of Flock (1964, 1965). Furthermore, the three major otolith organs of fishes (saccule, lagena, and utricle) were known to have different gross orientations in most fish species. Thus, the major mystery of possible directional hearing in fishes was thought to be solved, in principle, through the assumption that analyses of neural activity across cell arrays could reveal the axis of acoustic particle motion. This notion was called "vector detection" by Schuijf and Buwalda (1975).

Importantly, the additional assumptions here were that:

1. One end of the axis of acoustic particle motion pointed directly at the sound source.
2. Each auditory nerve fiber received input from only one hair cell or from a group of hair cells having the same directional orientation (an hypothesis of private directional channels).
3. This mode of stimulation was effective enough to operate at the sound levels usual for the species.

The first assumption was known to hold only for monopole sound sources (e.g., a pulsating source fluctuating in volume). However, van den Berg and Buwalda (1994) have pointed out that for any source order type (e.g., dipoles and higher-order sources), the axis of acoustic particle motion points directly radially (toward and away from the source) at each instant of a pressure waveform zero-crossing (= the pressure null). Thus, it is not certain that vector detection is useful for locating only monopole sources. The second assumption was not confirmed until the work of Hawkins and Horner (1981) on the directional response properties of saccular afferents in cod. These initial observations have been repeatedly confirmed in other species (e.g., Fay 1984, Fay and Edds-Walton 1997a; Lu et al. 1998). The third assumption of adequate sensitivity was tested indirectly in psychophysical experiments on sound detection by flatfishes without a swim bladder (Chapman and Sand 1974), indicating that displacement detection thresholds were as low as -220 dB re: 1 m (less than 0.1 nm) at the best frequency of hearing (near 100 Hz).

4.4 The 180° Ambiguity Problem

The notion of vector detection posed an important unsolved problem. This is that while a particle motion axis could be resolved by arrays of directional receivers, this solution could not determine which end of the axis pointed toward the source (i.e., specify the direction of sound propagation). This essential ambiguity has come to be known as the "180° ambiguity problem" and has dominated most theorizing and experimentation on directional hearing in fishes since the mid-1970s.

Schuijf (1975) and Schuijf and Buwalda (1975) conceived of a theoretical solution to this problem. In the simplest terms, a determination of the phase angle between acoustic particle motion and sound pressure could resolve this ambiguity. This can be intuitively understood at a very simple level as follows: Imagine an axis of particle motion that is from side to side. The monopole source could be oscillating from side to side either on the left or right of the receiving animal. However, if the sound is propagating from a source at the right, then leftward particle motions are preceded by rising pressure and leftward motions preceded by falling pressure. (The actual encoded phase angle between them is a function of source distance in the near field and also depends on the dynamics of the pressure and motion receivers the fish uses.) This "phase model" of directional hearing requires that both the sound pressure and particle motion waveforms be encoded at the periphery, and that appropriate central computations take place using useful representations of their phase or timing relations.

Schuijf and Buwalda (1975) went on to evaluate the phase model experimentally. Again using appetititive conditioning in a nearly free-field natural environment, they were able to condition cods to discriminate between sound sources directly in front and directly behind the animals. Furthermore, the directional choices could be reversed by manipulating the phase of sound pressure with respect to the phase of particle motion (180° phase shift) of a synthesized standing wave as the phase model predicted. This sort of experiment, repeated and extended several times later (e.g., Buwalda 1983; van den Berg and Schuijf 1983), represents the best evidence in support of the phase model for sound source localization by fishes.

A potential weakness of the phase model is its requirement that both sound pressure and acoustic particle motion be encoded separately at the periphery (or segregated somehow by central computations such as common-mode rejection [Fay and Olsho 1979; Buwalda et al. 1983]). In hearing generalist species with a swim bladder, this could be imagined as, for example, one set of hair cells (or otolith organs) oriented so as to receive reradiated or scattered particle motion from the swim bladder (indirect, or pressure-dependent stimulation), and another set somehow shielded from or insensitive to swim bladder signals that responded to direct particle motion stimulation. In this case, the assumption is that pressure-dependent input to the ears from the swim bladder is effective in hearing. There has been continuous speculation that in ostariophysine and other hearing specialist species, the lagena and utricle may also function as auditory organs (e.g., Wubbles and Schellart 1998). Since these probably do not receive strong (or any known) input from the swim bladder (Fay et al. 2002), but respond with great sensitivity to acoustic particle motion as inertial accelerometers (Fay 1984), the dual encoding assumption could hold, in principle. However, for species without a swim bladder (or equivalent) such as sharks and flatfish, and for hearing generalist species receiving negligible pressure-mediated input to the ears, this dual encoding assumption is less likely to be valid. In most hearing generalists lacking specialized pathways between the swim bladder and inner ears, it is sometimes assumed, but rarely demonstrated (see Chapman and Hawkins 1973 for data on cod), that the ears respond to displacements reradiated from the swim bladder. Since sharks, which lack a swim bladder, had been reported to be able to approach sound sources from distances as great as 200 m (e.g., Nelson 1965; Myrberg et al. 1972), how could the phase model apply to them?

Two possible answers to this question were provided by Schuijf and his colleagues. Schuijf (1981) suggested that a pressure-dependent phase reference for evaluating the phase of particle motion could be derived from the interference between direct, inertial ear stimulation and sound reflections from the water surface and bottom. Later, van den Berg and Schuijf (1983) demonstrated what they interpreted as sound pressure sensitivity in addition to particle motion sensitivity in the shark *Chiloscyllium griseum* in behavioral conditioning experiments using multiple sources to synthesize the sound fields. Based on those experiments, they suggested that pressure sensitivity could arise within the labyrinth owing to the presence of two flexible "windows" in the relatively rigid otic capsule (the oval window and the window to the endolymphatic duct) that could possibly release the pressure (a sort of "mobilization" hypothesis). These hypotheses have not been critically evaluated since that time.

Schellart and de Munck (1987) and de Munck and Schellart (1987) have provided a somewhat different possible solution to the 180° ambiguity problem. In their modeling and analysis of the fields impinging on otolith organs, they suggest two sources of input:, the direct inertial route and the indirect route due to reradiated fields from the swim bladder in species such as the cod having no specialized mechanical linkages between the swim bladder and the otolith organs. In this case, the authors point out that the interaction between these two fields will tend to produce elliptical displacement orbits of relative motion between the otoliths and their hair cell epithelia. These orbits could possibly play a role in encoding both the axis of acoustic particle motion and also solve the 180° ambiguity problem through sensing the rotation direction of the orbital motion. In this analysis, these authors suggested that the utricle is best suited as the receptor organ responsible for this encoding in the horizontal plane. However, there is no empirical evidence yet that the utricle is an auditory organ in fishes other than clupeids (herrings), and the additional prediction that this sort of encoding operates monaurally is inconsistent with empirical evidence that sound source localization in the horizontal plane requires binaural input. Behavioral conditioning studies have been carried out to indirectly evaluate the orbit model in the rainbow trout (*Salmo gairdneri*) by Schellart and Buwalda (1990). The results were equivocal, but details of their results suggested that binaural processing could theoretically contribute to sound source localization mechanisms based on processing the elliptical orbits of relative hair cell and otolith motion.

Rogers et al. (1988) presented a computational model for sound source localization that incorporates some of the elements of both the phase model of Schuijf (1975) and the orbit model of Schellart and de Munck (1987). The idea is essentially that pressure and multiaxis acceleration information are inherently contained in the totality of inputs from an otolith organ that responds to a combination of direct (acceleration) and indirect (swim bladder–mediated, proportional to sound pressure) inputs. Operations on matrix equations representing these inputs were shown to compute estimates of sound source location in both the near and far fields. This theory specifically predicts that pressure-dependent, swim bladder–mediated input to the ears, which alone is nondirectional and inadequate for sound source localization, is a necessary component for localization when combined with directional input from the ears' responses to direct acoustic particle motion. Here again, whether the fish is in the near field of a sound source or not is not critically important. It is necessary only that the otolith organs are activated in the direct mode by the kinetic components of underwater sound as well as by pressure-dependent input to the ears from the swim bladder or other gas bubble.

Most recently, Kalmijn (1997) has posited a novel mechanistic explanation for sound source localization in fishes. Focusing on approach or avoidance behaviors with respect to sound sources, Kalmijn has pointed out that a fish could approach any sound source accurately simply by swimming in a direction that maintained a constant angle with the locally determined axis of particle motion, which itself need not point to the sound source. This conception does not assume or explain a sound source localization decision based on sampling a sound field at one point in time. Rather, this is an ethological approach focusing on a mechanism for a specific behavior (possibly, both approach and avoidance). Note that for this sort of mechanism to work, the sound source must be assumed to be broadcasting nearly continuously for a relatively long period of time, and that the receiver must be assumed to be able to decide which direction along the pathway to take in approaching or avoiding the source. The behavior postulated could be evaluated, in principle, using a long-duration and attractive sound source (e.g., a male midshipman's advertisement call during the reproductive season), and a receptive animal (e.g., reproductively ready female) whose behavior could be tracked precisely with respect to the structure of the sound field at each point between the source and receiver.

5. Directional Acoustic Behaviors and Phonotaxis

Kleerekoper and Chagnon (1954) studied the behavior of the creek chub (*Semotilus a. atromaculatus*) in laboratory tanks in which a sound source was associated with food reinforcement. They found that the swimming pathways taken by the animals during sound presentation depended on the locations of sound sources in the experimental arena, and they concluded that the fish were "guided by fields of strongest intensity." Subsequently, Kleerekoper and Malar (1968) studied the movement patterns of carp (*Cyprinus carpio*) and sunfish (*Lepomis gibbosus*) during tone presentations in laboratory tanks. For both species, the presentation of a pure tone (2 kHz for carp, 700 Hz for sunfish) clearly resulted in a change in the swimming pathways and turning angles. Carp tended to move to tank areas of lower sound pressure level and seldom crossed into the fields of highest intensity when the sound was presented. However, no direct evidence was obtained that fishes located the sound sources in either of these studies.

Popper et al. (1973) studied behavior with respect to a sound source in the Hawaiian squirrelfishes (*Myripristus berndti* and *M. argyromus*) caged in an environment that was usual for the species. Both species responded to the playback of staccato and grunt vocalizations recorded from *M. berndti* by moving toward the sound source in a two-choice test. However, only sources at a 2-m distance (or less) were effective in eliciting this behavior; sources at a 2.9-m distance did not reliably elicit approach responses. Since this effect was shown to be independent of received sound pressure level, the authors concluded that source distance, per se, was an important factor in controlling these behaviors.

Leis et al. (2003) investigated the hypothesis that some larval marine fishes may orient with respect to reef sounds during searches for settlement sites at the end of their pelagic phase. Light traps were set near Lizard Island, Great Barrier Reef, Australia to capture larval fishes during the night. Some traps were equipped with sound sources that broadcast recordings of "reef sounds" typical for the area during the night. Pomacentrids, mullids, lethrinids, apogonids, and blennids constituted 95% of the trapped species. For all trap locations and dates, the number of pomacentrids caught by the reef noise traps significantly exceeded the number caught by silent traps. The effects for Mullidae, Apogonidae, and Blennidae were less consistent over locations and dates, but still, some significant differences were found. The authors concluded that some of these species are attracted to reef sounds at night from a distance of 65 m or less. It is reasonable to conclude that some sort of sound source localization is required for these behaviors.

In many fish species, males are known to signal their breeding territory locations through advertisement calls that attract females of the species (Fine et al. 1977). It is presumed, and sometimes has been demonstrated, that females are able to localize these sources using only the broadcast sounds as cues. Although there are reports of approaches to conspecifics and sound playbacks (phonotaxis) in many fish species (Fine et al. 1977), toadfish (family Batrachoididae) are the best studied (e.g., Gray and Winn 1961, Winn 1964 for *Opsanus tau*; Fish 1972). McKibben and Bass (1998) presented various synthesized sounds to plainfin midshipman (*Porchthys notatus*) from one of two loudspeakers near the center of a 4-m diameter cynindrical concrete tank (0.75 m deep) and observed the responses of females and males released within about 1 m from the loudspeakers. For continuous tones and harmonic complexes with a fundamental frequency near 100 Hz (at about 130 to 140 dB re: 1 μPa), gravid females were observed to exhibit a variety of behaviors that "usually began with a straight approach to one speaker." The authors concluded that the male's vocalization (a long-duration "hum" with a fundamental frequency of about 100 Hz) functions as a "call" that attracts gravid females that are ready for reproduction. These and other (McKibben and Bass 2001) studies on this species also represent some of the clearest evidence available that fishes are able to locate sound sources. Since these experiments were "closed-loop" in the sense that the sound continued during the phonotactic behavior, it is not known whether these animals were moving up an intensity gradient, or approached the source using another search strategy (e.g., the constant-angle mechanism proposed by Kalmijn 1997), or whether they had determined the source location at the time of initial release in the test arena. Further analyses of these behaviors using different types of sound sources and sound fields will help answer these questions.

Tavolga (1971, 1976) has documented an unusual example of the use of sound by a marine catfish (*Arius felis*) that strongly implies a capacity for sound source localization. Tavolga (1971) described the vocalizations of this species as pulses and trains of pulses with energy at frequencies between 100 and several kiloHertz. In preliminary experiments, animals were tested in a large laboratory test tank with visible and invisible plastic obstacles scattered about. The swimming fish had very few collisions with the obstacles. When the eyes were surgically removed, the vocalization quantity increased transiently, and the animals were still described as competent in avoiding obstacles. Subsequently, Tavolga (1976) introduced clear plastic barriers into the test arena for normal animals and animals that had been "muted" by surgically cutting the sonic muscles that deform the swim bladder for sound generation. All the muted animals behaved unusually with generally disoriented behavior and frequent collisions with the plastic barriers. Again, evidence was obtained that *Arius* could determine the presence and location of transparent plastic barriers using a sort of echolocation system. This is an interesting example of the use of vocal sounds to generally characterize the structure of the local environment.

6. Reflex Directional Orientation with Respect to Sound Sources

Many fishes produce fast escape responses (sometimes called fast startle or C-start responses) that are directional with respect to nearby sound sources (e.g., Moulton and Dixon 1967; Blaxter et al. 1981; Mueller 1981). For many of these fast responses, the M-cells of the medulla (Bartelmez 1915) are most likely responsible for initiating and directing the response. Mauthner cells are giant, reticulospinal cells that receive multimodal sensory input (Zottoli et al. 1995) and innervate motoneurons controlling contralateral body musculature. Beginning with the work of Furshpan and his colleagues (e.g., Furshpan and Furukawa 1962), the structures and physiological functions of M-cells have become an important example of how an identified single neuron of the vertebrate brain can create behavior. The paired M-cells (and other reticulospinal neurons) fire in a coordinated way to initiate rapid movement away from a local source of sound or other mechanical disturbance in some species of fish and larval amphibians. There is some evidence that primary afferent input from the saccule initiates the M-cell response through synapses directly on the M-cell lateral dendrite (e.g., Canfield and Eaton 1990). Directional decisions are likely made through synaptic processing mechanisms possibly involving both excitatory and inhibitory inputs from multiple sensory organs and at least one group of interneurons (i.e., passive hyperpolarizing potential cells: Korn and Faber 1975).

Since much of the work on M-cells has been done on goldfish, and since the goldfish saccule is known to be an auditory organ, M-cells have been understood as a mechanism for sound source localization. As such, the same questions that have arisen earlier in this chapter regarding source localization in a more general sense have been asked (and sometimes answered) of the M-cell system. The fundamental ones are:

1. What sensory organs or systems provide triggering and directional information to the M-cell?
2. How is the directional information encoded?
3. How are response thresholds and left-right directional decisions made given the 180° ambiguity problem?

Understanding this relatively simple neurobiological system offers the promise for understanding at least one neural circuit and processing strategy that could accomplish sound source localization. However, as noted above, intentional localization behaviors with respect to sound sources (e.g., phonotaxis, MAA discrimination) cannot be explained through the descending M-cell circuit, so it seems probable that the signal processing problems of source localization for fishes may have two independent circuit solutions within the fish auditory system.

How the M-cell system makes fast directional decisions is still a matter of speculation. Canfield and Eaton (1990) have shown that the M-cell excitation and firing requires input from the swim bladder via the saccule. This finding was initially surprising (but see Blaxter et al. 1981) since the swim bladder's input is mediated by sound pressure which, by itself, is nondirectional. However, an effective model that can account for both the initiation and directionality of the C-start must include the sound pressure signal polarity or phase in the computation (in accord with the phase model), so pressure information appears to be necessary (Guzik et al. 1999).

Eaton et al. (1995) have presented a connectionist model for this decision-making by M-cells that they characterize as an exclusive NOR ("NOT OR," XNOR) logical operation. A putative neural circuit was postulated that could implement this XNOR function in goldfish and other Otophysi. In this model, the M-cell receives direct excitatory input from saccular afferents of both positive and negative polarity, and the magnitude of swim bladder-mediated, pressure-dependent input from the saccules brings the M-cell membrane potential near threshold. At the same time, combinations of polarity-specific pressure inputs and direction-dependent displacement inputs combine via inhibitory interneurons (PHP cells) to provide the proper directionality (i.e., to solve the 180° ambiguity problem). In effect, this inhibition can pull the direct, pressure-mediated excitation just below M-cell spike threshold. This model is in accord with observations of strongly pressure-dependent excitatory postsynaptic potentials recorded intracellularly in M-cells (Canfield and Eaton 1990). Still unclear, however, is the source(s) of the directional displacement-sensitive input to the PHP cells. In principle, these could arise from the saccule, lagena, or utricle of the ear, or from the lateral line system. Also unclear at the moment are the sources of the pressure-dependent excitation in hearing generalist species and those lacking a swim bladder or equivalent. There are some indications that M-cell decision making may lead to different behaviors in different species, with visual cues weighted differently (Canfield and Rose 1996).

7. Physiological Studies on Directional Hearing in Fish

Neurophysiological investigations of directional hearing in fishes have focussed on the encoding of directional information in the afferents of the otolith organs and on the fates of these directional representations at various levels of the brainstem. The species investigated have been somewhat limited, including goldfish (*Carassius auratus*), toadfishes (*Opsanus tau* and *Porchthys notatus*), sleeper goby (*Dormitator latifrons*), rainbow trout (*Salmo gairdneri*), and Atlantic cod (*Gadus morhua*).

7.1 Directional Encoding at the Periphery

Single-unit studies on the peripheral encoding of directional information were first reported by Fay and Olsho (1979) and Fay (1981) for goldfish. Hawkins and Horner (1981) measured the first directional response patterns in recordings from the saccular and utricular nerve of the cod in response to whole-body oscillatory accelerations in the horizontal plane at frequencies between 63 and 250 Hz. Their major finding was that the response magnitude (expressed both in terms of spikes per cycle and phase locking) tended to vary according to a cosine-like function of vibration axis. This was significant because it indicated that each afferent studied represented the presumed directionality of a single hair cell or group of hair cells having the same directional orientation. In other words, each hair cell orientation appeared to have a private line to the brain, a requirement of the notion of "vector detection" assumed by Schuijf (1975) as the first stages of the phase model. For the saccule, the limited data set presented indicated that the best azimuthal axis of motion corresponded roughly with the horizontal-plane orientation of the saccular sensory epithelium and otolith. For utricular units, best azimuths varied widely, roughly in accord with the diversity of hair cell orientations over the (horizontal) surface of the utricular epithelium. The authors noted that utricular best sensitivity was similar to that of the saccule, suggesting that the utricle could possibly play a role in directional hearing. Finally, it was noted that the phase angle at which units synchronized varied widely among the units recorded and did not simply fall into two discrete groups, 180° out-of-phase with one another. Fay and Olsho (1979) and Fay (1981) also observed a nearly flat distribution of phase-locking angles among saccular and lagenar nerve units in goldfish. The phase model (and other related theories of directional hearing in fishes outlined above) assume that pressure and displacement "polarities" would be represented robustly in a bimodal distribution (180° separating peaks) of phase-locking angles, as predicted by anatomical hair cell orientation maps for otolith organs (e.g., Dale 1976; Platt 1977; Popper 1977). The fact that phase-locking angles do not cluster in such a simple and easily interpretable way (see also Fay and Edds-Walton 1997a for similar data on *Opsanus tau*) presents a problem for all current theories of sound source localization in fishes: Which neurons "represent" the phases of pressure or displacement waveforms that have to be compared to resolve the 180° ambiguity?

Studies on directional encoding in goldfish (Fay 1984; Ma and Fay 2002) and toadfish (Fay and Edds-Walton 1997a,b; Edds-Walton et al. 1999; Weeg et al. 2002) have used a three-dimensional electrodynamic "shaker" system to create whole-body translational accelerations varying in both azimuth and elevation. Figure 3.1 illustrates typical directional response patterns (DRPs) for saccular units of oyster toadfish. Data of this sort have led to the following generalizations:

1. Most saccular afferents respond in proportion to the cosine of the stimulus axis angle in azimuth and elevation, with a few exceptions (Fay and Edds-Walton 1997a). Thus, each afferent seems to represent the orientation of one hair cell, or a group of hair cells having the same directional orientation (Lu et al. 1998).
2. In the horizontal plane (azimuth), most saccular units respond best and with lowest thresholds to an axis angle that is approximately parallel with the saccular epithelium and otolith orientation in the head (see also Sand 1974).
3. In vertical planes, the diversity of best elevations among units corresponds qualitatively with the diversity of hair cell morphological polarizations on the saccular epithelium.
4. The best threshold sensitivity for otolithic afferents is very high; at 100 Hz, root-mean-square displacements that are effective in causing significant phase locking in the most sensitive afferents are about or below 0.1 nm. This is approximately the same amplitude of basilar membrane motion at behavioral detection threshold in mammals (Allen 1996).
5. Intracellular labeling studies indicate that anatomical hair cell orientation

front
0
-30
30
-60
60
left -90
90
-120
120
-150
150
180
dorsal
90
60
30
0 front
-30
-60
-90
Horizontal
Mid-Sagittal

maps do not quantitatively predict physiological directionality (Edds-Walton et al. 1999). This is probably due, at least in part, to the simplifications inherent in constructing two-dimensional map representations of three-dimensional structures. Thus, anatomical maps cannot substitute for physiological data in specifying the directional information transmitted to the brain over the auditory nerve.

6. As noted above, the phase angles of synchronization do not form simple bimodal distributions with peaks separated by 180°. These conclusions for toadfish (*Opsanus tau*) do not differ importantly from those based on similar work on the saccules of *Porichythys notatus* (Weeg et al. 2002), and on *Dormitator latifrons* (Lu et al. 1998).

Since best azimuths for the saccular afferents studied so far tend to cluster about the azimuthal angle in the head of the saccular epithelium and otolith (see also Sand 1974), the overall activation of each of the two saccules will tend to differ and will depend on the azimuth of the particle motion axis. Thus, azimuth angle could be computed by comparing the summed output of each saccule (e.g., through subtraction or common-mode rejection), but with a front–back ambiguity and two other ambiguous points corresponding to about plus and minus 90° (left–right) in *Opsanus tau* (Fig. 3.2). Since a relatively simple binaural comparison could compute azimuth, azimuthal localization in fishes could be a binaural process in fishes similar to that of terrestrial animals, with comparable ambiguities. This conclusion is consistent with the experiments of Moulton and Dixon (1967), Schuijf (1975), and Schuijf and Siemelink (1974) showing that the information from two intact labyrinths is necessary for the determination of sound source azimuth. Note, however, that in the case of fishes, binaural acoustic cues are not available; the binaural information derives from the directionality of the ears as they respond directly to acoustic particle motion. Fay and Edds-Walton (1997a) have observed that the phase angles at which the units synchronize to a tone stimulus are correlated with the differences in response magnitude (and effective sound excitation level) in nonspontaneous primary saccular afferents. This means that an interaural response phase or timing difference could play a role in represent-

FIGURE 3.1. Directional response functions (DRFs) for five saccular afferents from the left ear of one toadfish (*Opsanus tau*). Response magnitude is plotted as a function of stimulus axis in polar coordinates. (*Left*) DRFs in the horizontal plane (azimuth 0° = straight ahead). (*Right*) DRFs in the mid-sagittal plane (elevation 0° = straight ahead). For most afferents, DRFs were determined at several displacement levels. Response magnitude grows monotonically with signal level in 5-dB increments for panels with multiple functions. Signal levels range between 5 and 25 dB re: 1 nm displacement at 100 Hz. Note that the best axes in azimuth tend to cluster toward the left front-right rear axis while elevations are more diverse. Response magnitudes plotted are the z-statistics (vector strength squared times total number of spikes). (Unpublished data from Fay and Edds-Walton.)

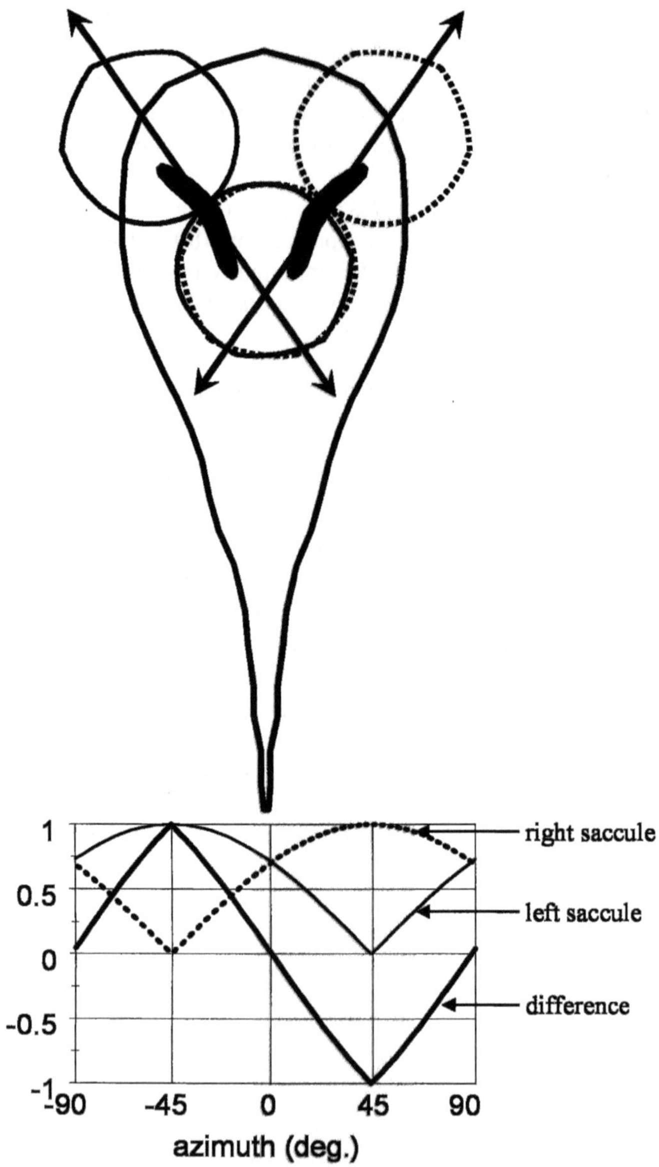

FIGURE 3.2. Simple model for using interaural differences in response magnitude to help determine the axis of acoustic particle motion in toadfish and other hearing generalist fishes. (*Top*) Each saccule is directional in the horizontal plane, roughly in the manner of a cosine function. *Arrows* indicate the best sensitivity of each saccule in the horizontal plane. (*Bottom*) The signed difference between the two differently oriented cosine functions (rectified) can represent the azimuth of the axis of acoustic particle motion. As is the case for other binaural vertebrates, ambiguities are present.

ing response magnitude, giving rise to a robust interaural response timing code for azimuth as well.

As discussed above, sound source azimuth could be represented in differences between the neurally coded outputs of the two ears. Coding for elevation seems to be a different matter, however. The elevation of a sound source (or the degree and axis angle of up–down acoustic particle motion) is represented within a sensory epithelium as the profile of activity across saccular afferents with different "best elevations" (see Fig. 3.1), as originally postulated by Schuijf (1975) in the "vector detection" process. It is interesting to note that there is a functionally similar hypothesis for determining elevation for human and other mammalian listeners; this is the hypothesis of processing the spectral profile (over the length of the cochlear epithelium) as shaped by the HRTF (e.g., Wightman and Kistler 1993). Thus, it is hypothesized that for both fishes and mammals, sound source elevation could be, in effect, mapped or coded as a monaural profile of excitation over the receptor organ surface.

The directional responses of auditory nerve units have also been studied for organs other than the saccule. Hawkins and Horner (1981) studied utricular units in the cod and found them to be most sensitive in the horizontal plane with substantially cosinelike DRPs. Fay (1984) surveyed lagenar and utricular as well as saccular units in goldfish. All three otolith organs had a similar distribution of displacement thresholds (lowest thresholds based on phase-locking near 0.1 nm, root-mean-square at 140 Hz) and essentially cosine-shaped DRPs. Lagenar units showed a wide distribution of best axes in elevation with a weak tendency to cluster in best azimuth along an axis parallel to the orientation of the lagenar epithelium in the horizontal plane. Most utricular units were most sensitive in the horizontal plane, in accord with the horizontal orientation of the utricular sensory epithelium. Lu et al. (2003) have studied lagenar DRPs in the sleeper goby. Surprisingly, many of the DRPs obtained deviated significantly from a cosine shape, showing more narrowly shaped DRPs than would be expected from hair cell directionality, and best thresholds that were somewhat higher than saccular units from the same species. More broadly shaped DRPs could be explained by excitatory convergence from hair cells having different directional orientations (Fay and Edds-Walton 1997a), but more narrowly shaped DRPs cannot be explained at present. The differences in sensitivity between lagenar and saccular units in the sleeper goby could possibly be related to the small size of the lagenar organ (characteristic of most hearing generalists).

7.2 Directional Representations in the Brain

The representations of directional acoustic information in the brain have been studied for *Carassius auratus* by Ma and Fay (2002), *Opsanus tau* by Edds-Walton and Fay, and for *Salmo gairdneri* by Wubbles, Schellart, and their colleagues. The major acoustic nuclei of the fish brainstem are the first-order descending octaval nucleus (DON) and the higher-order secondary octaval nuclei

(SON) of the medulla, and the torus semicircularis (TS) of the midbrain. Auditory responses of the SON, thalamic, and other forebrain auditory nuclei have been investigated, but the cells were not analyzed with respect to directional stimulation.

The great majority of single units recorded in the toadfish DON show simple directional preferences for the axis of whole-body translational acceleration. The maintenance of directionality in the DON (and in other auditory nuclei investigated) indicates that excitatory convergence from auditory neurons having different directional preferences tends not to occur in the brain; that is, directional selectivity originating at the periphery is maintained throughout the auditory brainstem. The axons of primary saccular afferents enter the DON anteriorally and project caudally throughout the rostrocaudal extent of the DON with collaterals heading medially (Edds-Walton et al. 1999). The sensitivity, frequency response, and phase locking of DON units are similar to those of saccular afferents, but the DRPs of most units tend to be more directionally selective than saccular afferents. This increased selectivity has been termed "sharpening" (Fay and Edds-Walton 1999; Edds-Walton and Fay 2003). Figure 3.3 shows typically sharpened DRPs from the brain of toadfish along with a graphical representation of a simple model mechanism that could account for sharpening (Edds-Walton and Fay 2003). The hypothesis is that a central cell receives excitatory input from one directional cell, and inhibitory input from another directional cell, both having cosine-like DRPs with different best axes in azimuth or elevation (Fay and Edds-Walton 1999). This excitatory–inhibitory convergence appears to be a common interaction in the auditory brainstem, and it inevitably would result in some degree of directional sharpening, depending on the best axes and weights associated with each input. Recordings from the TS of the midbrain (Fay and Edds-Walton 2001, Edds-Walton and Fay 2003) show similar unit sensitivity and frequency response as in the DON, but with dramatically reduced phase locking, and augmented directional sharpening (see Fig. 3.3).

Directional auditory responses were found both in the nucleus centralis (nominally, the "auditory" nucleus), and the nucleus ventrolateralis (nominally, the "lateral line" nucleus) of the TS in toadfish. In addition, many units recorded in both nuclei showed interactions of auditory and lateral line inputs (excitatory and inhibitory) (Fay and Edds-Walton 2001; Edds-Walton and Fay 2003). It is not known whether such bimodal interactions play a role in sound source localization; there are no major theories of source localization that require auditory–lateral line interactions. At the same time, however, source localization is likely a multimodal function (Braun et al. 2002), and the lateral line system could play an important role at short ranges (c.f. Weeg and Bass 2002).

In general, the distributions of best axes for brainstem auditory units are more widely distributed in azimuth and elevation than the same distributions for saccular afferents. Thus, the across-neuron or population representations of the axis of acoustic particle motion appear to be enhanced by excitatory–inhibitory interactions in the brainstem, particularly in azimuth. It is not known whether

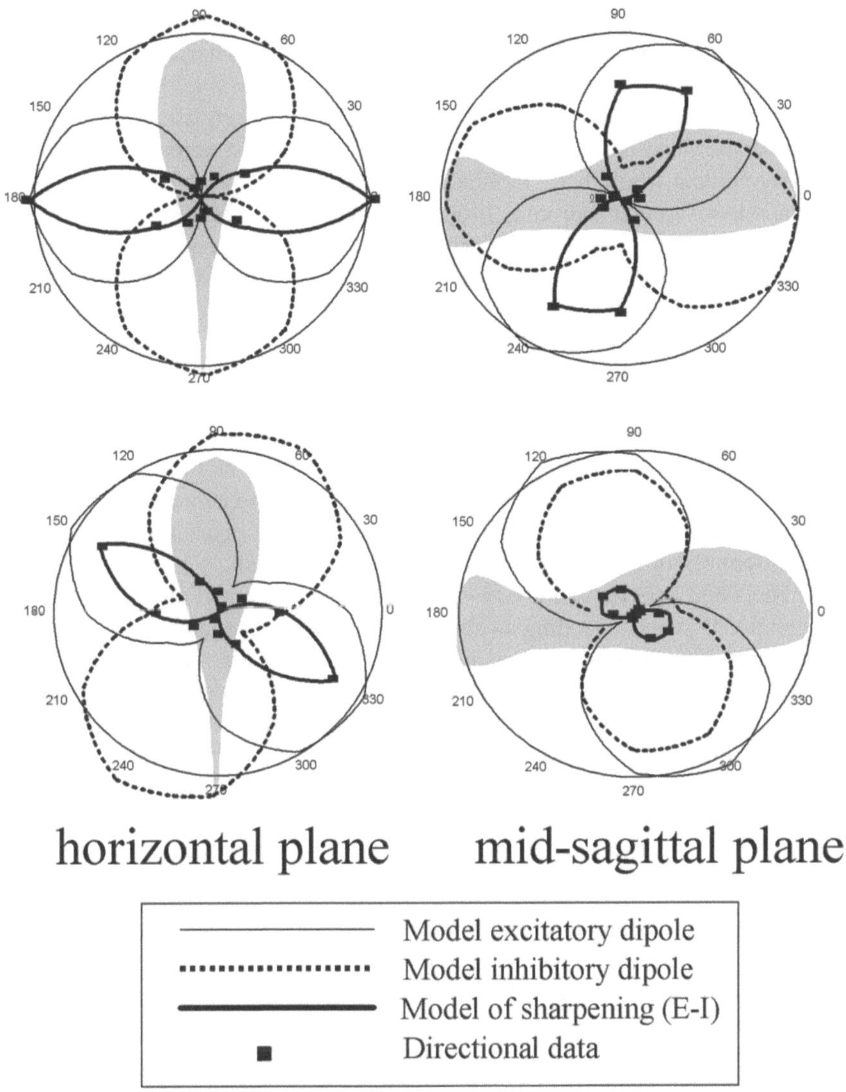

FIGURE 3.3. Demonstration of the sharpening model for two cells of the left torus semicircularis (C4-00, *top* and P2, *bottom*) in the horizontal (*far left*) and mid-sagittal (*middle*) planes of oyster toadfish. Stimulus levels were 25 and 20 dB re: 1 nm displacement at 100 Hz, respectively. Response magnitude is the z-statistic as in Fig. 3.1. One hypothetical excitatory input (*solid thin line*) and one hypothetical inhibitory input (*dashed line*) are shown for cell C4-00 (at 25 dB) and cell P2-01 (at 20 dB). The addition of the excitatory and inhibitory inputs would result in the sharpened directional response pattern shown with the *heavier solid line*. *Square symbols* are the data to be modeled (see Fay and Edds-Walton 1999).

this processing is based on binaural, monaural, or both types of neural interactions, but it is known that excitatory–inhibitory binaural interactions take place in the brainstem of the cod (Horner et al. 1980).

The directional characteristics of TS units also have been studied in goldfish, a hearing specialist (Ma and Fay 2002). In general, most units responded best to nearly vertical whole-body motion, in accord with the nearly uniform vertical orientation of saccular hair cells in goldfish and other Otophysi. Thus, excitatory inputs to the TS appear to be primarily, if not exclusively, from the saccule in goldfish. Nevertheless, deviations from cosine directionality among unit DRPs (i.e., sharpening) was also observed in the goldfish TS, and could be accounted for by simple excitatory–inhibitory interactions as in toadfish (see Fig. 3.3). This suggests that sound source localization in Otophysi, if it occurs at all (see Schuijf et al. 1977), may be based on computations taking place elsewhere in the ascending auditory system where lagenar or utricular inputs could be used to help resolve the axis of acoustic particle motion in a population code comprised of a wide distribution of best axes among neurons. In any case, the representation of acoustic particle motion appears at present to be organized quite differently in the midbrains of toadfish and goldfish.

Wubbels and Schellart and their colleagues have presented a series of studies on directional sound encoding in the midbrain (torus semicircularis or TS) of the rainbow trout (*Oncorhynchus mykiss*). This species is a hearing generalist and was assumed to receive both direct motion as well as reradiated, pressure-dependent motion inputs from the swim bladder to the ears under normal listening conditions. Fish were stimulated in neurophysiological studies by whole-body acceleration at various angles in the horizontal plane using a vibrating platform that could be rotated to any angle (Schellart et al. 1995). Several important observations on directional encoding were made, including the following:

1. Some units were classified as directional (about 44%), and some nondirectional (Wubbels and Schellart 1997).
2. Directional units were described as roughly mapped in the TS with the medial TS containing rostrocaudal orientations and the lateral TS containing all possible orientations (Wubbles et al. 1995).
3. Based on individual electrode tracks, the TS was described as having a columnar organization with similar best axes of horizontal motion tending to be constant within vertical columns (Wubbels et al. 1995; Wubbels and Schellart 1998).
4. Some phase-locked units had phase angles of synchronization that did not vary with the stimulus axis angle (except for the expected 180° shift at one angle around the circle), while others showed a phase shift that varied continuously with stimulus angle over 360° (Wubbels and Schellart 1997).

Wubbels and Schellart concluded that these and other results strongly supported the phase model. Further, they speculated that the rostrocaudally oriented units of the medial TS were channels activated by swim bladder–dependent motion input, while the diversely oriented units of the lateral TS represented direct motion input to the otolith organs. The utricle was hypothesized to be the most important otolith organ supplying the direct motion-dependent input because of its horizontal orientation. The authors further speculated that the units with synchronization angles independent of stimulus direction represented pressure-dependent swim bladder inputs while the units with variable synchronization phase angles represented direct motion inputs. Wubbels and Schellart (1997) then concluded that "the phase difference between the(se) two unequivocally encodes the stimulus direction (0 to 360°)" (i.e., solves the 180° ambiguity problem). This conclusion could be strengthened by a mechanistic explanation for the direction-dependent variation in synchronization angle shown by some units and by a testable model for the final step that solves the 180° ambiguity.

8. Summary and Conclusions

1. There are multiple observations on the acoustical behaviors of many fish species that strongly suggest the capacity for sound source localization. Most of these observations take place within the near field of sound sources and likely depend on the response of one or more otolith organs to acoustic particle motion.
2. The question of whether localization takes place in the near or far fields is no longer important because we have learned that processing in the near field does not imply that the lateral line system must be responsible for the behavior. The otolith organs can respond directly to acoustic particle motion in both fields given sufficient amplitude.
3. Most conditioning and psychophysical studies on the discrimination of sound source location provide evidence consistent with the hypothesis that fishes are able to locate sound sources in a way analogous to localization capacities of humans and other tetrapods, both in azimuth and elevation. However, most of these studies fail to unequivocally demonstrate that fishes can actually locate the sources in space (i.e., "know" where the sources are).
4. An explanation for sound source localization behavior at the level of Mauthner cells and other reticulospinal neurons are relatively successful for reflex orientation behaviors but cannot serve to explain conditioning and psychophysical observations on sound source localization.
5. There are several, related theories of sound source localization by fishes. All present theories postulate that the process begins with the determination of the axis of acoustic particle motion by processing the profile of activity over an array of peripheral channels that directly reflect diverse hair cell and receptor organ orientations ("vector detection").
6. Neurophysiological studies on units of the auditory nerve and brainstem are consistent with the notion of vector detection and show that most brainstem cells maintain and enhance the directionality originating from hair cell directionality at the periphery. However, goldfish and other Otophysi present a clear

problem for this view because there is little or no variation of hair cell directionality in the saccule and at the level of the midbrain. This has led to the speculation that Otophysi use other otolith organs in addition to the saccule for encoding the axis of acoustic particle motion. At present, it is still unclear whether or not goldfish and other Otophysi are able to locate sound sources.

7. Vector detection leaves an essential "180° ambiguity" as an unsolved problem (Which end of the axis points to the source? In what direction is the sound propagating?). The "phase model" of directional hearing has been moderately successful in solving this ambiguity as part of a more complex theory, and in experiments deriving from the theory. However, the 180° ambiguity is not the only essential ambiguity that occurs in sound source localization throughout the vertebrates. It is not certain that immediate auditory processing, alone, must be able to solve this problem for appropriate behavior to occur with respect to sound sources.

8. While the phase model is successful in a general sense, it is difficult to apply in several important cases (i.e., for fishes lacking swim bladders, and for Otophysi) where effectively independent representations of the particle motion and pressure waveforms are required by the theory but are not evident in the ear or through more peripheral structures.

9. Additional problems for vector detection and the phase model are that the axis of acoustic particle motion points directly at the source only for monopole sources, and that clear and unambiguous representations of waveform phase have not been observed in auditory nerve units (distributions of phase-locking angles tend to be uniform).

10. While there are behavioral and electrophysiological observations that are consistent with sound source localization in fishes, there are no examples of localization capacities in a single species that have a complete and satisfying description and theoretical explanation. Sound source localization in fishes remains incompletely understood.

Acknowledgments. The writing of this chapter was supported by research grants from the NIH (NIDCD) to R. Fay. Thanks to Peggy Edds-Walton for editorial suggestions, comments, and additions that have improved the manuscript. Thanks to Arthur Popper for editorial comments. Thanks also to my colleagues in the study of hearing in vertebrate animals for their attention to the question of sound source localization in fishes.

References

Allen J (1996) OHCs shift the excitation pattern via BM tension. In: Lewis ER, Long GR, Lyon RF, Narins PM, Steele CR, Hecht-Poinar E (eds), Diversity in Auditory Mechanics. Singapore: World Scientific, pp. 167–175.

Bartelmez GW (1915) Mauthner's cell and the nucleus motorious tegmenti. J Comp Neurol 25:87–128.

Blaxter JH, Denton EJ, Gray JAB (1981) Acousticolateralis system in clupeid fishes. In: Tavolga WN, Popper AN, Fay RR (eds), Hearing and Sound Communication in Fishes. New York: Springer-Verlag, pp. 39–56.

Braun C, Coombs S, Fay R (2002) Multisensory interactions within the octavolateralis systems: What is the nature of multisensory integration? Brain Behav Evol 59:162–176.

Bregman AS (1990) Auditory Scene Analysis. The Perceptual Organisation of Sound. Cambridge, MA: MIT Press.

Buwalda RJA, Schuijf A, Hawkins AD (1983) Discrimination by the cod of sounds from opposing directions. J Comp Physiol A 150:175–184.

Canfield JG, Eaton RC (1990) Swim bladder acoustic pressure transduction initiates Mauthner-mediated escape. Nature 347:760–762.

Canfield JG, Rose GJ (1996) Hierarchical sensory guidance of Mauthner-mediated escape response in goldfish (*Carassius auratus*) and cichlids (*Haplochromis burtoni*). Brain Behav Evol 48:137–156.

Chapman CJ (1973) Field studies of hearing in teleost fish. Helgolander Meeresunters 24:371–390.

Chapman CJ, Hawkins AD (1973) A field study of hearing in the cod, *Gadus morhua* L. J Comp Physiol A 85:147–167.

Chapman CJ, Johnstone ADF (1974) Some auditory discrimination experiments on marine fish. J Exp Biol 61:521–528.

Chapman CJ, Sand O (1974) Field studies of hearing in two species of flatfish, *Pleuronectes platessa* (L.) and *Limanda limanda* (L.) (Family Pleuronectidae). Comp Biochem Physiol 47:371–385.

Cherry EC (1953) Some experiments on the recognition of speech, with one and with two ears. J Acoust Soc Am 25:975–979.

Dale T (1976) The labyrinthine mechanoreceptor organs of the cod (*Gadus morhua* L. (Teleostei: Gadidae). Norw J Zool 24:85–128.

de Munck JC, Schellart NAM (1987) A model for the nearfield acoustics of the fish swim bladder and its relevance for directional hearing. J Acoust Soc Am 81:556–560.

de Vries HL (1950) The mechanics of the labyrinth otoliths. Acta Oto-Laryngol 38: 262–273.

Dijkgraaf S (1960) Hearing in bony fishes. Proc Roy Soc B 152:51–54.

Eaton RC, Canfield JG, Guzik AL (1995) Left-right discrimination of sound onset by the Mauthner system. Brain Behav Evol 46:165–179.

Edds-Walton P, Fay RR (2003) Directional selectivity and frequency tuning of midbrain cells in the oyster toadfish, *Opsanus tau*. J Comp Physiol 189:527–543.

Edds-Walton PL, Fay RR, Highstein SM (1999) Dendritic arbors and central projections of auditory fibers from the saccule of the toadfish (*Opsanus tau*). J Comp Neurol 411: 212–238.

Enger PS, Hawkins AD, Dand O, Chapman CJ (1973) Directional sensitivity of saccular microphonic potentials in the haddock. J Exp Biol 59:425–434.

Fay RR (1981) Coding of acoustic information in the eighth nerve. In: Tavolga W, Popper AN, Fay RR (eds), Hearing and Sound Communication in Fishes. New York: Springer-Verlag, pp. 189–219.

Fay RR (1984) The goldfish ear codes the axis of acoustic particle motion in three dimensions. Science 225:951–954.

Fay RR, Edds-Walton PL (1997a) Directional response properties of saccular afferents of the toadfish, *Opsanus tau*. Hear Res 111:1–21.

Fay RR, Edds-Walton PL (1997b) Diversity in frequency response properties of saccular afferents of the toadfish (*Opsanus tau*). Hear Res 113:235–246.

Fay RR, Edds-Walton PL (1999) Sharpening of directional auditory responses in the descending octaval nucleus of the toadfish (*Opsanus tau*). Biol Bull 197:240–241.

Fay RR, Edds-Walton PL (2001) Bimodal units in the torus semicircularis of the toadfish (*Opsanus tau*). Biol Bull 201:280–281.

Fay RR, Olsho LW (1979) Discharge patterns of lagenar and saccular neurones of the goldfish eighth nerve: displacement sensitivity and directional characteristics. Comp Biochem Physiol 62:377–386.

Fay RR, Coombs SL, Elepfandt A (2002) Response of goldfish otolithic afferents to a moving dipole sound source. Bioacoustics 12:172–173.

Fine M, Winn H, Olla B (1977) Communication in fishes. In: Sebeok T (ed), How Animals Communicate. Bloomington: Indiana University Press, pp. 472–518.

Fish JF (1972) The effect of sound playback on the toadfish. In: Winn HE, Olla BL, (eds), Behavior of Marine Animals, Vol. 2. New York: Plenum, pp. 386–434.

Flock Å (1964) Structure of the macula utriculi with special reference to directional interplay of sensory responses as revealed by morphological polarization. J Cell Biol 22:413–431.

Flock Å (1965) Electron microscopic and electrophysiological studies on the lateral line canal organ. Acta Oto-laryngol Suppl 199:1–90.

Furshpan EJ, Furukawa T (1962) Intracellular and extracellular responses of the several regions of the Mauthner cell of the goldfish. J Neurophysiol 25:732–771.

Gray GA, Winn HE (1961) Reproductive ecology and sound production of the toadfish, *Opsanus tau*. Ecology 42:274–282.

Guzik AL, Eaton RC, Mathis DW (1999) A connectionist model of left-right discrimination by the Mauthner system. J Comp Neurosci 6:121–144.

Harris GG, van Bergeijk WA van (1962) Evidence that the lateral line organ responds to near-field displacements of sound sources in water. J Acoust Soc Am 34:1831–1841.

Hawkins AD, Horner K (1981) Directional characteristics of primary auditory neurons from the cod ear. In: Tavolga WN, Popper AN, Fay RR (eds), Hearing and Sound Communication in Fishes. New York: Springer-Verlag, pp. 311–328.

Hawkins AD, Sand O (1977) Directional hearing in the median vertical plane by the cod. J Comp Physiol 122:1–8.

Heffner RS, Heffner HE (1982) Hearing in the elephant (*Elephas maximus*): absolute sensitivity, frequency discrimination, and sound localization. J Comp Psychol 96:926–944.

Hirsh IJ (1948) The influence of interaural phase on interaural summation and inhibition. J Acoust Soc Am 20:536–544.

Horner K, Sand O, Enger PS (1980) Binaural interaction in the cod. J Exp Biol 85:323–331.

Kalmijn AJ (1997) Electric and near-field acoustic detection, a comparative study. Acta Physiol Scand 161 (Suppl 638): 25–38.

Kleerekoper H, Chagnon EC (1954) Hearing in fish with special reference to *Semotilus atromaculatus atromaculatus* (Mitchill). J Fish Res Bd Can 11:130–152.

Kleerekoper H, Malar T (1968) Orientation through sound in fishes. In: de Reuck AVS, and Knight J (eds), Hearing Mechanisms in Vertebrates. London: J & A Churchill, pp. 188–201.

Korn H, Faber DS (1975) An electrically mediated inhibition in goldfish medulla. J Neurophysiol 38:452–471.

Leis JM, Carson-Ewart BM, Hay AC, Cato DH (2003) Coral-reef sounds enable nocturnal navigation by some reef-fish larvae in some places and at some times. J Fish Biol 63: 724–737.

Lu Z, Song J, Popper AN (1998) Encoding of acoustic directional information by saccular afferents of the sleeper goby, *Dormitator latifrons*. J Comp Physiol A 182:805–815.

Lu Z, Xu Z, Buchser WJ (2003) Acoustic response properties of lagenar nerve fibers in the sleeper goby, *Dormitator latifrons*. J Comp Physiol A 189:889–905.

Ma W-L, Fay RR (2002) Neural representations of the axis of acoustic particle motion in the nucleus centralis of the torus semicircularis of the goldfish, *Carassius auratus*. J Comp Physiol A 188:301–313.

McKibben JR, Bass AH (1998) Behavioral assessment of acoustic parameters relevant to signal recognition and preference in a vocal fish. J Acoust Soc Am 104:3520–3533.

McKibben JR, Bass AH (2001) Effects of temporal envelope modulation on acoustic signal recognition in a vocal fish, the plainfin midshipman. J Acoust Soc Am 109: 2934–2943.

Moulton JM, Dixon RH (1967) Directional hearing in fishes. In: Tavolga WN (ed), Marine Bio-acoustics, Vol. II. New York: Pergamon Press, pp. 187–228.

Mueller TJ (1981) Goldfish respond to sound direction in the Mauthner-cell initiated startle behavior. Ph.D. dissertation, University of Southern California, Los Angeles.

Myrberg AA, Ha SJ, Walewski S, Banbury JC (1972) Effectiveness of acoustic signals in attracting epipelagic sharks to an underwater sound source. Bull Mar Sci 22:926–949.

Nelson DR (1965) Hearing and acoustic orientation in the lemon shark *Negaprion brevirostris* (Poey) and other large sharks. Ph.D. dissertation, University of Miami.

Platt C (1977) Hair cell distribution and orientation in goldfish otolith organs. J Comp Neurol 172:283–297.

Popper AN (1977) A scanning electron microscopic study of the sacculus and lagena in the ears of fifteen species of teleost fishes. J Morphol 153:397–418.

Popper AN, Salmon A, Parvulescu (1973) Sound localization by the Hawaiian squirrelfishes, *Myripristis berndti* and *M. argyromus*. Anim Behav 21:86–97.

Pumphrey RJ (1950) Hearing. Symp Soc Exp Biol 4:1–18.

Rakerd B, Hartmann WM (1985) Localization of sound in rooms. II: The effects of a single reflecting surface. J Acoust Soc Am 78:524–533.

Reinhardt F (1935) Uber Richtungswharnehmung bei Fischen, besonders bei der Elritze (*Phoxinus laevis* L.) und beim Zwergwels (*Amiurus nebulosus* Raf.). Z Vergl Physiol 22:570–603.

Rogers PH, Popper AN, Cox M, Saidel WM (1988) Processing of acoustic signals in the auditory system of bony fish. J Acoust Soc Am 83:338–349.

Sand O (1974) Directional sensitivity of microphonic potentials from the perch ear. J Exp Biol 60:881–899.

Schellart NAM, Buwalda RJA (1990) Directional variant and invariant hearing thresholds in the rainbow trout (*Salmo gairdneri*). J Exp Biol 149:113–131.

Schellart NAM, de Munck JC (1987) A model for directional and distance hearing in swimbladder-bearing fish based on the displacement orbits of the hair cells. J Acoust Soc Am 82:822–829.

Schellart NAM, Wubbels RJ, Schreurs W, Faber A (1995) Two-dimensional vibrating platform in nm range. Med Biol Eng Comp 33:217–220.

Schuijf A (1975) Directional hearing of cod (*Gadus morhua*) under approximate free field conditions. J Comp Physiol A 98:307–332.
Schuijf A (1981) Models of acoustic localization. In: Tavolga WN, Popper AN, Fay RR (eds), Hearing and Sound Communication in Fishes. New York: Springer-Verlag, pp. 267–310.
Schuijf A, Buwalda RJA (1975) On the mechanism of directional hearing in cod (*Gadus morhua*). J Comp Physiol A 98:333–344.
Schuijf A, Hawkins AD (1983) Acoustic distance discrimination by the cod. Nature 302: 143–144.
Schuijf A, Siemelink M (1974) The ability of cod (*Gadus morhua*) to orient towards a sound source. Experientia 30:773–774.
Schuijf A, Baretta JW, Windschut JT (1972) A field investigation on the deiscrimination of sound direction in *Labrus berggylta* (Pisces: Perciformes). Netherl J Zool 22:81–104.
Schuijf A, Visser C, Willers A, Buwalda RJ (1977) Acoustic localization in an ostariophysine fish. Experientia 33:1062–1063.
Tavolga WN (1971) Acoustic orientation in the sea catfish, *Galeichthys felis*. Ann NY Acad Sci 188:80–97.
Tavolga WN (1976) Acoustic obstacle detection in the sea catfish (*Arius felis*). In: Schuijf A, Hawkins AD (eds), Sound Reception in Fish. Amsterdam: Elsevier, pp. 185–204.
van Bergeijk WA (1964) Directional and nondirectional hearing in Fish. In: Tavolga WA (ed), Marine Bioacoustics. London: Pergamon Press, pp. 269–301.
van Bergeijk WA (1967) The evolution of vertebrate hearing. In: Neff WD (ed), Contributions to Sensory Physiology, Vol 2. New York: Academic Press, pp. 1–49.
van den Berg AV, Buwalda RJA (1994) Sound localization in underwater sound fields. Biol Cybern 70:255–265.
van den Berg AV, and Schuijf A (1983) Discrimination of sounds based on the phase difference between the particle motion and acoustic pressure in the shark *Chiloscyllium griseum*. Proc Roy Soc Lond B 218:127–134.
von Frisch K, Dijkgraaf S (1935) Can fish perceive sound direction? Z Vergl Physiol 22:641–655.
Weeg MS, Bass AH (2002) Frequency response properties of lateral line superficial neuromasts in a vocal fish, with evidence for acoustic sensitivity. J Neurophysiol 88: 1252–1262.
Weeg M, Fay RR, Bass A (2002) Directional response and frequency tuning in saccular nerve fibers of a vocal fish, *Porichthys notatus*. J Comp Physiol 188:631–641.
Wightman F, Kistler D (1993) Sound localization. In: Yost WA, Popper AN, Fay RR (eds), Human Psychophysics. New York: Springer-Verlag, pp. 155–192.
Winn HE (1964) The biological significance of fish sounds. In: Tavolga WN (ed), Marine Bioacoustics. New York:Pergamon Press, pp. 213–231.
Wubbels RJ, Schellart NAM (1997) Neuronal encoding of sound direction in the auditory midbrain of the rainbow trout. J Neurophysiol 77:3060–3074.
Wubbels RJ, Schellart NAM (1998) An analysis of the relationship between the response characteristics and toporgaphy of directional- and non-directional auditory neurons in the torus semicircularis of the rainbow trout. J Exp Biol 201:1947–1958.
Wubbles RJ, Schellart NAM, Goossens JHHLM (1995) Mapping of sound direction in the trout lower midbrain. Neurosci Lett 199:179–182.
Zottoli SJ, Bentley AP, Prendergast BJ, Riell HI (1995) Comparative studies on the Mauthner cell of teleost fish in relation to sensory input. Brain Behav Evol 56:151–164.

4

Directional Hearing in Nonmammalian Tetrapods

Jakob Christensen-Dalsgaard

1. Introduction

The nonmammalian tetrapods—amphibians, reptiles and birds—are a diverse assembly of animals with body mass ranging from below 1 g to more than 100 kg and adapted to almost any habitat on Earth. Apart from being tetrapods, these animals do not form a natural group. However, they share an important functional characteristic—a tympanic ear with a single auditory ossicle—and, as will be outlined below, the limitations of this monossicular ear may impose common constraints on the directional hearing of these species. Another shared constraint in all vertebrates is the general, conserved organization of inner ear and central auditory pathways.

The focus of this chapter is the origin of directional hearing, the general principles of directionality of the monossicular ear, and the special characteristics of directional hearing in the different groups. The main thesis is that the primitive condition of the ear in all groups is one in which the tympana are air coupled and that a pressure-sensitive ear represents a later (derived) specialization. Also, following the current view of the independent evolution of tympanic hearing in these groups, the differences in the organization of neural processing of directional hearing are outlined. This chapter does not attempt to review the older litterature on directional hearing in detail, since it has been covered in excellent reviews (e.g., Fay and Feng 1987; Eggermont 1988; Klump 1988, 2000).

1.1 Origin of the Monossicular Ear

The earlier view of the evolution of tetrapod hearing was based on the general similarity of the tympanic ears of tetrapods and stated that tympanic hearing emerged early in the tetrapods (or even before the tetrapods, van Bergeijk 1966) and was conserved in the lineages leading to recent amphibians and amniotes (Goodrich 1930). However, this view was challenged by Lombard and Bolt (1979) and Bolt and Lombard (1985), who provided evidence from the morphology of the middle ear in recent amphibians and their tetrapod ancestors, leading to the conclusion that tympanic hearing had evolved independently in anurans (frogs and toads) and in the amniotes. Studies on fossil tetrapods have shown that a tympanic middle ear is not a primitive characteristic of tetrapods (Clack 1993) and that even the amniote ancestors probably did not have a tympanic ear (Clack 1997; Manley and Clack 2004). Therefore, the informed consensus today is that the columellar–tympanum connection has emerged independently at least five times, that is, in the lines leading to amphibians, turtles, lepidosaurs (lizards and snakes), archosaurs (crocodiles and birds), and mammals, and that the inner ear (but not the auditory organs!) and middle ear bone (columella/stapes) is homologous in the tetrapods (Lombard and Bolt 1979; Clack 1997; Manley and Köppl 1998).

In this light, the tympanic ears of all groups are independently derived traits, and, furthermore, many of the similarities of the tympanic ears in tetrapods are probably caused by convergent evolution. Also, it is not self-evident anymore that all the central auditory nuclei are homologous in the tetrapods beyond the basic homology as octaval nuclei (McCormick 1999). An obvious, but still important, point to note is that none of the extant groups can be regarded as representing the ancestral condition of any of the others.

1.2 Origin of Directional Hearing

Unfortunately, directional hearing is not linked to any specific morphological character and therefore it cannot be traced in the fossil record. It would be tempting to link the emergence of directional hearing to the emergence of the tympanic ear, but this would be incorrect, since also atympanic ears can show directionality. For example, frogs show enhanced, nontympanic directionality at low frequencies (see Section 3.3.4). Similarly, ancestral tetrapods, even if atympanic, could have had a crude (low-frequency) directional hearing, since the hair cells in their sensory maculae would encode the direction of vibrations of the skull induced by sound: stimulation along the hair cell's axis produces maximal responses with 180° phase difference for stimulation parallel and antiparallel to the hair cell's orientation (see Jørgensen and Christensen-Dalsgaard 1997b and Section 3.3.4). Thus, binaural amplitude and phase comparisons would probably already be useful to sharpen the directional response, and some of the neuronal substrate subserving directional hearing could already have been in place from the early tetrapods. However, the emergence of tympanic hearing changed directional hearing by (1) increasing sensitivity, (2) extending the frequency range, (3) enabling the auditory system to use time-of-arrival and intensity difference cues, and (4) enabling a new directional mechanism by acoustical coupling of the eardrums. Therefore, the emergence of tympanic hearing is an important landmark in the evolution of directional hearing.

The anurans (frogs and toads), the only amphibians that have a tympanic membrane, probably emerged in the Triassic. In the amniote lineages, tympanic hearing also emerged during the Triassic (Clack 1997; Manley and Köppl 1998; Manley and Clack 2004). It has been speculated that this timing coincides with the evolution of sound-producing insects (earliest orthopterans date from the Permian; Hoy 1992) and that the evolutionary push for high-frequency hearing occurred in small insectivores of the different amniote lineages. If this hypothesis is true, localization of sounds associated with prey organisms would also have been a major driving force in the initial evolution of the tympanic ear that was later exploited by secondary adaptations for sound communication in the anurans and some of the amniote lineages.

2. General Properties of the Monossicular Ear

2.1 Structure of the Ear

The structure of the auditory periphery in a representative anuran, lizard, and bird is shown schematically in Figure 4.1. Anurans and lizards show the same general configuration in that they have middle ear cavities that are connected through the mouth cavities by relatively large, permanently open Eustachian tubes, but the anuran head (and body) is generally much more transparent to sound than the head and body of the other groups (see Section 3.3). In birds

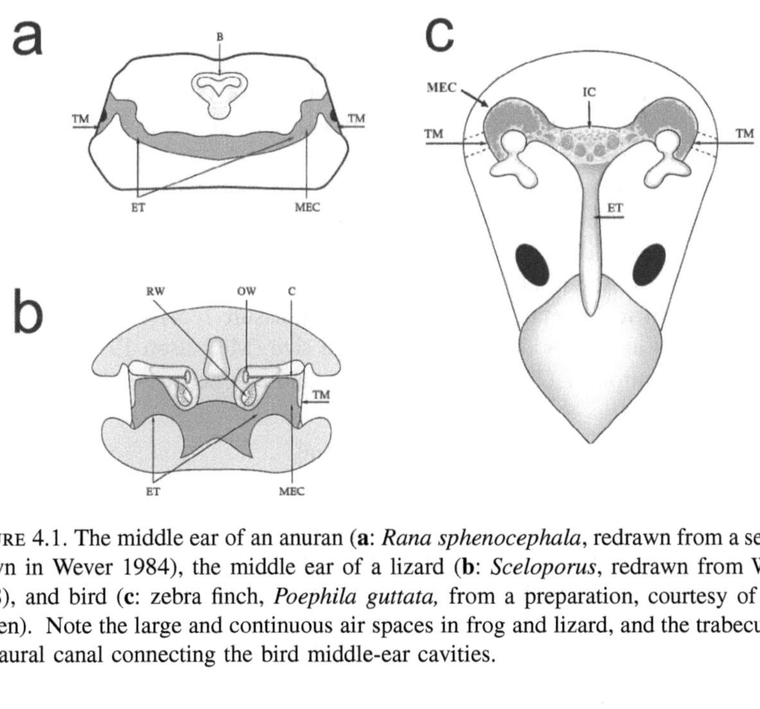

FIGURE 4.1. The middle ear of an anuran (**a**: *Rana sphenocephala*, redrawn from a section shown in Wever 1984), the middle ear of a lizard (**b**: *Sceloporus*, redrawn from Wever 1978), and bird (**c**: zebra finch, *Poephila guttata*, from a preparation, courtesy of O.N. Larsen). Note the large and continuous air spaces in frog and lizard, and the trabeculated interaural canal connecting the bird middle-ear cavities.

and crocodiles, the middle ear cavities are connected by an interaural canal that is connected to the mouth cavity through a common Eustachian tube (Kühne and Lewis 1985). Also, the avian ear has a relatively long external ear canal.

The anurans, reptiles, and birds all have a single auditory ossicle, the columella (homolog to the mammalian stapes) and an extracolumella between columella and eardrum. In reptiles and birds, the extracolumella is an incompletely ossified and complex structure with several processes, whereas in anurans it is a lump of soft cartilage. In all of these species, the columella–extracolumella link has been shown to have an essential function in the impedance matching of the ear by creating a mechanical lever (Manley 1972, 1990; Jørgensen and Kanneworff 1998).

2.2 *Pressure and Pressure-Difference Receivers*

Tympanic ears can be divided in two functional categories. Ears in which the tympanum lines a closed middle ear cavity are pressure receivers and nondirectional, since they respond to sound pressure, a scalar. An example of a pressure receiver ear is the mammalian ear, in which the Eustachian tubes are narrow and usually closed, resulting in functionally closed and separated middle ear cavities. In pressure receiver ears, directional information is extracted by the central nervous system (CNS) using binaural comparisons of the inputs, such as interaural time differences (ITDs) and interaural level differences (ILDs). In mammals, the duplex theory proposes that localization of low- and high-frequency sounds depends on the ITD and ILD, respectively (Wightman et al. 1987). ITD includes both interaural phase differences (IPDs) and onset time differences (OTDs). IPD is by far the most salient cue to direction at low frequencies in humans (Blauert 1997), and a localization ability based on time differences declines with frequency as would be expected by the decline in phase locking by the auditory fibers. However, at high frequencies, in which neural phase locking is decreased, OTD is probably an important cue for localization of more complex sounds, such as amplitude-modulated sounds, based on timing differences of the sound envelope.

The problem for a small animal in using a pressure receiver is that both the maximal ITDs and ILDs depend on the head size (Michelsen 1994). The simplest approximation of ITD, assuming that sound can penetrate the head, is

$$\mathrm{ITD} = \frac{2r}{c} \cdot \sin\theta$$

where r is the head radius, c is sound velocity, and θ the angle of sound incidence. If—more realistically—sound is assumed to propagate along the head perimeter, the expression becomes

$$\mathrm{ITD} = \frac{r}{c} \cdot (\sin\theta + \theta)$$

(Woodworth and Schlosberg 1962; Blauert 1997; see also Klump 2000 for more accurate approximations). For an animal with a head diameter of 2 cm, for example, the maximal ITD (contra- or ipsilateral sound incidence) is 58 μs using the simplest approximation (76 μs assuming propagation along the perimeter). For a 10° sound incidence angle, the ITD would be reduced to 10 μs. ILDs depend on the diffraction by the head and body of the animal. As a rough rule of thumb, diffraction effects are seen when the dimensions of an object is larger than 1/10 of the wavelength (Larsen 1995), for example, 3.4 cm at 1 kHz, where the wavelength is 34 cm (but note that close to this limit the effects will be very small, and robust effects are seen only for objects larger than 1/4 of a wavelength). Therefore, for most frogs, reptiles and smaller birds measurable diffraction effects are restricted to higher frequencies (above 4 kHz).

In contrast, the principle of the pressure gradient or pressure difference receiver ear is that binaural interaction takes place on the tympanum itself. Sound reaches both sides of the tympanic membrane, and the driving force for membrane motion is proportional to the instantaneous pressure difference between the two sides. Obviously, membrane motion depends on phase as well as on amplitude differences between the two sides of the membrane. The pressure gradient receiver is directional, because the phase shift between sounds reaching the two sides of the membrane is directional, and both ILD and ITD cues are larger than in a pressure receiver ear. However, the directivity (the physical directional characteristics of the receiver) is very frequency dependent. At very low frequencies, the phase difference between direct and indirect sound will be small at any direction of sound incidence, so the vibration amplitudes of the membrane will be small. At high frequencies, the phase difference between direct and indirect sound exceeds 360°; hence, the phase cues become ambiguous.

Any ear in which the two tympana are coupled through Eustachian tubes or interaural canals is potentially a pressure difference receiver. However, for the ear to exhibit any significant directionality, the sound from the contralateral ear must reach the ipsilateral ear with little excess attenuation. Evidently, if there is no diffraction or interaural attenuation, so that the amplitudes of direct and indirect sound are equal, the pressure difference will range from 0 (when direct and indirect sound is in phase) to twice the level of direct sound (when the two sides are 180° out of phase). If the indirect sound pressure is 0.5 that of direct sound pressure p, then the pressure difference ranges from $0.5p$ to $1.5\ p$ and, generally, the smaller the indirect component, the smaller the directionality (ITD as well as ILD; see Klump 2000) generated by the pressure difference receiver. However, at high frequencies where sound is diffracted around the head of the animal, the sound amplitudes reaching the shaded ear can be so small that they are comparable in amplitude to sound reaching the ear via internal pathways, even though attenuation through the internal pathways is considerable (Michelsen 1998).

Any tubelike structure such as the Eustachian tubes will exhibit frequency-dependent attenuation depending on its length and thickness, and this will limit the frequency range within which the receiver is directional. In contrast to the ideal pressure difference receiver, which is just a suspended membrane, in real-world ears the eardrums are connected by tubes and cavities, and the ears behave like a combination of a pressure and pressure-difference receiver (a pressure–pressure difference receiver; Fay and Feng 1987).

2.3 Acoustical Models

The directivity index V for a simple receiver consisting of a membrane, a cavity, and an second sound entry can be modeled by

$$V = 20 \log \frac{1 + B\cos\theta}{1 + B}$$

$$B = \frac{\Delta L}{c \cdot C_A \cdot R_A}$$

(Beranek 1986) where B is a dimensionless constant, ΔL is the distance between the eardrums, c is the speed of sound, C_A is the compliance, and R_A is the resistance of the cavity. With a large interaural resistance, B approaches zero (omnidirectional pressure receiver). Conversely, for small resistances (large B) the directivity will approach $V = 20 \log \cos\theta$, producing a figure-of-eight directionality with a null for sound coming from frontal and caudal directions.

More realistic model calculations have been based on electrical network analog of the auditory periphery of frogs and birds (Fletcher 1992; see Fig. 4.2). For an initiation into such network modeling, the reader is referred to Fletcher and Thwaites (1979) and Fletcher (1992). In brief, any acoustical system in which acoustical flow is one dimensional (such as propagation in tubes and through membranes) will have an analogous electrical circuit in which cavities correspond to capacitors, tubes to inductances (at low frequencies; for high-frequency approximations see Fletcher 1992) and sound absorbers to resistances. At high frequencies, a key assumption inherent in the electrical analogy—that the acoustical elements can be treated as lumped elements—is no longer valid. Therefore, as a rule of thumb all elements must be smaller than 0.5 * wavelength (Morse 1948), that is, 3.4 cm at 5 kHz, which means that the models are applicable only to low-frequency hearing in small animals. These kinds of models are of course strongly simplified; nonetheless, with realistic parameters they make it possible to evaluate the contributions of Eustachian tubes and mouth cavity volume to the measured response. As will be noted below, some of the network models fail to give a reasonable fit to the observed data, usually because some of the basic assumptions are violated, for example, that the sound entrances are not localized, but distributed as in the frogs. In these instances, more sophisticated models are needed.

An even more simplified model of the avian auditory periphery than the network analog was developed by Calford (1988). Here, the difference between direct sound and indirect sound, delayed and attenuated during propagation

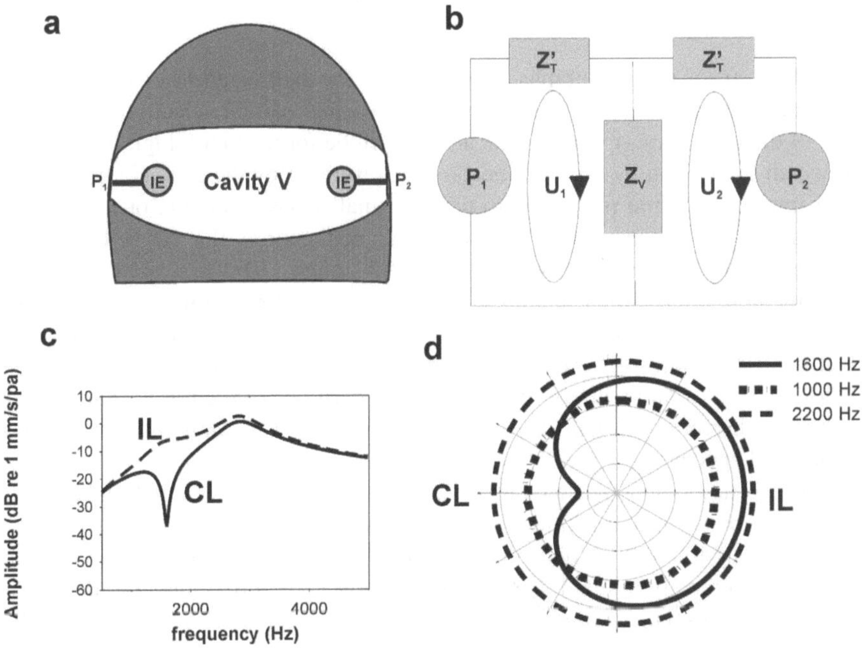

FIGURE 4.2. An example of an analog model of a very simple (lizardlike) ear with two eardrums connected by a large cavity (**a**). The equivalent electrical diagram of such an ear is shown in (**b**). Sound only enters via the tympana (p_1 and p_2) delayed by arrival-time differences, and is filtered by impedances of tympanum (Z'_T) and middle ear cavity (Z_v) before reaching the other tympanum. With 'realistic' parameters the vibration amplitude of the tympana are highly directional in a limited frequency range, as shown by the spectra (**c**) and polar plots (**d**). The parameters used are based on measurements from a lizard. (From Fletcher 1992; Christensen-Dalsgaard and Manley 2005; and unpublished data.)

through the interaural canal, was calculated. The advantages of this model are its conceptual simplicity and that no parameters need to be estimated (the only parameters entering the equations are dimensions of the interaural canal and canal attenuation; both can be measured with relative ease). The model is important, because it allows calculation of the additional delays caused by the indirect pathway. The disadvantage of the model is that the phases of direct and indirect sound are not realistic, since the complex impedances associated with inductance of the interaural canal and Eustachian tube and with the capacitance of the mouth cavity that would generate a (frequency-dependent) phase shift of the indirect signal are neglected. It may be advantageous to use the models discussed by Fletcher (1992, see e.g., pp. 208–212), since they are based on a realistic, if simplified, acoustical description of the system.

2.4 Optimization of Directionality

It follows from the paragraphs above that in the optimization of the auditory periphery of a small animal for directional hearing, natural selection can follow one of three courses. Either (1) the animal will be forced to use high frequencies at which diffraction by the head becomes measurable by the ears, (2) the animal will improve its time resolution to use the small ITDs available, or (3) by manipulating the dimensions of the interaural connections the ear will become inherently directional in a limited frequency range. The first solution is the mammalian, where small mammals extend their frequency range of hearing to frequencies at which diffraction produces useful directional cues (Heffner and Heffner 1992), but also used by some bird species, most notably the owls (see below). The second solution is also used by barn owls (Section 5.2) and probably by other bird species. The third solution is used by anurans, reptiles, and some bird species.

There is an obvious reason why nonmammalian tetrapods cannot extend their frequency range to that of mammals of comparable size. As shown by Manley (1972, 1990) for reptiles and birds the ear is impedance matched by insertion of the incompletely ossified extracolumella between eardrum and columella. At low frequencies vibrations of the eardrum is transferred with little loss to the columellar footplate. At higher frequencies, however, the transfer breaks down because the extracolumella–columella link flexes, and a further high-frequency limitation is that the tympanic membrane vibration tends to break up in higher vibration modes at higher frequencies (Manley 1972). In these modes, nodes can form at the extracolumellar attachment, effectively limiting the sound input to the inner ear. The high-frequency limit of the nonmammals does not exceed 12 kHz, and for most species sensitivity drops around 5 kHz. Here, the wavelength is 6.8 cm, so for the smallest animals (head size around 1 cm) diffraction might just be measurable.

2.5 The Pressure Difference Receiver: Primitive or Derived?

Consider the emergence of a tympanic ear in a generalized ancestral tetrapod. This animal would have a mouth cavity and a columella probably functioning as a structural element (see Clack 1997). It is likely that the first eardrums would have been especially compliant areas of skin covering fenestrations in the skull and contacted by the columella. The most compliant areas would connect directly to the mouth cavity. Thus, there would be little obstruction of sound reaching the internal surface of the eardrum, and, hence, the ear would in a certain frequency range function as a pressure difference receiver. In contrast, the pressure receiver in the terrestrial tetrapods is considerably more complicated, since it depends on a specialized structure, that is, an enclosed cavity behind the tympanic membrane. Therefore, the pressure receiver ear is probably a derived condition and the pressure difference receiver in the amphibians and reptiles reflects the primitive condition—not necessarily as a special adaptation for directional hearing, but rather as a byproduct of having an ear that is coupled to the respiratory structures. What, then, are the selection pressures that can lead to the pressure receiver?

It is evident that some undesirable effects will result from the ears being coupled to the respiratory and food-intake pathways. One effect is that breathing will produce noise that is very efficiently coupled to the eardrum. However, it may be of equal consequence that the frequency and directional characteristics of the ear are quite variable. Inflation of the lungs and changes of the volume of the mouth cavity will change the characteristics of the ear (Rheinlaender et al. 1981, Jørgensen et al. 1991), and this instability of the ear directivity may pose a problem for the processing of directional information by the CNS. Frog calls tend to be at frequencies at which the directionality of their ear is not maximal, but stable, probably because at the frequencies where the directionality is maximal it is also quite variable (Jørgensen 1991; Jørgensen and Gerhardt 1991; see below). Thus, a step in the evolution of a pressure receiver could be to isolate the ear from the respiratory pathway (see also Clack 1997). The development of a separate interaural canal in the archosaurs (crocodiles and dinosaurs including birds) can be seen as a step in this direction.

Another disadvantage of the pressure difference receiver is that the two ears are functionally converted to one directional ear, since they are coupled. Thus, monaural directional cues generated by diffraction cannot be used. Such cues could aid the segregation of sound components from multiple sources, for example, in distinguishing between one sound source located equidistantly from the ears and two sound sources stimulating each ear equally (auditory streaming, Bregman 1990).

Obviously, an advantage of the pressure-difference receiver is that it produces a directional response at low frequencies, whereas an unspecialized ear may have few directional cues. However, the drawback is that the directionality is strongly frequency dependent. Consequently, the useful frequencies for sound localization may lie in a restricted frequency range. In the context of sound communication, the animal can place its signals within the operational range of the pressure-difference receiver. However, for animals that rely on passive hearing the sound emitted by important sources (such as high-frequency rustling noises made by prey) may lie outside of the useful frequency range. For example, it has been speculated that an evolutionary push for the development of mammalian high-frequency hearing could have been that insect prey increased the frequencies of their communication sounds into the ultrasonic range (Clack 1997). The resulting selection pressure to detect and localize such sounds would lead to improved high-frequency hearing, to a reliance on diffraction cues, and to further the functional isolation of the two ears.

3. Directional Hearing in Amphibians

Directional hearing in amphibians has only been studied in anurans (the frogs and toads). Both the urodeles (salamanders) and the caecilians lack a tympanic ear (although the columella is present), but they may still possess directional hearing comparable to the low-frequency extratympanic hearing of anurans (see Section 3.3.4). Apart from a recent study showing that the marbled newt, *Triturus marmoratus*, will perform phonotaxis toward sympatric anuran choruses (Diego-Rasilla and Luengo 2004), far less is known about these groups than about the anurans.

3.1 Behavioral Studies of Frog Directional Hearing

In almost all species of anurans, males produce advertisement calls to attract females, and the female's identification and localization of the advertisement call is a prerequisite for successful mating. Furthermore, given that the female incurs increased predation risks during her phonotactic approach, it is a reasonable assumption that natural selection should act to shorten the phonotactic approach by maximizing her efficiency in localizing conspecific calls.

It is natural, therefore, that the main focus of behavioral studies of frog directional hearing has been on localization of the mating call, especially since the only robust sound localization behavior in anurans thus far has been observed in phonotaxis. As pointed out by Gerhardt (1995), the problem with this "naturalistic" approach is that behavioral studies on anurans are difficult to compare with psychophysical experiments using conditioning in other animal groups, because there is no way to test the localizability of nonattractive signals (Klump 1995). However, it has proved to be difficult to condition anurans to acoustic stimuli. Food conditioning does not seem to work with acoustic stimuli. So far, the only quantitative results have been obtained with aversive conditioning (Elepfandt et al. 2000) and reflex modification (Megela-Simmons et al. 1985). None of these methods seem to work very robustly in frogs and they have not been applied to directional hearing studies.

In earlier phonotaxis experiments, frogs (usually gravid females) were placed in an arena and the localization path toward a loudspeaker continuously playing the advertisement call was recorded. Not all frog species work equally well in such a setup, but some of the hylid treefrogs have a very robust and consequently well-studied phonotactic behavior. It is unfortunate, though, that the ranid "laboratory" grass frogs *Rana temporaria* and *R. pipiens*, on which the bulk of physiological experiments have been performed, do not exhibit very robust phonotaxis.

A study of the sound localization behavior in two treefrog species, *Hyla cinerea* and *H. gratiosa* (Feng et al. 1976) showed that gravid females could accurately locate the sound source (a loudspeaker broadcasting the mating call continuously at a level comparable to a calling male—86 dB SPL in 2m distance). The phonotactic paths were shown, but the accuracy was not quantified.

Furthermore, the frogs were unable to locate the sound source when one eardrum was covered with vaseline (Feng et al. 1976). In this case, the frogs would turn in circles toward the unoccluded ear, indicating that interaural comparison (acoustical—by the pressure difference receiver—and neural) is necessary for a normal phonotactic response. The accuracy of phonotaxis toward the mating call was quantified in *H. cinerea* by Rheinlaender et al. (1979). The average jump error was 16.1°, but the head orientation error after head scanning movements was smaller (mean 8.4°), as was the jump error after scanning (11.8°), suggesting that scanning improves the localization accuracy. Later, azimuthal localization accuracy was quantified in the species *H. versicolor* (mean jump error angle 19.4°, Jørgensen and Gerhardt 1991), *Hyperolius marmoratus* (mean jump error angle 22.0°, Passmore et al. 1984), and, interestingly, in the small dendrobatid *Colostethus nubicola* (mean jump error angle 23.2°, Gerhardt and Rheinlaender 1980). In other words, azimuthal localization accuracy is remarkably similar in the species studied (around 20°), including in the very tiny *C. nubicola*. However, in this species the small head width (5 mm) is probably compensated for by the high-frequency advertisement call (5 to 6 kHz).

All of the studies discussed above have been so-called closed-loop experiments, in which sound is emitted continuously (Rheinlaender and Klump 1988). In closed-loop experiments the frogs can locate the sound source using lateralization by scanning movements of the head until the ears are equally stimulated, or even simply by following the pressure gradient (by moving in the direction of increasing sound level). In contrast, true angle discrimination must be investigated in open-loop experiments, in which the sound is switched off after the frog has made an orienting response (Klump 1995). Such brief sounds are not attractive in all frog species, and angle discrimination has so far been demonstrated only in the barking treefrog, *H. gratiosa*, that does respond to single sound pulses. Head orienting and jump error angles are 21.2° and 24.6°, respectively (Klump and Gerhardt 1989).

The role of head scanning for localization acuity is a matter of current debate. As mentioned above, head scanning improved localization in *Hyla cinerea* and *Hyperolius marmoratus*. However, in the open-loop study of *Hyla gratiosa*, the localization acuity without scanning was comparable to the acuity in the other species. Furthermore, lateral scanning movements were not observed in *Hyla versicolor* (Jørgensen and Gerhardt 1991).

Many arboreal frog species locate conspecifics calling from elevated sites, that is, they have to localize sound in elevation as well as in azimuth. Localization of sound in elevation was first demonstrated in *Hyla cinerea* (Gerhardt and Rheinlaender 1982). The mean jump error angle of *Hyperolius marmoratus* in a three-dimensional grid (closed-loop) was 43.0°, that is, approximately twice as large as the error angle in a two-dimensional grid (Passmore et al. 1984). In *Hyla versicolor*, the mean three-dimensional error angle was 23° (excluding vertical climbs; with vertical climbs the error angle was 36°), close to the azimuthal error angle of 19.4° (Jørgensen and Gerhardt 1991). The localization of elevated sound sources is still difficult to explain, since mechanisms such as binaural comparisons for azimuthal localization cannot be invoked. The frequency response of the eardrum (see below) seems to vary systematically with elevation (Jørgensen and Gerhardt 1991), but the frog needs a reference to utilize this cue. Vertical head scanning movements would be a possible way to compare auditory responses at different elevation angles, but such movements are not reported for *H. versicolor*. However, the frogs make quick orientation movements in response to sound onset (Jørgensen and Gerhardt 1991), and such movements might enable the frogs to compare different elevation angles. Another possibility is that the frog has some kind of acoustic memory enabling a comparison of sequential responses at different elevation angles. In both cases the elevation angle determination should work only in a closed-loop experiment. Thus, it would be interesting to investigate three-dimensional phonotaxis in an open-loop experiment.

While most of the studies reviewed so far have dealt with localization of single sources, in the real world frogs face the problem of localizing and discriminating in the presence of several sound sources, whether they be masking noise or calling males emitting sound more or less synchronously in a chorus. For example, female *H. gratiosa* were attracted and made accurate phonotactic movements toward a chorus of calling males at least 160 m away (Gerhardt and Klump 1988). The detection of the advertisement call in noise by *H. cinerea* females was shown to improve, but only 3 dB or less, with angular separation of masker and target when the separation was 45° or 90° (Schwartz and Gerhardt 1989). Other angles were not tested, but if the spatial release from masking reflects the directionality of the auditory system, an effect of angular separation should be expected at least down to the 20° found in the phonotaxis experiments.

The ability to separate simultaneously calling males has been investigated in *Hyperolius marmoratus* (Passmore and Telford 1981). Here, neither phonotactic paths nor duration of the phonotactic approach was affected by simultaneous playback of the mating call from two speakers placed 0.5 m apart (corresponding to an angular separation of approximately 35° at the release point of the frog). In a clever experiment, female *H. versicolor* was presented with advertisement calls emitted from either adjacent or spatially separated speakers (Schwartz and Gerhardt 1995). The calls were time shifted so that calls from speaker pairs overlapped, thereby obscuring the temporal pattern (in fact, changing it to the temporal pattern of *H. chrysoscelis*). The test was whether the females would choose the spatially separated pair over the adjacent pair, and it was shown that females would choose pairs separated by 120°, but not by 45°. Even at 120° the preference could be counteracted by dropping the level of one of the adjacent speakers by 3 dB. Compared to neurophysiological data (midbrain multiunit recordings) that showed a 9 dB release from masking for a 120° angular separation, the behavioral performance seems to be relatively poor. One reason may be that the behavioral experiments do not measure directionality as such, but rather female choice—not whether the sounds presented are localizable, but also whether they are attractive. For example, as mentioned by the authors, the female performance could be offset by a preference for closely spaced calling males. However, an alternative interpretation is that processing of sound from multiple sources degrades the localization accuracy, indicating that the separation of sounds emitted simultaneously from multiple sources (i.e., auditory streaming, Bregman 1990) should be difficult for the frog, maybe as a result of the acoustical coupling of the two ears.

3.2 Structure of the Frog Ear

A schematic diagram of the frog ear is shown in Figure 4.3 (see Lewis and Narins 1999 for a full review of amphibian ear structure). In brief, the two large middle-ear cavities are bounded by a tympanum and coupled through the mouth cavity by wide, permanently open Eustachian tubes (see Fig. 4.1a). Vibrations of the tympanum are coupled to the inner ear through the middle ear bone, the columella. The inner ear is encased in the otic capsule that has two major openings, the round and oval window. The columellar footplate sits in the oval window, and uniquely to the amphibians, a second movable element is inserted in the oval window. This is the operculum, which is connected to the scapula through the opercularis muscle. Vibrations generated by the columellar footplate or operculum at the oval window travel through the otic capsule to the other pressure release window, the round window.

Three inner-ear organs can be regarded as acoustic sensors: the otolithic sacculus primarily responds to low-frequency vibrations (BF 40 to 80 Hz), but can

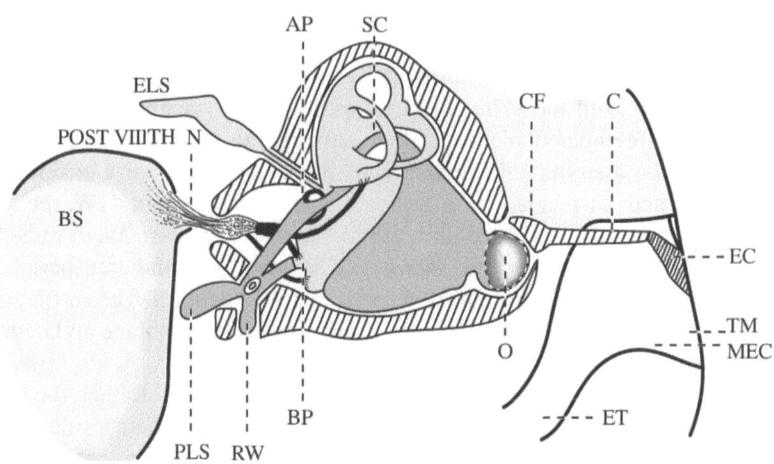

FIGURE 4.3. A diagram of the frog ear seen in transverse section at the level of the eardrum. Note that the operculum (O) is drawn in simulated 3-D (extending out of the plane of the figure). BS, brainstem; ELS, endolymphatic sac; PLS, perilymphatic sac; Post VIIIth n, posterior branch of the VIIIth nerve; SC, semicircular canals. (Redrawn and altered from a diagram by Frishkopf and Goldstein 1963. © 1963 American Institute of Physics; reprinted by permission.)

also be stimulated by intense sounds (Christensen-Dalsgaard and Narins 1993). The amphibian papilla responds to low-frequency sound (from below 100 Hz up to 1200 Hz) and vibrations, while the basilar papilla responds to high-frequency sounds (above approximately 1.5 kHz).

The amphibian papilla (AP) is by far the most complicated of the auditory organs. It contains a large sensory macula in which the best frequencies of the sensory hair cells are organized tonotopically. In contrast, the basilar papilla (BP) is probably a simple resonator that is tuned to a single best frequency, which is usually the higher frequency component of the advertisement call. The AP and BP are located close to the round window and, according to recent measurements of the acoustic flow resulting from columellar vibrations in *R. catesbeiana* (Purgue and Narins 2000), the acoustic flow is directed through the perilymphatic and endolymphatic spaces and diverges according to stimulus frequency. The frequency dependence of the acoustic flow is such that the BP contact membrane is maximally stimulated above 1100 Hz, whereas the AP contact membrane displays a peak for frequencies below 500 Hz.

3.3 Biophysics of Directional Hearing in Anurans

Understanding directional hearing in the anurans is complicated, since sound enters the frog ear by multiple pathways: through tympana, the lungs, the mouth cavity, and the nares as well as through extratympanic pathways. The following paragraphs will characterize each of these inputs.

3.3.1 The Tympanic Input

Anurans lack external ear structures and external ear canals and the tympana are located flush with the skin. In most species, the tympanic membrane is a relatively undifferentiated piece of skin, although in the aquatic clawed frog (*Xenopus laevis*) the tympanic "membrane" is a cartilaginous disk covered with normal skin and suspended in a delicate membranous frame (Wever 1985, Christensen-Dalsgaard and Elepfandt 1995). In the Southeast Asian ranid frog (*Amolops jerboa*), the tympanic membrane is very thin and transparent and clearly differentiated from normal skin (personal observation). The cartilaginous extracolumella is attached to the center of the tympanic membrane and connects it to the columella.

The columella is not driven like a piston by the membrane. Rather, the ventral edge of the columellar footplate is firmly connected to the otic capsule, and the columella rotates around this fulcrum, producing a lever ratio of approximately 6 (Jørgensen and Kanneworff 1998; Werner 2003). The rotational instead of translational movement of the columella has the consequence that inward movement of the tympanum results in outward movement of the columellar footplate, contrary to the motion in amniotes. A key element in the mechanism is that the inward movement of the eardrum is converted to a downward displacement of the distal end of the columella. This happens because the soft extracolumella slides down relative to the eardrum during inward movement of the eardrum (Jørgensen and Kanneworff 1998). However, the weakness of this mechanism probably is the same as described for reptiles and birds whose middle ear transduction also rely on a flexible extracolumella (Manley 1990). At high frequencies, the coupling between extracolumella and columella decreases, and so does the efficiency of transmission of eardrum vibrations to the inner ear.

Another factor that may limit the high-frequency transmission by the middle ear is the changes in the vibration pattern of the eardrum at high frequencies. At low frequencies, the membrane will usually vibrate in the fundamental mode in which all parts of the eardrum move in phase. However, at higher frequencies the eardrum vibration tends to break up into higher modes where parts of the eardrum move 180° out of phase (Jørgensen 1993; Purgue 1997) and the sound radiated to the internal pathways therefore may cancel. Also, in these modes, the site of attachment of the extracolumella may move very little.

3.3.2 The Lung Input

The lungs of several species of frogs vibrate as a simple resonator in the sound field with a characteristic frequency set by the lung volume (Jørgensen 1991) and hence by the size of the frog, and a corresponding low-frequency peak can be seen in the eardrum vibration spectrum (Narins et al. 1988; Jørgensen et al. 1991; Ehret et al. 1993). Furthermore, Jørgensen et al. (1991) showed that the low-frequency peak in *Eleutherodactylus coqui* could be diminished by shielding the lungs (Fig. 4.4). How sound is coupled from the lungs to the middle ear cavity is not clear. The pathway from the lungs to the middle ear cavity is obstructed by the glottal slit. During most of the respiratory cycle the glottis is closed and the lungs inflated. The glottis is open only briefly during the respiratory cycle when the lungs are emptied and refilled with air. Opening of the glottis leads to instant deflation of the lungs. Therefore, the efficient transfer of sound during the brief glottis-open periods is probably not very important in the normal function of the ear. Moreover, when the glottis is closed sound is transferred efficiently from the lungs to the middle ear cavity (Jørgensen et al.

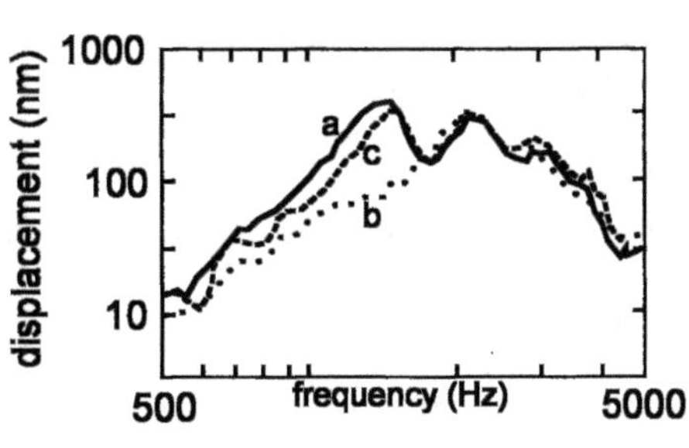

FIGURE 4.4. The lung input to the ear of *Eleutherodactylus coqui*. The figure shows eardrum vibration spectra measured by laser vibrometry before (*a*) and after (*b*) loading the body wall of awake frogs with Vaseline. Curve *c* shows the response after removal of the Vaseline. The low-frequency peak corresponds to the frequency of body wall vibrations. (From Jørgensen et al. 1991. © 1991 Springer-Verlag.)

1991). This idea was corroborated by the finding that blocking the glottal slit in *Xenopus* reduced the lung input to the tympanum (Christensen-Dalsgaard and Elepfandt 1995).

However, it has also been proposed that sound is transferred from the lungs to the ear through the round window (Hetherington and Lindquist 1999). The round window is isolated by a layer of muscle and connective tissue from the respiratory tract.

Another proposed pathway is through endolymphatic channels that connect directly to the inner ear (Narins et al. 1988). At present sound entry through both pathways are hypothetical and the importance of them needs to be verified by experiment. Finally, the relatively large pulsations of the lungs will generate vibrations in the adjacent tissue that may be coupled to the inner ear (for example, through the scapula and opercularis muscle), even in the absence of specialized structures.

3.3.3 Mouth Cavity and Nares

In earlier studies and models of anuran directional hearing, in addition to the tympana, the nares were regarded as the main point of entry for sound into the mouth cavity (Fletcher and Thwaites 1979; Palmer and Pinder 1984). However, blocking the nares with grease does not affect ear directionality (Aertsen et al. 1986). Furthermore, Vlaming et al. (1984) showed that the effect of stimulating inside the mouth cavity is almost identical to contralateral stimulation and suggested that sound can enter the mouth cavity through the mouth floor with little attenuation. However, it is also evident that opening the mouth obscures the coupling between the two ears and changes the directionality of the ear (Feng 1980; Feng and Shofner 1981; Vlaming et al. 1984). Thus, the mouth floor cannot be totally transparent to sound. A partial solution to this discrepancy could be that the lung input was not known at the time of the experiments of Vlaming et al. (1984). Hence, at least part of the sound entering the mouth cavity could have done so through the lung–glottis pathway described above. Rheinlaender et al. (1981) showed that altering the mouth cavity volume by inserting molds changed the directivity of the ear. They speculated that the frog could actively change the directionality by changing mouth cavity volume. However, their molds only allowed connection between the Eustachian tubes (and in one experiment the nares), so the mouth floor input or lung input was blocked. Thus, the increased directionality could also have resulted from an increased interaural transmission, because the mouth input was blocked. The idea that the frogs would be able to actively change the directionality is attractive, but probably unlikely, since the directional cues generated would be variable and thus difficult to process by the CNS (the same argument as for the variable directionality generated by the lung input; see Section 3.4).

3.3.4 The Extratympanic Input(s)

Neurophysiological experiments (Lombard and Straughan 1974; Wilczynski et al. 1987) showed that the anuran ear is remarkably sensitive at low frequencies, where the eardrum shows very little sensitivity. Wilczynski et al. (1987) compared auditory nerve fiber thresholds in frogs stimulated by a coupler and by free-field stimulation with the eardrum shielded by the coupler housing. They showed that the thresholds for the two stimulation routes were similar up to 1 kHz. Also, the directionality of the low-frequency fibers is pronounced, in contrast to the small directionality measured at the eardrum (Feng 1980; Wang et al. 1996; Jørgensen and Christensen-Dalsgaard 1997a; see Section 3.5.1). Furthermore, some anurans have secondarily lost parts of the middle ear, so they do not have a functional tympanic ear (Jaslow et al. 1988; Hetherington and Lindquist 1999), yet most of these species communicate by sound. Recordings from their auditory system shows responses to acoustic stimulation with thresholds that are elevated compared to tympanate species, but not more than approximately 20 dB in the low-frequency range (Walkowiak 1980). Hence, extratympanic sensitivity obviously is important for low-frequency sensitivity and directionality. The main characteristics (inferred from neurophysiological studies) are: (1) the sensitivity is maximal in the frequency range of 100 to 400 Hz, (2) the extratympanic directionality has a figure-of-eight characteristic with a frontal null, and (3) the phase difference between ipsi- and contralateral stimulation approaches 180° (see Section 3.6.1).

The origin of the extratympanic input is still unknown, but several studies have attempted to assign it to different acoustical pathways. Most interest has centered on the operculum, a movable cartilaginous element inserted in the oval window. The operculum is connected to the scapula by the opercularis muscle. It has been proposed that the operculum could be implicated in extratympanic sensitivity, since the low-frequency sensitivity decreased after section of the opercularis muscle (Lombard and Straughan 1974). Eggermont (1988) speculated that the opercularis complex may have a resonance frequency around 2 to 300 Hz and be acted upon by sound entering through the eardrum and through the mouth cavity. Conversely, Christensen-Dalsgaard et al. (1997) reported that laser vibrometry measurements from the operculum show that it vibrates 20 dB less than the columellar footplate when stimulated with free-field sound. Moreover, the peak vibration frequencies of opercular vibrations were around 1200 to 1500 Hz. It should be noted, however, that they had to expose the operculum for the laser vibrometry measurements, which may conceivably have changed its frequency response.

Christensen-Dalsgaard and Narins (1993) proposed sound-induced substrate vibrations as the origin of the extratympanic sensitivity. However, it was later shown that the low-frequency sensitivity is essentially unchanged when the sound-induced vibrations are canceled (Christensen-Dalsgaard and Jørgensen 1996). Other putative extratympanic pathways, as yet unconfirmed by experiments, may be sound entering via the round window (Hetherington and Lindquist 1999) or via endolymphatic pathways (Narins et al. 1988). Christensen-Dalsgaard et al. (1997) reported that removal of the tympana also affects the low-frequency sensitivity, in contrast to the earlier observations of Lombard and Straughan (1974). The effect of detympanation is largest at high frequencies, but can be measured down to 150 Hz. This puzzling observation shows that the extratympanic sensitivity may be quite closely coupled to tympanic hearing. Jørgensen and Christensen-Dalsgaard (1997b) proposed an alternative model for extratympanic hearing, where the inner-ear fluids are set in motion by a combination of bone conduction and differential motion of otic capsule and columella. The directionality of such a system would result from the fact that the inner-ear fluids will show maximal vibrations when the head is displaced along the axis of the pressure release windows. A hair cell oriented along this axis would show figure-of-eight directivity and a maximal phaseshift of 180° (stimulation parallel and antiparallel to the hair cell's orientation). Frog VIIIth nerve fibers show well-defined best axes of sensitivity to vibration in three dimensions (Brandt and Christensen-Dalsgaard 2001). Manipulations of the system by severing the opercularis muscle or removing the tympanum will change the impedances as seen from the inner ear and may conceivably affect the bone-conduction pathways (e.g., by "shunting" vibrations through the operculum or columellar footplate and reducing the effective stimulus for the sensory cells).

3.4 Measurements of Eardrum Directionality

Eardrum directionality has been measured in several anuran species: *Rana esculenta* (Pinder and Palmer 1983); *R. temporaria* (Vlaming et al. 1984); *Hyla cinerea* (Michelsen et al. 1986); *Eleutherodactylus coqui* (Jørgensen et al. 1991); *R. temporaria, H. versicolor, H. chrysoscelis*, and *H. gratiosa* (Jørgensen 1991); and *Bufo calamita* (Christensen and Christensen-Dalsgaard 1997). In all species, the frequency response of eardrum vibration stimulated with free-field sound has a bandpass characteristic with one or two peaks, where the low-frequency peak usually corresponds to the frequency of lung resonance (Section 3.3.2).

The eardrum vibration spectrum varies systematically with sound direction. Generally, the resulting polar plots (Fig. 4.5) are ovoidal with a maximal difference of 6 to 10 dB between ipsi- and contralateral sound incidence, but at very low and very high frequencies, eardrum vibration amplitude as well as directionality decreases. Around the lung resonance frequency, eardrum directionality is small, as is the directionality of lung vibrations (Jørgensen 1991; Jørgensen et al. 1991). However, directionality is maximal at frequencies between the two peaks. If this directionality maximum was exploited by the frogs, the call frequencies would be expected to be placed in this frequency region, but that is not the case in any of the species investigated. Jørgensen and Gerhardt (1991) tested whether female *H. versicolor* had improved localization abilities when using these intermediate frequencies and concluded that sound localization was poorer at the intermediate frequencies, at which eardrum directionality is maximal, probably because the directionality at these frequencies is also quite variable and affected by small changes in the inflation of the lungs.

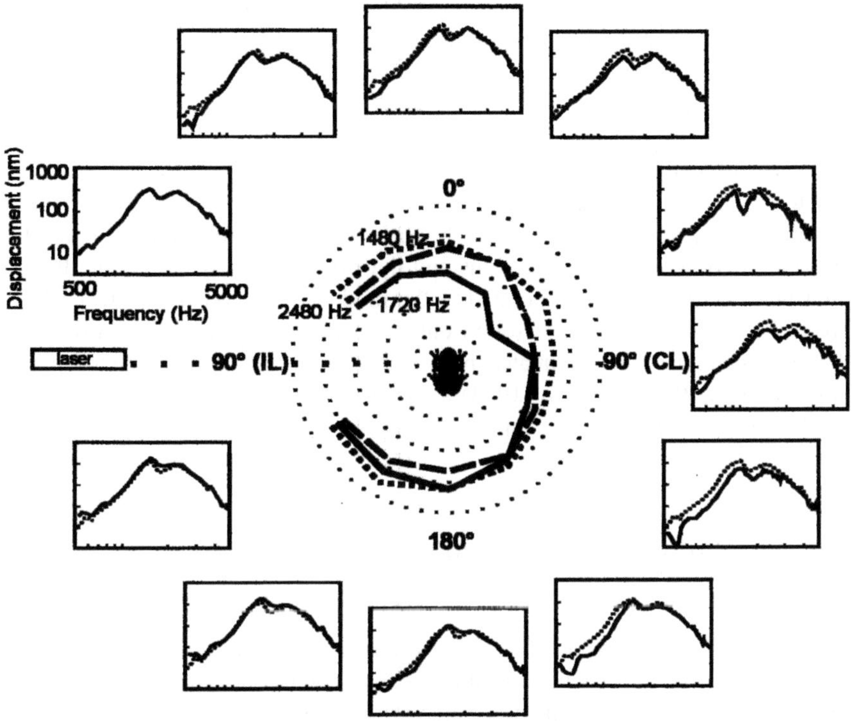

FIGURE 4.5. Directional response of the eardrum in *E. coqui* measured by laser vibrometry. The figure shows polar plots at three frequencies (5 db/circle), and the *inset* figures are vibration spectra taken at different directions. In each figure, the IL spectrum is shown as a reference (thin line). (Redrawn from Jørgensen et al. 1991. © 1991 Springer-Verlag.)

3.5 Models of Anuran Eardrum Directionality

The common characteristic of all current models of the anuran acoustic periphery (Fletcher and Thwaites 1979; Palmer and Pinder 1984; Aertsen et al. 1986) is that they rely on electrical analog models (see Section 2.3). Also, the tympanic inputs and the mouth cavity are modelled similarly in all models. The models differ, however, in the inputs. For example, a crucial element in the model by Palmer and Pinder (1984) is that the input to the mouth cavity is through the nares, that are given a tubelike radiation impedance. In contrast to this, Aertsen et al. (1986) ascribe the mouth cavity input to a delocalized, general transparency of the mouth floor to sound. Aertsen et al. incorporate the directional characteristics of the extratympanic pathway as known from neurophysiology experiments in their model and obtain, not surprisingly, a nice fit to this part of the experimental data.

All models published so far fail to give a reasonable fit to the experimental data if realistic parameters are used. The model by Palmer and Pinder (1984) predicts a figure-of-eight shaped directivity at low frequencies. This directivity is observed in neural recordings, but not in eardrum vibrations and is probably a property of the extratympanic pathway. The model of Aertsen et al predicts generally higher directionality at low frequencies. However, the eardrum vibration measurements (e.g., Jørgensen 1991; Jørgensen et al. 1991) show highest sensitivity and directionality at and between the peak frequencies and very little directionality and sensitivity at low frequencies. One explanation for the discrepancy between model predictions and the experimental data may be that the lung input needs to be incorporated in the models, but another serious problem probably is that some of the sound entrances are not very well localized and therefore cannot be approximated by a single input with a well-defined phase. For example, Vlaming et al. (1984) showed that sound enters via most of the head region. Also, the lung input would essentially cover a large area of the dorsum.

3.6 Neurophysiology of Anuran Directional Hearing

3.6.1 The Auditory Nerve

Afferent nerve fibers from the inner-ear organs are collected in the eighth or auditory nerve that enters the dorsal part of the brainstem. The number of fibers innervating the amphibian papilla ranges from 141 (*Ascaphus*) to 1548 (*R. catesbeiana*); those innervating the basilar papilla range from 31 in *Ascaphus* to 392 in *R. catesbeiana* (Will and Fritzsch 1988). In directional hearing studies, the auditory nerve must be exposed from the dorsal side to allow the animal to sit in a normal posture and to avoid decoupling the ears by opening the mouth. This type of experiments have only been undertaken in two relatively large ranid species: *R. pipiens* (Feng 1980; Feng and Shofner 1981; White et al. 1992; Schmitz et al. 1992; Wang et al. 1996; Wang and Narins 1996) and *R. temporaria* (Jørgensen and Christensen-Dalsgaard 1997a,b). Comparative studies, especially on frogs that have a well-studied phonotactic behavior (e.g., hylids), are thus badly needed.

In the following, neural coding of direction by spike rate and spike timing will be discussed. It should be realized from the outset that this separation is somewhat artificial, since spike rate and spike timing are linked through phenomena such as intensity-latency trading, in that spike latency decreases monotonically with stimulus level, whereas spike rate usually shows a monotonic increase (Feng 1982; Jørgensen and Christensen-Dalsgaard 1997b). Furthermore, in the central processing of auditory nerve information in the frogs, there is at present no evidence of separate time and intensity pathways such as reported, for example, for barn owls (see Section 5.2.4). Rather, in the CNS both response timing and response strength are integrated, for example, by inhibitory interneurons in the DMN where the output depends both on input timing and strength (see Section 3.6.2).

3.6.1.1 Spike-Rate Coding of Sound Direction

Single-cell recordings from the anuran auditory nerve using free-field stimulation have shown that the auditory fibers have two types of directional responses (Feng 1980; Jørgensen and Christensen-Dalsgaard 1997a; see Fig. 4.6). For low-frequency stimulation, a polar plot of spike rates shows a figure-of-eight directivity pattern with a frontal "null," that is, very low sensitivity to sound coming from the frontal and caudal directions, and equally high sensitivity to sound from ipsi- and contralateral directions. The axis of least sensitivity is tilted relative to the frog's symmetry axis. For high-frequency stimulation, the directivity pattern is ovoidal with the highest sensitivity for sound coming from the ipsilateral direction. The directivity pattern of a fiber depends on its char-

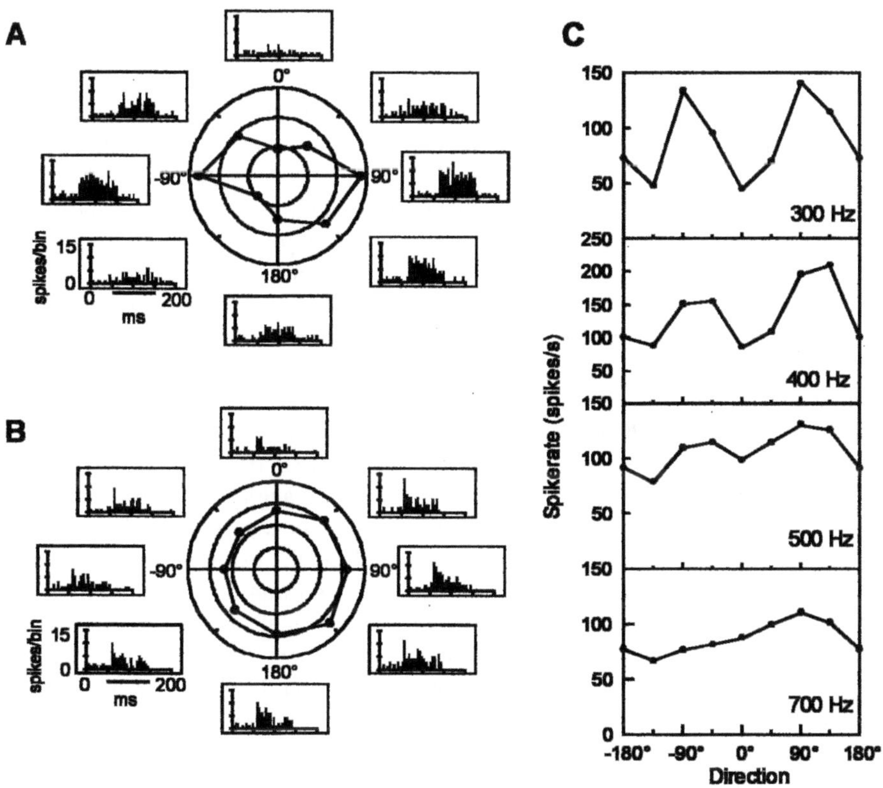

FIGURE 4.6. Response of VIIIth nerve fibers in *Rana temporaria* to directional stimulation. (**A**) Response of a low-frequency fiber stimulated at BF (300 Hz). Note the "figure-of-eight" response with low sensitivity at frontal directions. (**B**) Response of a BP fiber stimulated at its BF (1500 Hz). Here, the response is ovoidal. In (**C**) the response of a fiber stimulated at different frequencies shows that the response changes systematically with frequency. (From Jørgensen and Christensen-Dalsgaard 1997a. © Springer-Verlag.)

acteristic frequency (CF, the frequency where the cell is most sensitive) and not on stimulus frequency (Feng 1980; Jørgensen and Christensen-Dalsgaard 1997a). Accordingly, tuning curves change shape with stimulus direction (White et al. 1992).

For all fibers, the directionality depends strongly on stimulus intensity. Since almost all fibers have a relatively narrow dynamic range (median 20 dB) and saturating rate-level curves (Christensen-Dalsgaard et al. 1998), at high stimulus intensities the directionality will also show saturation and therefore decrease. Conversely, at low stimulus levels the response from some of the directions will be below threshold. Note, however, that this limited dynamic range is not necessarily a problem in the processing of directional information, since the thresholds of fibers span a range of approximately 60 dB, and, furthermore, cells with different CFs will be recruited at high stimulus intensities. When the spike rates are recalculated as equivalent decibel values by reading the levels corresponding to the measured spike rates off the fiber's rate-level curve (measured with ipsilateral stimulation) (Feng 1980), the resulting directivity is the directivity of the entire acoustic periphery and can be compared to the directivity of the tympanum such as Fig. 4.5. For the low-frequency fibers the maximal differences between ipsi- and contralateral stimulation in equivalent dB is 15 dB in *R. temporaria* (Jørgensen and Christensen-Dalsgaard 1997a) and 1 to 8 dB in *R. pipiens* (Feng 1980). For high-frequency fibers, the maximal directional difference is 10 dB in *R. temporaria* and 5 to 10 dB in *R. pipiens*. The high-frequency directivity is directly comparable in shape and magnitude to the directivity of the eardrum. However, at low frequencies, where the eardrum shows little directivity, the nerve fibers show the highest directionality and a figure-of-eight directivity pattern that is not found in the eardrum measurements. Here, the nerve fiber directivity undoubtedly reflects the directionality of the extratympanic pathway. Simultaneous single cell recordings and laser vibrometry measurements in *R. pipiens* auditory nerve fibers showed that 55% of the fibers show some degree of extratympanic directionality (Wang et al. 1996). Interestingly, however, in detympanated frogs the low frequency directionality is also changed, suggesting that detympanation also affects the extratympanic pathway (Christensen-Dalsgaard et al. 1997). When interaural transmission in *R. pipiens* is reduced by filling the mouth cavity, directionality at all frequencies decreases and the directivity patterns of the auditory fibers are ovoidal (Feng and Shofner 1981; see Fig. 4.7, second row). It may be surprising that blocking the mouth cavity also changes the directivity pattern at the low, extratympanic frequencies. However, it should be realized that filling the mouth cavity not only blocks interaural transmission. By blocking sound access to the middle ear cavity, the ear is converted to a pressure receiver, and this changes the frequency response of the eardrum and its directionality. After covering the contralateral eardrum, the directionality and directivity pattern of the low-frequency fibers was unchanged, but for mid- and high-frequency fibers directionality decreased and the directivity pattern changed to omnidirectional (Fig. 4.7, third row). When the frog's mouth was forced open, an increased directionality and a figure-of-eight

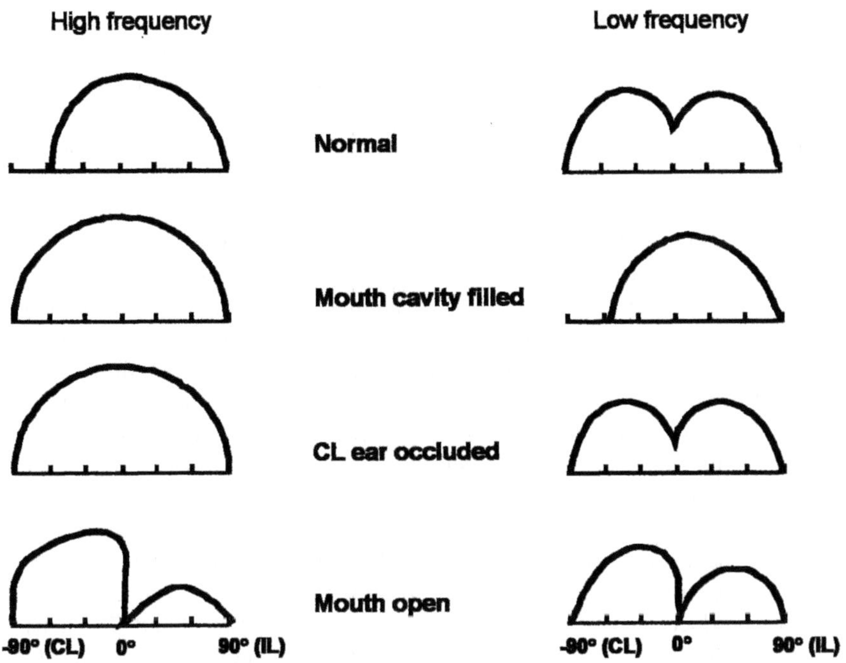

FIGURE 4.7. Directional response of *R. pipiens* auditory fibers after manipulations of the auditory periphery. (Reprinted from Feng and Shofner 1981. © 1981 with permission from Elsevier.) See text for details.

directivity pattern was found at all frequencies (Feng and Shofner 1981; Fig. 4.7, bottom row). Conceivably, opening the mouth changes the characteristics of the ear to that of a pure pressure-difference receiver, probably caused by easy access of sound to both sides of the eardrums. Interestingly, this result again suggests that the mouth floor is not transparent to sound (see above). Taken together, Feng and Shofner's experiments are consistent with the view that low-frequency directionality essentially is extratympanic, whereas directionality at higher frequencies is produced by the acoustics of the coupled middle-ear cavities and respiratory pathway resulting in combined pressure–pressure-difference receiver directivity.

3.6.1.2 *Spike Timing Coding of Sound Direction*

Response timing in auditory nerve fibers of *R. pipiens* and *R. temporaria* depends on the direction of sound incidence (Schmitz et al. 1992; Jørgensen and Christensen-Dalsgaard 1997b). Schmitz et al. investigated the directionality of phase locking and showed that preferred phase, but not vector strength (i.e., the degree of phase locking), varied systematically with sound direction. Polar plots of the phase differences showed an ovoidal directivity, and the directionality of phase locking decreased with fiber CF. Contralateral stimulation always produced a phase lead relative to ipsilateral stimulation. The magnitude of the phase lead was, however, quite variable (150° to 360° in the 200 to 300-Hz range). Jørgensen and Christensen-Dalsgaard (1997b) found qualitatively similar results showing a phase lead for contralateral stimulation in 84 of 86 neurons (the remaining two showed a phase lead for ipsilateral stimulation), but a much more homogeneous distribution of phase leads. They report a mean phase shift of 140° at 200 to 300 Hz decreasing with frequency to 100° at 600 to 700 Hz. These phase shifts correspond to time shifts of 2 ms at 200 Hz and 0.5 ms at 700 Hz (see Fig. 4.8). In contrast, the maximal timeshift resulting from arrival time differences at the two eardrums is only 60 μs (assuming a 2-cm interaural distance). Measurements of the directional phase shift at the eardrum show a maximal difference of 60 to 100°. At higher frequencies, therefore, the phase shifts of the fibers are largely caused by the directionality of the frog ear. At 200 to 300 Hz, however, the phase shift is caused by the extratympanic pathway (see Section 3.4.4). Spike latencies also show systematic changes with sound direction with a difference of up to 2 ms between ipsi- and contralateral stimulation. The latency difference is probably caused by directional changes in stimulus intensity (time-intensity trading). Both the directional latency and phase changes produce large interaural time differences. Jørgensen and Christensen-Dalsgaard (1997b) calculated that a hypothetical binaural neuron in the CNS that compared inputs from two auditory fibers with equal directional characteristics would register systematic variation in interaural time differences with direction with a range of ± 1.6 ms.

These directional effects are only seen at frequencies to which the auditory fibers show robust phase locking (below 500 to 600 Hz). However, the auditory fibers also show phase locking to the envelope of amplitude modulated (AM) stimuli (Christensen-Dalsgaard and Jørgensen, in preparation). The time shift of the spikes again varies systematically with direction, but surprisingly, there is now a phase lead for stimuli from ipsilateral directions. The time shifts (up to 3 ms) are comparable to those produced by phase locking to the carrier and

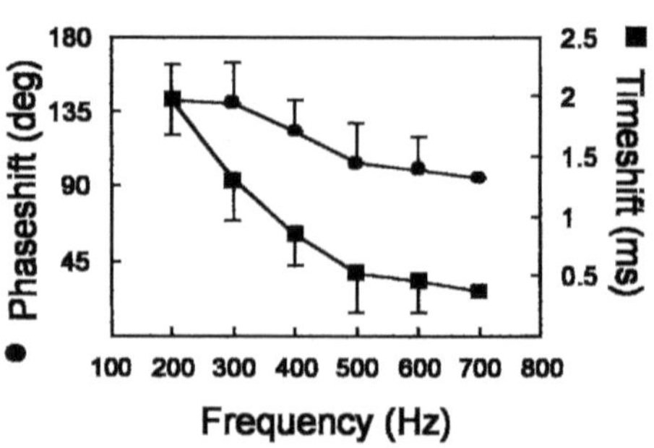

FIGURE 4.8. Ipsilateral–contralateral phase shifts in *R. temporaria* auditory fibers. The two curves show phase shifts (*circles*) and time differences calculated from the phase shifts (*squares*). Note the decline in phase difference with frequency, but also that the phase shift at the higher frequencies is much larger than expected from the head size. (From Jørgensen and Christensen-Dalsgaard 1997b. © Springer-Verlag.)

are independent of AM rate and carrier frequency. Similar directional time shifts are found for stimulation with the advertisement call, but not for stimulation with AM noise. If the AM time shifts were caused by time-intensity trading, so the effects were due to directional changes in stimulus intensity, the time shifts should be smaller, with high AM rates producing shorter rise–fall times and should be independent of the carrier (tone/noise). Since this is not the case, the time shifts are generated by an unknown mechanism, perhaps related to short-term adaptation during the AM stimulus.

3.6.2 Processing of Directional Information in the Dorsal Medullary Nucleus (DMN)

The first auditory nucleus, the DMN (see Fig. 4.9) (also called the dorsolateral or dorsomedial nucleus), is also the first stage in the processing of directional information (for a review of central auditory processing, see Feng and Schellart 1999). DMN has traditionally been homologized with the mammalian cochlear nucleus, but it is now realized that the "homology" is as octaval nucleus (Will 1988; McCormick 1999) and does not imply similarity in processing of auditory stimuli, given the independent origin of tympanic hearing in the two groups. Also, in contrast to its mammalian counterpart (the cochlear nucleus) the DMN is innervated by commissural fibers from the contralateral DMN (Feng 1986; Will 1988). Anatomical studies of the DMN have shown that the nucleus is heterogeneous in that it has six different cell types (Feng and Lin 1996), although it does not exhibit the clear subdivisions found in its amniote counterparts (Will 1988). So far, nothing is known about the location and morphology of the binaural cells in the DMN. Dichotic stimulation (where the ears were uncoupled by opening the mouth) of neurons in the DMN in *Rana catesbeiana*

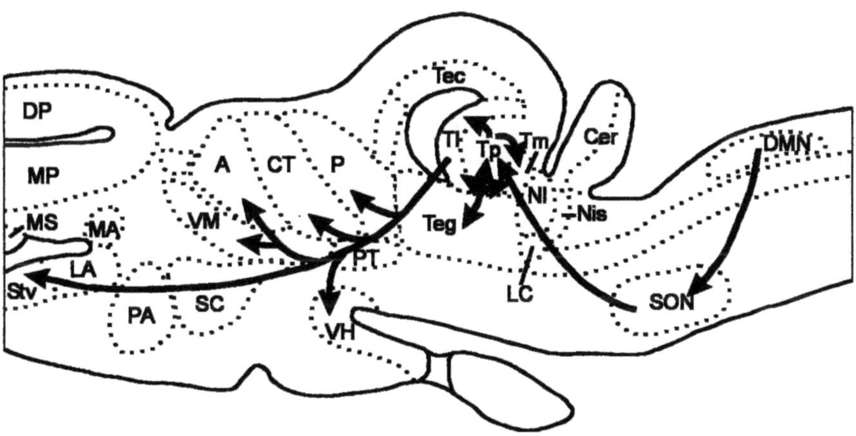

FIGURE 4.9. Diagram of the ascending auditory connections in the anuran brain. (Redrawn from Endepols et al. 2000. © 2000, with permission from Elsevier.)

(Feng and Capranica 1976) showed that approximately half of the cells studied were monaural and that most of these cells were excited by the ipsilateral ear. Of the binaural cells, most were EI cells (excitatory = inhibitory, meaning that they were excited by one ear and inhibited by the other). In most of these cells, the contralateral ear was excitatory. The EI cells were sensitive to interaural time differences of 150 μs and interaural level differences of 3 dB. Recently, binaural cells in *R. temporaria* have been studied using both closed-field and free-field stimulation (Christensen-Dalsgaard and Kanneworff 2005; Kanneworff and Christensen-Dalsgaard, in preparation). A subset of cells is inhibited by a combination of ITD and ILD (Fig. 4.10a). ITD responses are always seen as inhibition, and the cells probably respond to IPD (the ITD response is repetitive with the stimulus cycle). Interestingly, the responses are similar to recent data from mammals (Brand et al. 2002) where ITD sensitivity is generated by precise, fast inhibition. Using closed-field stimulation, it is possible to separate neural interaction from acoustical interaction resulting from coupling of the ears. However, it may be difficult to relate the results to natural, free-field stimulation. In a pilot study of free-field responses of DMN neurons, Christensen-Dalsgaard and Kanneworff (2005) report that the directionality in many cases was not much different from the directionality of VIIIth nerve fibers (Fig. 4.10b, c). However, most low-frequency neurons (Fig. 4.10b) showed ovoidal directivity in contrast to the figure-of-eight directivity of the auditory nerve fibers. Also, some of their high-frequency neurons (Fig. 4.10c) show increased directionality that probably is caused by inhibition. Such a sharpening probably is caused by the EI neurons. Note that the minimal ITD where inhibition in EI neurons was observed in Feng and Capranica's study (1976) was only 150 μs. As stated above, in a free sound field the directional interaural time difference found in the auditory nerve fibers can be much larger (up to 2 ms latency differences; Jørgensen and Christensen-Dalsgaard 1997b). If the latency difference of the DMN neurons to contra- and ipsilateral stimulation is 1 to 2 ms (reported by Feng and Capranica 1976, for the EE [excitatory-excitatory] neurons) the response of an EI neuron could range from total inhibition (inhibitory side leads) to a shortened excitatory response (excitatory side leads), depending on inhibitory and excitatory strength. Conversely, the EE cells that receive excitatory input from both ears probably do not increase the directionality compared to the auditory nerve fibers, unless they are coincidence detectors that have so far not been reported from the anuran

FIGURE 4.10. Responses of neurons in the DMN of *R. temporaria* to dichotic stimuli (**A**) and free-field stimulation (**B**, **C**). (**A**) is an ITD–ILD response area; the number of spikes elicited at each combination is indicated by a grayscale code. This low-frequency neuron is inhibited at ITDs from 0.4 to 1 ms, IL leading. (**B**, **C**) shows the response of two DMN neurons to free-field sound. The thin line is the average response of VIIIth nerve fibers at the same best frequencies (**B**: 300 Hz, **C**: 1500 Hz), indicating a sharpening of directional sensitivity already at this stage. (From Christensen-Dalsgaard and Kanneworff 2005. © 2005, with permission from Elsevier.)

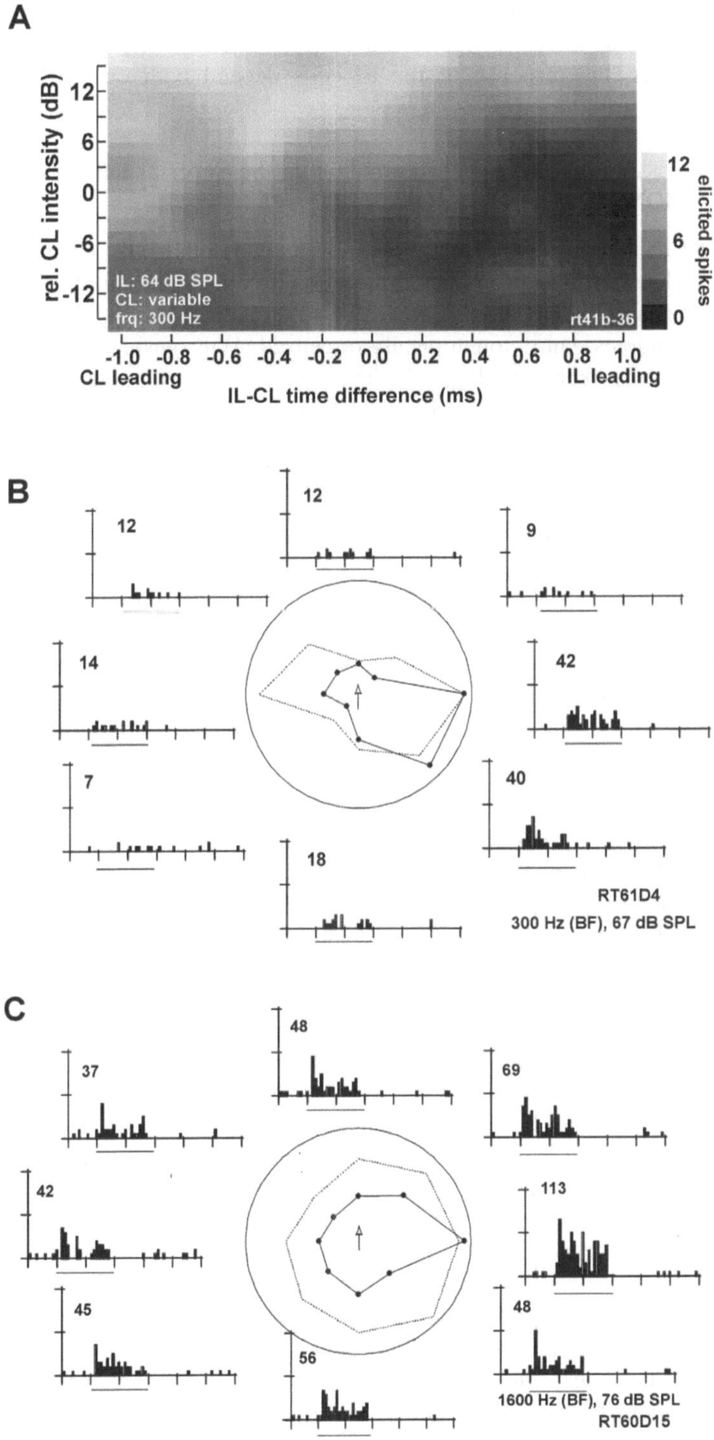
A
rel. CL intensity (dB)
12
6
0
-6
-12
IL: 64 dB SPL
CL: variable
frq: 300 Hz
rt41b-36
12
6
0
elicited spikes
-1.0 -0.8 -0.6 -0.4 -0.2 0.0 0.2 0.4 0.6 0.8 1.0
CL leading
IL leading
IL-CL time difference (ms)
B
12
12
9
14
42
7
40
18
RT61D4
300 Hz (BF), 67 dB SPL
C
48
37
69
42
113
45
48
56
1600 Hz (BF), 76 dB SPL
RT60D15

DMN. Note also that, in contrast to avian and mammalian auditory systems, there is no evidence of a segregation of time and intensity pathways at this stage. Rather, the response of the EI cells depend both on level and time difference.

3.6.3 Processing of Directional Information in the Superior Olivary Nucleus (SON)

The SON of anurans (Fig. 4.9) receives projections bilaterally from the dorsolateral nuclei with most prominent projections from the contralateral DMN (Will 1988). It has no subdivisions comparable to the lateral and medial olivary nucleus of amniotes, but is tonotopically organized (Wilczynski 1988). The SON has traditionally been homologized with the amniote SON, but as for the DMN, the homology is disputed (Will 1988; McCormick 1999), as is the homology of the SON within the amniotes (Carr and Code 2000). Only one study of directional processing in the SON has been published (Feng and Capranica 1978). Here, responses of SON neurons in *H. cinerea* to dichotic stimulation were investigated. A little less than half of the cells were binaural, and most of these were EI cells. Generally, the directional response characteristics of these cells are very similar to those of the DMN cells. From the limited data available there is no indication of a sharpening of the directionality or a separation in time and intensity pathways as reported for the SON of birds and mammals, but further investigations of the directional processing in the SON are obviously needed.

3.6.4 Processing of Directional Information in the Torus Semicircularis (TS)

The anuran midbrain auditory center TS in anurans is homologous to the inferior colliculus (IC) of mammals and birds and to the torus semicircularis in reptiles (Carr and Code 2000). It is subdivided into five nuclei of which three—the principal (T_p), magnocellular (T_m) and laminar nuclei (T_l)—are auditory (see Fig. 4.9). The principal nucleus receives most of the inputs from the caudal brainstem nuclei, for example, direct projections bilaterally, but predominantly from the contralateral DMN and from the ipsilateral SON. The arborizations of cells in the principal nucleus are small and projections are found mainly within the torus (Luksch and Walkowiak 1998). This nucleus is tonotopically organized. The magnocellular and laminar nuclei receive most of their ascending projections from the thalamus and have descending projections to the DMN and SON (Feng and Lin 1991). The cells in these nuclei differ in their projection patterns, one cell type in each having mainly intrinsic projections within the TS (Luksch and Walkowiak 1998). To summarize, the principal nucleus is the input layer of the TS, whereas the other nuclei serve audiomotor functions, the laminar nucleus probably being the main output station of toral auditory processing (Luksch and Walkowiak 1998). (Note, however, that the organization of the TS in the aquatic frog *Xenopus laevis* is apparently totally different. Here, the laminar nucleus receives all ascending projections from the DMN, and the principal and magnocellular nuclei receive projections from the lateral line nucleus [Edwards and Kelley 2001]). In contrast to the paucity of data from the lower auditory stations in the CNS, the accessibility of the TS has generated a wealth of neurophysiological data on processing of directional information. Kaulen et al. (1972) made single-cell recordings from TS with dichotic (coupler) stimulation in "the frog"and found that most of the cells were monaural. Approximately 40% were binaural, and half of these were EE cells. Of the rest, almost all were EI cells (ipsilateral inhibitory). Unfortunately, in these experiments as in some of the later studies, the mouth was closed during the recordings, so the ears were coupled acoustically. This probably leads to an overrepresentation of EE cells (Epping and Eggermont 1985), but could also influence the responses of EI cells. With free-field stimulation, most cells in the TS show direction-dependent changes in firing rates and/or latencies (Feng 1981). The units were distributed over the principal, magnocellular and laminaris nuclei and showed two major classes of directional responses. Most (with CFs ranging from 135 to 2100 Hz) had ovoidal directional responses, usually with highest sensitivity from the contralateral side. The rest (CFs ranging from 295 to 1950 Hz) had "V-shaped" (i.e., figure-of-eight) directional responses. Note that these types of directional responses are also found in auditory nerve fibers (see Section 3.5.1). In auditory nerve fibers, however, the directivity pattern is frequency specific (V-shaped at low and ovoidal at high frequencies), and V-shaped directional TS responses at high frequencies reflects additional processing by the CNS. Mellsen and Epping (1990), using closed-field dichotic stimulation in *R. temporaria* found that almost all units were binaural. Of the binaural units, most were EI units with BFs uniformly distributed between 100 and 3000 Hz and most sensitive to IIDs from -4 to 4 dB. Forty percent of the units showed intensity-invariant responses. Gooler et al. (1993) showed that tuning curves of single neurons in the TS (free-field stimulation) varied systematically with sound direction; the tuning curves were broader with contralateral than with ipsilateral stimulation. Similarly, the isointensity frequency response showed a narrower bandwith for ipsilateral than for contralateral stimulation (Xu et al. 1994). That these effects are due to neural interactions, especially ipsilateral inhibition mediated by γ-aminobutyric acid (GABA), has recently been shown in a series of elegant experiments (Zhang et al. 1999). However, it is also suggested that binaural interactions probably takes place in lower stations in the brainstem, maybe also by GABA-mediated inhibition, or more likely by faster inhibitory transmitters such as glycine. While most of these studies deal with ILDs, it should be noted that there is apparently not a clear division of time and level processing in the anuran auditory pathway. Rather, as discussed for the auditory nerve and DMN, directional phase and level changes are coupled. Furthermore, the effects of inhibition probably are intensified by the large timeshifts that accompany directional changes in level in the auditory nerve. Approximately half of the units in the TS showed intensity-invariant responses to click stimulation, and most ITD-selective units showed a well-defined latency of the response to the excitatory, contralateral stimulus (Melssen and Epping 1992). The inhibitory, ipsilateral stimulus depressed activity within a time window of a few milliseconds following excitation. Whether there is a spatiotopic organization of the TS neurons (i.e., an auditory space map) is controversial. Pettigrew et al. (1981) found spatiotopy in multiunit recordings from the TS of *R. temporaria* and *Rana esculenta*. They also proposed the simple explanation that if the best direction of a neuron varied systematically with frequency, this would produce a spatial map because of the tonotopicity in TS. This idea was corroborated by field potential recordings in *B. marinus* by Pettigrew and Carlile (1984) suggesting that the optimal stimulus angle changed with frequency. However, other studies (Wilczynski 1988) have shown that the tonotopicity in TS is not very pronounced, and furthermore, Pettigrew and Carlile probably included areas outside the TS in their data. To conclude, it is probably safe to say that no robust spatiotopy is found in the anuran TS and that the data are compatible with a model in which sound direction (encoded in the spike rate of individual TS neurons; Feng 1981) is processed in separate frequency channels (since the TS neurons are tuned) and encoded by neuronal ensemble activity. At the population level the directional information is sharpened by inhibition in the TS that will produce strong lateralization cues.

3.6.5 The Processing of Directional Information in the Forebrain

Next to nothing is known about directional processing in the forebrain. Lesion experiments have shown that female toads (*B. americanus*) will perform phonotaxis after complete removal of the telencephalon and dorsal diencephalon (Schmidt 1988). However, Walkowiak et al. (1998) showed that phonotaxis in *H. versicolor* was affected by lesions in the septum (MS, see Fig. 4.9) and the striatum (Stv) and abolished completely by lesions in the preoptic area (PA), but apparently unaffected by lesions in the dorsomedial pallium (DP, MP). Lesions of the thalamus (A, C, P) did not affect phonotaxis, whereas even small lesions in the torus produced a degraded phonotactic response (Endepols et al. 2003). In summary, the forebrain lesions seem to affect only the initiation or control of the phonotactic response. The experiments do not permit any evaluation of whether directional hearing as such is degraded (as would, e.g., lesion experiments showing that the frogs showed phonotaxis, but that localization accuracy was reduced), but show that all the processing necessary for sound direction determination likely occurs in the TS. The question then is how pattern recognition (i.e., mating call identification) and localization are integrated in the TS (apparently, there are no specialized centers for spatial hearing and pattern recognition), and, even more fundamentally: How is the directional information "read out"? Obviously, the contralateral inhibition found in the TS (Zhang et al. 1999) can generate a robust lateralized response in a "winner takes all" fashion. However, such a simple lateralization does not explain the behavioral results showing true angle discrimination (Klump and Gerhardt 1989). The absence of any robust spatiotopy as well as the generally distributed nature of frequency representation in the TS might suggest that directional information is processed in separate frequency bands (since the directional input is strongly frequency dependent, see above; note also that all the directionally sensitive TS cells are tuned) and integrated with pattern recognition subsystems to generate a direction-specific (but probably distributed) excitation pattern. Each excitation pattern could then elicit a pattern of muscle activity turning the frog in the relevant direction. Whether the frog would move or not would then be controlled by inhibition (i.e., forebrain structures). Interestingly, a simulation experiment on visual orientation in salamanders showed that a relatively small network consisting of 300 neurons (100 optic tectum [OT] neurons with "coarse coding," i.e., large receptive fields, 100 interneurons, 100 motoneurons, and 4 muscles) and incorporating neuroanatomical and physiological features of the salamander visual brain can be trained to localize moving prey (Eurich et al. 1995). The model does not incorporate a motorneuron map. Rather, all neurons participate in coding of prey location, and the distributed coding of the tectal neurons is transformed directly into a distributed activation of the muscles. It remains to be shown whether phonotaxis in anurans can be explained by similar models. What is needed is probably simultaneously recordings from many neurons under directional stimulation.

4. Directional Hearing in "Reptiles"

The reptiles do not form a "natural" taxonomic group, since they are amniotes united by primitive characteristics (i.e., a paraphyletic group; see Manley 2004). Thus, crocodiles are more closely related to birds than to the other reptile groups, and turtles and tortoises are as distantly related to other reptiles as to mammals. This section concentrates on lacertids (the lizards), since the (few) data available on reptile directional hearing have been obtained in this group.

4.1 Behavioral Investigations of Lacertid Directional Hearing

Only in one case has a behavioral use of directional hearing been demonstrated in any lizard (or reptile). It was shown that Mediterranean geckos (*Hemidactylus tursicus*) will intercept calling crickets and also perform phonotaxis toward a speaker playing cricket songs (carrier frequency 6.6 kHz; Sakaluk and Belwood 1984). Interestingly, the data suggest that the behavior is acquired, since only adults show a significant phonotactic response. The members of one lacertid family, the true geckos (Gekkonidae) are highly vocal, but no phonotaxis (or indeed any clear responses) to call playbacks have been shown so far. Investigations of hearing using conditioned responses to sound in the reptiles have met with as little success as in the anurans (Manley 2000). However, one experimental approach seems to work, namely that lizards open their eyes in response to sounds (Berger 1924). An audiogram from *Tiliqua rugosa* based on this paradigm matches the neural audiogram reasonably well (Manley 2000). This approach has so far not been applied to directional hearing studies, but it would be interesting in the future to investigate, for example, directional release from masking or habituation.

4.2 The Lacertid Ear

Lizards do not have an external ear, although some species have an external ear canal, while in other species the eardrum is flush with the surrounding skin. The tympanic membrane is usually delicate and clearly distinct from normal skin and is usually convex. The single auditory ossicle, the columella, is connected to the eardrum by an extracolumella, that is generally not strongly ossified and with up to four fingerlike processes (Wever 1978; Manley 1990; Saunders et al. 2000). The extracolumella is probably essential for the impedance matching of the ear by being one arm in a second-order lever system with a lever ratio of approximately 3 in *Gekko gecko* (Manley 1990) and 2 in an eublepharid and a pygopodid gekko species (Werner et al. 1998). An essential feature of the lever system is that there is a flexible connection between extracolumella and columella. At low frequencies (below 4 kHz) the extracolumella pivots as a stiff rod around a fulcrum and pushes the columella. As pointed out by Manley (1990), the system is less efficient at high frequencies, because the energy is lost in flexion at the extracolumella–columella joint and thus poorly transmitted. A limited high-frequency response thus appears to be an inherent drawback of the design.

4.3 Biophysics of Lacertid Directional Hearing

Very little information exists on lizard directional hearing. Wever (1978), noting the very wide Eustachian tubes, suggested that the ear of some lizards could operate as a pressure-difference receiver. However, most of the earlier studies of the eardrum response were made using closed field stimulation. Preliminary data from a free-field investigation of midbrain auditory neurons in *G. gecko* is reviewed by Manley (1981), who together with co-workers found highly directional units in the torus semicircularis. These units exhibited ovoidal directivity with activity almost completely suppressed at (mostly) ipsilateral angles. However, as Manley (1981) pointed out, the responses are probably both due to neural inhibition and acoustical interaction. Recently, Christensen-Dalsgaard and Manley (2005) have studied the directional characteristics of the tympanum in four lizard species stimulated with free-field sound. The tympana of all species showed bandpass characteristics and a remarkable directivity (Fig. 4.11a). In some of the animals, the difference between ipsi- and contralateral stimulation exceeded 25 dB in the frequency range from 1 to 3 kHz, and the directivity is dependent on acoustical coupling of the eardrum. In this frequency range, sound shadowing hardly contributes to the difference. The directivity pattern of the eardrum is ovoidal and highly asymmetrical around the midline (i.e., with a

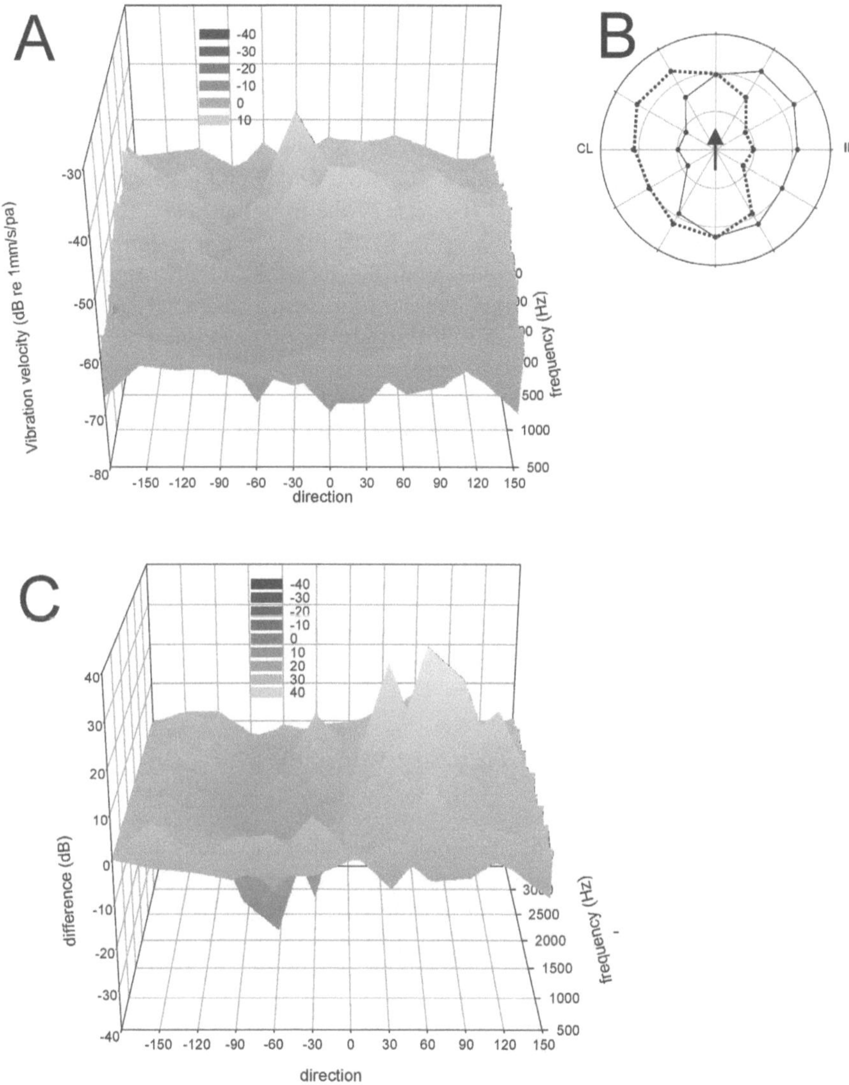

FIGURE 4.11. Directivity of eardrum vibrations in the iguanid *Ctenosaura* measured with laser vibrometry and free-field stimulation. The three-dimensional plots in (**A**) and (**C**) show the response as a function of frequency and direction (vibration amplitude is indicated by a grayscale), and each *horizontal line* corresponds to a polar plot. The eardrum has its maximal sensitivity and directionality around 2 kHz, but as shown in (**A**), contralateral sensitivity is generally depressed. A special feature of the directivity is that it is highly asymmetrical with the highest sensitivity in the IL frontal quadrant, as shown in the polar plot (**B**). If the inputs from the two ears are subtracted, the asymmetry produces a considerable sharpening of the directivity (**C**). Here, the reflection along the midline is subtracted from the response. (Redrawn from Christensen-Dalsgaard and Manley 2005 and unpublished data.)

large difference between, e.g., 30° ipsilateral and 30° contralateral, Fig. 4.11b). Any mechanism that performs binaural subtraction (as, e.g., an EI neuron) will exaggerate this directivity. A simplified model of the output of such a neuron is shown in the interaural vibration amplitude difference (IVAD) plot (Fig. 4.11c), where a mirror image of the directivity plot is subtracted from itself (Jørgensen et al. 1991). Note that the shape of the IVAD plot is generally similar to the eardrum directivity, but that (because of the asymmetrical directivity) the directionality is much sharper, with up to 40 dB difference between 30° ipsilateral and 30° contralateral. In conclusion, the fact that all investigated lizard species essentially show a similar, pressure-difference characteristic and furthermore, that the characteristic reflects a primitive organization of the periphery (i.e., that the middle ear cavities are almost continuous with the pharynx) suggests that a pressure difference characteristic and the associated low-frequency directionality is a feature of most lizard ears. The larger species, for example, some of the iguanids and varanids, should be able to exploit ILDs generated by diffraction and also have large ITDs resulting from arrival time differences at the ears. Consequently, it could be expected that some of these species would have developed uncoupled, pressure sensitive ears during the course of evolution, but that does not seem to be the case; also in the larger species (such as *Iguana iguana*) the middle ear cavities are connected through wide Eustachian tubes (G.A. Manley, personal communication).

4.4. Neuroanatomy and Neurophysiology of Lizard Directional Hearing

Apart from the study on TS neurons mentioned above (Manley 1981) the processing of directional information in the lizard (or reptile) CNS has not been studied. The same divisions of the cochlear nucleus (CN) (i.e., in a nucleus angularis and magnocellularis) as in the birds have been described (see review in Carr and Code 2000). In birds, the division in nucleus angularis and magnocellularis reflect a functional division of time and intensity processing, at least in the barn owl, and it is hypothetized that the nuclei in reptiles should serve a similar functional division (Carr and Code 2000). At least in the alligator lizard (*Gerrhonotus multicarinatus*) there is anatomical evidence that two types of auditory afferents (low-frequency tectorial and high-frequency free standing fibers) project differently in the cochlear nucleus (Szpir et al. 1990). Endbulb terminations were found only in the tectorial fibers and only in the magnocellular nucleus. This finding should be noted, since endbulb swellings with the associated, very efficient synaptic transmission is a characteristic of cochlear nucleus cells in the time coding pathway in birds and mammals. It should also be noted, however, that most lizard nucleus magnocellularis cells are small to medium-sized and therefore may not be functionally equivalent to the avian magnocellularis cells, even if the nuclei are homologous (which by no means can be assumed). Furthermore, it could be argued that the special characteristics of the pressure difference receiver and the high directionality of the periphery in the lizards would necessitate a different central processing, with emphasis, for example, on EI cells that could sharpen the directionality considerably (as shown by the IVAD plots above). Therefore, it would be of considerable interest to investigate the directional processing in the CNS of lizards. Physiological data from the CN (reviewed in Manley 1981) show that the CN in *G. gecko* (and probably also in other lizards) is tonotopically organized. All neurons are tuned, many have primary-like (i.e., phasic-tonic) responses to sound, but very phasic responses also are common. The anatomical data from *Iguana iguana* (Foster and Hall 1978) and *Varanus exanthematicus* (ten Donkelaar et al. 1987) show that the earliest stage of binaural interaction probably is at the level of the trapezoid body or SON (the trapezoid body was included in the SON by ten Donkelaar et al 1987) that receives projections from both ipsi- and contralateral nucleus angularis. Note, however that the *I. iguana* SON apparently lacks structures similar to the MSO in mammals and the nucleus laminaris (NL) in crocodiles and birds (Foster and Hall 1978). From the SON, bilateral projections to the TS have been found in both lizard species, where highly directional cells have been found in *G. gecko*, as outlined above (Manley 1981).

5. Directional Hearing in Birds

For a general review of directional hearing in birds, the reader is referred to Knudsen (1980) and to a recent review in this series (Klump 2000). The aim of the present section is to provide a counterpoint to that review by focusing on the biophysics of directional hearing and especially the evidence for pressure difference/ pressure sensitivity of the avian ear.

5.1 Biophysics of Directional Hearing in Birds

The structure of the avian ear is similar to the lizard ear. Birds usually have no external auditory structures (with the exception of some of the owls, see below). However, an ear canal is always present, but short (2 to 7 mm, Saunders et al. 2000). The single ossicle (columella) is connected to the eardrum via an especially complex extracolumella with three processes, which probably improves the lever ratio of the ear, but probably also limits the high-frequency sensitivity of the ear, depending on the flexibility of the columella–extracolumella connection (Manley 1990).

In birds, arising from an archosaur–dinosaur lineage, the ancestral condition probably is that the middle ears are partially isolated from the respiratory pathway, but connected via an interaural canal that is also found in crocodilians (Wever 1978) and probably in nonavian dinosaurs including *Tyrannosaurus rex* (Larsen and Pettigrew 1988; J.D. Pettigrew, personal observation). Reflecting this ancestral condition, all birds have an interaural canal and the eardrums of all birds are therefore to some extent coupled acoustically. To what extent the acoustic coupling produces usable pressure-difference receiver directivity has been debated extensively. The evidence is reviewed in the following paragraphs.

5.1.1 Directivity of the Auditory Periphery

Studies of directivity roughly fall into two groups. One group of researchers has shown that the directionality is greater than expected from diffraction effects and have compared the directivity to that of a pressure-difference receiver. Another group of researchers found generally small directionality that could result from diffraction and assumed that the ears are functionally uncoupled pressure receivers.

5.1.1.1 Evidence for Pressure-Difference Receiver Operation of the Auditory Periphery

The pioneering studies on bird directional hearing were performed by Schwartzkopff (1950, 1952), who found that the bullfinch (*Pyrrhula pyrrhula*) auditory periphery showed higher directionality than expected from diffraction, but, surprisingly, that this directionality did not change on ear occlusion. Consequently, he concluded that the ears operated as independent pressure receivers. Coles and co-workers (Coles et al. 1980; Hill et al. 1980) showed pronounced directionality of the quail (*Coturnix coturnix*) auditory periphery. Hill et al. (1980) measured sound transmission through the quail head using inserted microphones and found less than 5 dB interaural canal transmission attenuation at frequencies below 5 kHz. At higher frequencies, attenuation increased above 20 dB. From the attenuation data, they predicted strongly asymmetrical directivity with 10- to 20-dB directional difference in the 1- to 4-kHz range. Cochlear microphonics in anesthetized quail showed a variety of directivity patterns, cardioid at lower frequencies and figure-of-eight shaped at high frequencies. The directivities were altered when one eardrum was blocked. Larsen and Popov (1995) found very similar results using laser vibrometry and sound diffraction measurements from quail. They reported an enhancement of interaural delay of 40 μs and an interaural canal attenuation of 6 dB at 1 kHz. Interaural coupling was also inferred by Calford (1988) from a study of frequency selectivity in the IC of nine different bird species. All species except owls exhibited a poorly represented frequency range, which was correlated with the dimensions of their interaural canal. Model calculations based on the interaural canal dimensions showed that the "missing" frequency ranges corresponded to frequency regions in which tympanic directionality generated by interaural coupling was poor. The proposed model, which was based on the addition of direct and indirect sound components (delayed by propagation time and attenuated by the interaural canal), was subsequently used to calculate interaural delays, which were shown to be frequency dependent and, especially at low frequencies, much larger than travel-time delays. Pettigrew and Larsen (1990) reported that neurons in the IC of the plains-wanderer (*Pedionomus torquatus*) showed very directional responses to low-frequency sound (300 Hz) and ascribed the directionality to the large interaural canal in this species. Similarly, Hyson et al. (1994) measured bilateral cochlear microphonics in the chicken and found larger ITDs (up to 200 μs at low frequencies) than expected from travel time differences. They also report up to ±30% (±3dB) directional change of IID (relative to frontal stimulation), and effects of contralateral tympanum occlusion at low, but not on high frequencies, and conclude that interaural coupling enhances the sound localization cues. Finally, Larsen et al. (1997) showed that the interaural coupling and normal operation of the tympanum of budgerigars was very dependent on the intracranial air pressure (ICA) (Fig. 4.12). This is a very important finding, because the ICA tends to decrease in anesthetized birds, unless they are vented. The result is that tympanic vibrations are impeded (the tympanum is sucked inwards), and interaural coupling decreases by around 20 dB and tympanal directivity by 6 dB or more in nonvented compared to vented birds.

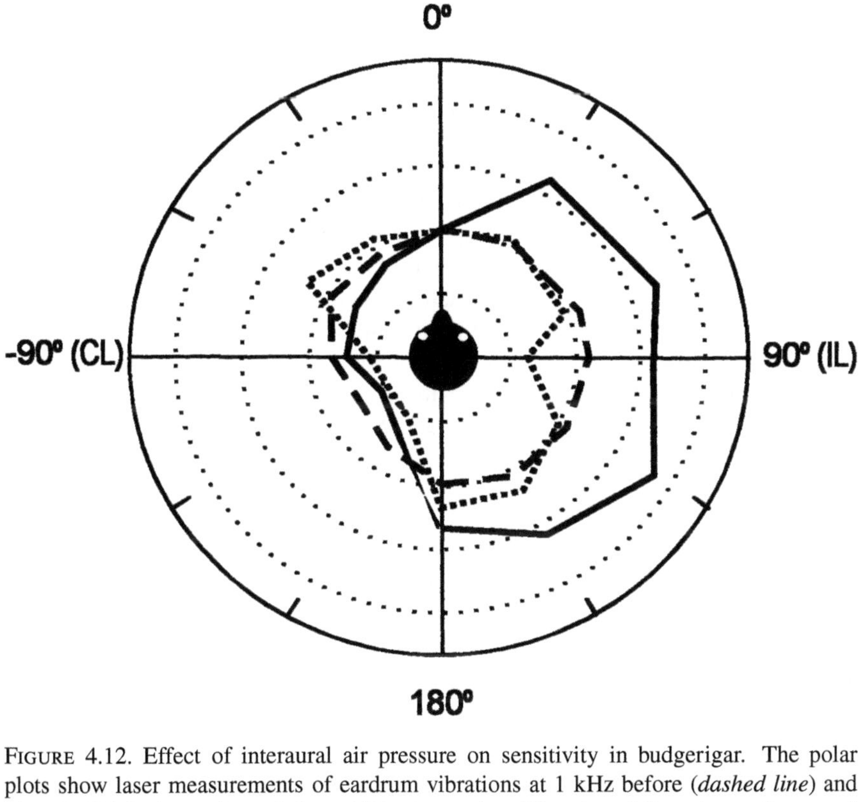

FIGURE 4.12. Effect of interaural air pressure on sensitivity in budgerigar. The polar plots show laser measurements of eardrum vibrations at 1 kHz before (*dashed line*) and after (*solid line*) venting of the middle-ear cavity. The *dotted line* shows the sound pressure. Note the large difference in directivity and sensitivity before and after venting. Scale: 3 dB/unit circle. (Redrawn from Larsen et al. 1997 © World Scientific Publishers, reprinted by permission.)

5.1.1.2 Experiments Showing Small Effects of Interaural Coupling

Owls have large interaural canals, and earlier studies (Payne 1971) assumed that acoustical coupling of the ears would contribute to the directionality. However, Moiseff and Konishi (1981) recorded from monaural cells in the "cochlear nucleus" of barn owls and showed that interaural attenuation (measured as the difference between the unit's threshold to ipsi- and contralateral stimulation via couplers) increased with frequency from 13 dB at 3.5 kHz to 63 dB at 7 kHz. Measurements with probe microphones also showed large interaural attenuation at higher frequencies. Thus, at behaviorally relevant frequencies, the ears would be functionally isolated pressure receivers. In contrast, Coles and Guppy (1988) report that directionality in the barn owl, measured by cochlear microphonics, exceeds the directionality produced by the external ear and suggest that interaural coupling is important for the directionality also at high frequencies. The reason for the discrepancy between the measurements of Moiseff and Konishi (1981) and Coles and Guppy (1988) is not clear. However, microphonic recordings are inherently unstable and the level can fluctuate during the time needed to measure directional sensitivity in the entire frontal hemifield. Therefore, it would be important to know about the reproducibility of the directional patterns, but this information is not given in the paper by Coles and Guppy. In contrast, threshold comparisons are probably much more reliable. Also, since head-related transfer functions generate virtual space stimulation in the owl with identical responses to free-field sound and virtual stimulation (Keller et al. 1998; Poganiatz et al. 2001) interaural coupling cannot be very important. Rosowski and Saunders (1980) used cochlear microphonics to measure interaural transmission in the chicken and found virtually no attenuation by the interaural canal, but 15- to 20-dB attenuation by the tympanum. With such an impedance mismatch by the tympanum, the directionality should be negligible, which contrasts with Hyson et al.'s (1994) data from the chicken. However, the level of the microphonics measured by Rosowski and Saunders are much lower than the levels measured by Hyson et al. (1994), suggesting that the ears had not been working optimally. Thus, the 15- to 20-dB attenuation is probably too high. Lewald (1990) investigated the directionality of cochlear microphonics in anesthetized pigeons and found less than 5 dB effects of interaural sound transmission and small effects of blocking the contralateral ear. Perhaps the most convincing case for a bird with uncoupled ears was presented by Klump and Larsen (1991) in their work on the starling. They used laser vibrometry to measure free-field directional characteristics of the eardrum in anesthetized starlings and showed that the largest directionality when corrected for sound diffraction by the animal was 3.4 dB (average 1.13 dB). These results in anesthetized birds were paralleled by cochlear microphonics from awake birds. To summarize, there is solid evidence that some species (quail, chicken, and budgerigar) have enhanced peripheral directionality caused by interaural coupling. However, it is also evident that there is considerable species variation, and that some species (barn owl, starling) probably have functionally separate ears. Most importantly, perhaps, the results of Larsen et al. (1997) have reopened the field, since many of the earlier measurements on anesthetized animals need to be redone with appropriate venting of the middle ear cavities.

5.2 Directional Hearing in Barn Owls and Other Birds

The barn owl is a model organism within the field of neuroethology and directional hearing studies. The reason for this is that the barn owl is extremely specialized for directional hearing (Payne 1971; Wagner 2002) and exhibits a very robust sound localization behavior associated with prey capture. This behavior and the associated neurophysiology has lent itself to rigorous, careful laboratory studies for three decades, with the consequence that probably more is known about sound localization in the barn owl than in any other organism. Any discussion of bird directional hearing, therefore, would be clarified by investigating similarities and dissimilarities between directional hearing in the barn owl and that of other birds, summarized in Table 4.1. The field has been extensively reviewed, however, and is only outlined briefly here. The reader is referred to Klump (2000), Wagner (2002) and, especially, to reviews by Konishi (1973, 2000) and Knudsen (2002).

5.2.1 Sound Localization Behavior

Early experiments (reviewed in Payne 1971) demonstrated that barn owls can locate prey accurately using acoustical cues (i.e., the rustling noises made by

TABLE 4.1. Specializations of the owl localization pathway.

Feature	Reference
Extended high-frequency hearing	Konishi (1973); Köppl et al. (1993)
Very low thresholds (−20 dB SPL)	Konishi (1973)
Asymmetrical auditory periphery	Konishi (1973)
Sound-reflecting facial ruff	Konishi (1973)
Increased phase-locking properties of auditory-nerve fibers (up to 9 to 10 kHz)	Köppl (1997)
Longest basilar papilla	Smith et al. (1985)
Overrepresentation of high-frequency range in basilar papilla ("auditory fovea")	Köppl et al. (1993)
Large size and convoluted shape of NA	Köppl (2001)
Multiple layering of NL	Carr and Konishi (1990)
Kv 3.1 channels in NM that reduces the duration of action potentials	Parameshwaran et al. (2001)
Segregation of time and intensity pathways	Sullivan and Konishi (1984)
Small receptive fields and ordered spatiotopic representation in the ICC	Knudsen and Konishi (1978a,b)
Sharp spatial tuning in ICx	Wagner (1993)
Spatial map in ICx and SCC	Knudsen (1982)

the prey). The accuracy of striking the target depends on the stimulus bandwidth; a 4-kHz noise band centered on 7 kHz was most efficient (Konishi 1973). Later, Konishi, Knudsen and co-workers used a setup in which the turning angle of perched owls was measured (Knudsen et al. 1979). These behavioral studies have shown that the orientation error of the barn owl for noise is approximately 5°, but around three times as large for tones (Knudsen and Konishi 1979). Furthermore, occluding the right ear produced an orientation bias downwards and to the left; occluding the left ear produced an orientation bias upwards and to the right. Finally, removing the ruff feathers of the facial mask only disturbed localization in elevation. Recent studies using a pupillary dilation response have found minimum audible angles of approximately 3° (Bala et al. 2003).

The performance of other birds in sound localization tasks vary among the investigated species, but are not as acute as in the barn owl. Early experiments on bullfinches (Schwartzkopff 1950) showed minimal angular resolution of 25°, and minimal resolution angles of 20° in Great tits (*Parus major*) were reported by Klump et al. (1986). The authors suggested that the ITD cues generated by comparing the inputs from two uncoupled ears (18 μs at 25°, approximately twice the minimal ITD in barn owls) are sufficient to explain the localization accuracy. Park and Dooling (1991) reported minimal resolution angles (noise stimulation) of 17° and 25°, respectively, in budgerigar (*Melopsitaccus undulatus*) and canary (*Serinus canarius*), at 1 and 2 kHz, the minimal resolution angles were larger. Outside the owls, the smallest minimal resolution angles were found in other aerial predators like the marsh hawk (*Circus cyaneus*) (2°) and red-tailed hawk (*Buteo jamaicensis*) (8 to 10°) (Rice 1982; see also Klump 2000, Table 6.1). Note, however, that a study of the passerine bird *Pipilo erythrophtalmus* (Towhee) showed much higher acuity than that of other songbirds: 5 to 9° azimuth as well as accurate estimation of distance (Nelson and Stoddard 1998); the authors suggest that birds would use vision at short distances and therefore might not attend to auditory stimuli.

5.2.2 External Ear in the Owls

The auditory periphery of barn owls and some other owl species (Volman and Konishi 1990) has a very special feature that is unique among the tetrapods. The ears are asymmetric, that is, the left and right ear opening differ in size and/or placement. The function of the ear asymmetry is to enable the owl to locate sound accurately in elevation. For an animal that has two symmetrical ears, the region of equal intensity for all frequencies is a plane aligned with the medial plane of the animal, that is, no binaural comparisons can resolve sound elevation. For the asymmetrical ears of the owl, however, the iso-intensity plane is a complex contour that changes with sound frequency: the plane is almost horizontal at high frequencies and almost vertical at lower frequencies (below 6 kHz). For the lower frequencies, the ongoing time difference varies systematically with azimuth because of the differences in travel distance to the two ears. Thus, (low-frequency) ITD and (high-frequency) IID define the two axes of a coordinate system (Moiseff 1989), and a particular location is uniquely defined in elevation and azimuth by its low-frequency ITD and high-frequency IID (which explains why the localization of tones is inherently more difficult for the barn owl than the localization of noise). Furthermore, some of the owls have a well-developed feathered facial mask or ruff that is a parabolic sound reflector. The ruff of the barn owl amplifies some frequencies by more than 10 dB (Volman and Konishi 1990). It should be noted that the effect of the asymmetrical ear and of the facial ruff is only seen at frequencies above 3 to 4 kHz.

5.2.3 The Auditory Papilla and Auditory Nerve

The bird basilar papilla or cochlea is an elongated structure containing the sensory epithelium of hair cells covered by a tectorial membrane. The papilla of the barn owl is longer than the papilla of all other birds—11 mm (Köppl et al. 1993) compared to 3.1 mm in pigeon and 3.7 mm in chicken (Gleich and Manley 2000)—which reflects the extended high-frequency hearing (Smith et al. 1985; Fischer et al. 1988). The auditory papilla is tonotopically organized, and in the barn owl papilla the high frequency range (5 to 10 kHz) is highly overrepresented in an "acoustic fovea" (Köppl et al. 1993), whereas the low-frequency part of the papilla is comparable to that of other birds (Gleich and Manley 2000). The auditory nerve contains an ordered array of fibers tuned to different CFs, but the directional characteristics of avian auditory nerve fibers have not been studied. Avian fibers phase-lock to relatively high frequencies; 4 kHz in the starling (Gleich and Narins 1988) and up to 10 kHz in the barn owl (Köppl 1997). The physiological basis of the specialization for high-frequency phase locking is at present unknown.

5.2.4 Nucleus Angularis (NA) and Magnocellularis (NM)

In the barn owl, information on spike timing and spike rate becomes first segregated in the auditory nuclei in the sense that NA cells show weak phase locking, large dynamic range and high rate-level slopes, whereas NM cells phase lock up to 9 kHz, and their response is almost independent of the intensity of the stimulus (Sullivan and Konishi 1984). The strong phase locking in the NM cells is inferior compared to the auditory nerve input (Köppl 1997). Anatomically, four major classes of cell types have been described from the barn owl NA, classified by the branching pattern of their dendrites *in* (two morphological types), *across*, and *vertical* to the isofrequency plane (Soares and Carr 2001). The same cell types are found in the chicken NA and the morphological types also have distinct physiological properties (Soares et al. 2002). A recent study has identified five distinct physiological response types in the NA of the barn owl, and the nucleus is probably not dedicated to sound level processing (Köppl and Carr 2003). However, some of the NA cells innervate the posterior nucleus of the ventral lateral lemniscus (VLVp), the first site of binaural computation of ILD in the owl (Manley et al. 1988; Mogdans and Knudsen 1994). The NM cells have very characteristic, regular branching patterns sending off collaterals that enter the NL at equidistant points along the tonotopic band (Carr and Konishi 1990). The NM cells have morphological specializations for relaying time information efficiently, notably large and calyx-shaped synapses (endbulbs of Held) and specialized potassium channels that reduce the duration of action potentials (Carr et al. 2001; Parameshwaran et al. 2001). Functionally, NM neurons in the owl are delay lines that, together with the contralateral NM neurons, interface on NL neurons. The NM cells are tonotopically organized and project to specific isofrequency bands in the NL (Carr and Konishi 1990). In other birds and crocodilians, only the contralateral NM neurons are delay lines, whereas the ipsilateral NM neurons have a fixed delay (Overholt et al. 1992; Kubke et al. 2002; see also Fig. 4.13).

5.2.5 Nucleus Laminaris

As a second-order auditory nucleus, the NL is equivalent to the MSO of mammals. NL cells have large, oval cell bodies and short dendrites (Carr and Konishi 1990) and discharge when action potentials arriving from the ipsi- and contralateral NM neurons coincide in time (Carr and Konishi 1990). The cells are often described as ITD-sensitive, but they really encode IPDs, since stimulus cycle time ITD shifts will produce identical coincidences, and the output of NL cells do show multiple peaks at ITD intervals corresponding to the cycle time of the stimulus. As the NL neurons are stimulated by appropriately delayed coincident action potentials from the IL and CL NM, they project an ordered array of IPDs to the IC. NL is hypertrophied in owls compared to other birds (chicken; see Fig. 4.13), where the cells form a single layer of bipolar neurons (Carr and Konishi 1990; Joseph and Hyson 1993), probably the primitive avian pattern (Kubke et al. 2002). However, also in other birds the mechanism of coincidence detection has been reported (Joseph and Hyson 1993). The hypertrophy of the owl NL occurs during a “second phase” in development; the first phase produces the plesiomorphic NL and is similar in chicken and owl (Kubke et al. 2002). Probably the owl NL reflects specialization for high-frequency IPD processing (Köppl 1997; Kubke et al. 2002) or just improves the sensitivity by parallel calculation of IPD in many cells (Carr and Code 2000).

With the current focus on ITD processing by fast (glycinergic) inhibition in the mammalian MSO (Brand et al. 2002), it may be worthwhile to note that avian NL neurons do receive inhibitory inputs, but the inhibition is relatively slow (GABAergic), controlled from the SON and most likely used to control the sensitivity of the NL cells (Brückner and Hyson 1998; Carr et al. 2001).

5.2.6 The Posterior Nucleus of the Ventral Lateral Lemniscus (VLVp or LLDp)

The VLVp receives projections from the NA and process ILD. The physiological properties of cells in the VLVp have only been studied in the barn owl (Manley et al. 1988; Mogdans and Knudsen 1994). The cells receive excitatory inputs from the CL NA and inhibitory inputs from the IL NA. The strength of

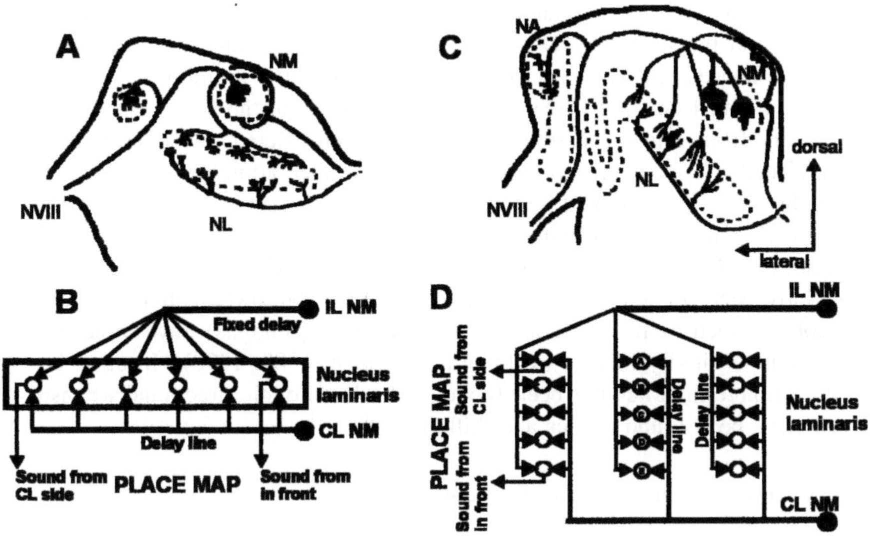

FIGURE 4.13. Comparison of NL ITD coding in chicken (**A**, **B**) and barn owl (**C**, **D**). In both chicken and owl, the NL detects coincidence between inputs from the IL and CL nucleus magnocellularis and conforms to a Jeffress model (**B**, **D**). In the chicken (schematic cross section in **A**), the NL is only a monolayer, and the IL magnocellularis input has a fixed delay. In the barn owl (schematic cross section in **C**), the NL is hypertrophied with several rows of coincidence detectors and a delay line input from both IL and CL NM. (Redrawn from Kubke and Carr 2000. © 2000, with permission from Elsevier.)

inhibition received from the IL side decrease with depth in the nucleus, and consequently, a map of ILD is generated in the nucleus (Manley et al. 1988). The cells have chopperlike discharges and show level-dependent discharge patterns (so the response to a weak monaural tone can be distinguished from the response to a loud binaural tone) (Mogdans and Knudsen 1994).

5.2.7 The Inferior Colliculus (IC)

The avian IC is homologous to the IC in mammals and to the torus semicircularis in fish, amphibians, and reptiles. In the literature, it is also called nucleus mesencephalicus dorsalis (MLD) or torus. The IC is divided into an external nucleus (ICX) and a central nucleus consisting of a lateral shell (ICCls), central core (ICCc) and medial shell (ICCms) (Carr and Code 2000). The ICCc receives projections both directly and indirectly from the NL, and the ICCms and ICCls from the NA and is still tonotopically organized. The ILD and ITD pathway converges in the ICX, where the information is combined across frequency to produce spatial receptive fields, ITD giving the azimuthal and ILD the elevational component (Takahashi et al. 1984). Also, the IPD information is converted to true ITD by comparing IPD responses across frequencies, since they will align at the shortest ITD (Wagner et al. 1987; Konishi 2000). The space-sensitive neurons in ICX have an excitatory center and an inhibitory surround (Knudsen and Konishi 1978b). The receptive fields of the space-sensitive neurons vary from 7 to 40° azimuth and 23° to "unrestricted" elevation (Knudsen and Konishi 1978a) and are organized into a spatiotopic map. The ICX is probably also an important site for plasticity and instruction by visual input, since owls that are reared with prisms show changes in the ITD coding in ICX, but not in ICC (Gold and Knudsen 2000; Knudsen 2002). The other owl species that have been studied have comparable ICX maps, but owls with a symmetrical periphery, such as the great horned owl (*Bubo virginianus*) have broad or no elevation sensitivity in ICX (Volman and Konishi 1989, 1990). In other birds, a comparable map has not been found in the IC, though directionally sensitive cells have been found (Coles and Aitkin 1979).

5.2.8 Optic Tectum (Superior Colliculus)

The OT or the superior colliculus is bimodal, and most cells in the barn owl OT respond to both visual and auditory stimuli. The cells show space-specific responses independent of the type of sound or its intensity. The width of the receptive field is narrow, and lesions in the map produce behavioral deficits (but the owls can recover) (Wagner 1993). Also, the auditory and visual maps are generally aligned. It has been shown both in behavioral experiments (Gold and Knudsen 2000) and neurophysiological experiments (Hyde and Knudsen 2000) that the visual tectal neurons control the auditory map by projecting an instructional signal back into the IC in juvenile owls (Knudsen 2002). Pigeons (*Columba livia*) also have a map of auditory space in the OT that is aligned with the visual map (Lewald and Dörrscheidt 1998). The units are very broadly tuned, however (width 43 to 166° azimuth, 16 to 186° elevation).

6. Summary and Perspectives

The main point of the present review is that directional hearing in the recent tetrapods reflects the independent origin of tympanic hearing in the different groups (but note that even the primitive, nontympanic tetrapods can have had a crude directional hearing based on sound-induced vibrations of the skull, similar to the directional hearing of their fish ancestors). Furthermore, in the three groups reviewed here, the primitive organization of the auditory periphery is probably one where the middle ears are coupled, potentially producing pressure-difference receiver directionality. That the directionality of anurans and lacertids is produced by acoustical interaction of the two ears is beyond dispute. However, in some bird species, at least, evolution has led to functional isolation of the two ears, most notably in auditory specialists like the barn owl and relatives. Whether this functional isolation is caused by selection against the coupled ear (which has disadvantages; see Section 2.5) or is just caused by adaptation that shifts sensitivity to higher frequencies where transmission through the interaural canal happens to be attenuated, is unknown at present. In most of the other birds studied, however, the degree of interaural coupling is disputed at present. Large values of interaural attenuation (20 dB or more) have been reported and this would mean that the ears were functionally isolated. However, from simple considerations of the acoustics of a system with two receivers coupled by a pipe (Fletcher 1992) it is difficult to accept that these large attenuations should apply throughout the entire frequency range. Also, recent findings showing that the acoustics of the periphery is highly sensitive to anesthesia, causing the Eustachian tubes to close (Larsen et al. 1997), necessitate a reevaluation of the older data, preferably in combination with realistic models based on electrical analogs. On the other hand, it is likely that bird groups other than the owls have evolved functional isolation of the ears, and hence the mechanisms of directionality should be expected to vary among different bird groups.

Very recently, it has been shown that the neural processing underlying acute ITD sensitivity in mammals is very different from the processing in barn owls. In mammals binaural interaction is probably mediated by fast and precise inhibition of neurons in the MSO, and ITD is encoded by the spike rate rather than by the activity of characteristic cells in an ordered array of best ITDs (Brand et al. 2002). Similarly, binaural interaction based on inhibition is also found in the frog DMN (Christensen-Dalsgaard and Kanneworff 2005). Thus, apparently, the Jeffress-type coincidence detectors found in the avian NL are unique in the tetrapods (Grothe 2003). The processing of directional hearing probably reflects independent, convergent evolution in the different groups of tetrapods based on ancestral features. Also in the tetrapod ancestors, bilateral neural comparison and computation of the relative level of excitation on the two sides of the animal could have been important, so that neural mechanisms subserving bilateral interaction may be a fundamental and plesiomorphic feature that can be coopted by evolving auditory systems.

The central processing of directional information apparently varies among the three groups. In the well-characterized sound localization pathway of birds, intensity and time cues are segregated at the level of the first auditory nuclei with a specialized structure, NL, for ITD processing using delay lines. In barn owls, spatiotopy in the IC and in the OT has been demonstrated. Anurans have binaural interaction already at the level of the first auditory nucleus, apparently no clear segregation of time and intensity pathways, and no robust spatial map. In the lacertid reptiles, binaural interaction starts like in the archosaurs at the level of the SON. However, some lacertids, at least, lack specialized structures comparable to the MSO in mammals and NL in archosaurs that function in temporal (ITD) processing. While essentially nothing is known of the central processing of directional information in lizards, it could be argued that a useful directionality could result from binaural difference (EI) cells, since the eardrum directivity is markedly asymmetrical and the directionality therefore will be enhanced by such cells (Fig. 4.10b, c).

The point is that the neural computation of directionality with a pressure-difference input could be radically different from the computation using a pressure-sensitive ear such as the barn owl's. For example in the first case, the directionality (IID as well as ITD) will be strongly frequency dependent, so it would be sensible to compare binaural inputs in separate frequency channels and compute a direction for each of them. Since the ITD would be strongly frequency dependent, it might not be advantageous to have a specialized time pathway. To address these questions, much more comparative work will be needed in the future.

The organization of the sound localization pathways and the amount of neural tissue devoted to the processing of sound direction in the species reviewed here probably accurately reflects the magnitude of selection pressures for directional hearing. While it is obviously always of some advantage to be able to determine the direction to a sound source, the extreme selection pressure for directional hearing in a barn owl that has to locate moving prey has led to specialization for directional hearing at every stage of the auditory pathway. In contrast, the magnitude of selection pressures for frog directional hearing is less easy to assess. Even if a female frog has to locate a calling male to reproduce—so there is a selection pressure for directional hearing—much of the burden of communication is on the sender, which has to call nonstop for hours, and whose communication calls are probably tailored to be maximally audible and localizable (Wilczynski et al. 2001). Furthermore, in some frog species, mate localisation will probably be largely carried out using nonacoustic cues, and the selection pressure for a sharpened directional hearing in those species is unknown, as in the nonvocal lizards. A related major question in anuran hearing is the extent to which the auditory system is dedicated to the processing of conspecific calls. Specialization of the auditory system for communication might be suggested by the ubiquity of call communication within the anurans and the virtual absence of response to noncommunication sound, but the fact that nongekkonid lizards, that do not communicate by sound or show any robust behavioral response to sound, have very sensitive and directional ears, should caution us: A major and primitive function of audition in the vertebrates might not be related to communication, but rather simply to the necessity of collecting information about changing features of the environment.

Acknowledgments. I thank Geoffrey A. Manley for comments on the manuscript and Nina Krogh for the drawings in Figures 4.1 and 4.3.

Abbreviations

A	anterior thalamic nucleus
AM	amplitude Modulated
AP	amphibian papilla

BP	basilar papilla
C	columella
Cb	cerebellar nucleus
Cer	cerebellum
CF	characteristic frequency
CL	contralateral
CN	cochlear nucleus
CoF	columellar footplate
CT	central thalamic nucleus
DMN	dorsal medullary nucleus
DP	dorsal pallium
EC	extracolumella
EE	excitatory-excitatory
EI	excitatory-inhibitory
ET	Eustachian tube
GABA	γ-aminobutyric acid
HRTF	head-related transfer function
IC	inferior colliculus
ICA	intracranial air pressure
ICc	inferior colliculus, core
ICx	inferior colliculus, external nucleus
IL	ipsilateral
ILD	interaural level difference
IPD	interaural phase difference
ITD	interaural time difference
IVAD	interaural vibration amplitude difference
LA	lateral amygdala
LC	locus coeruleus
LPv	lateral pallium, ventral portion
MA	medial amygdala
MEC	middle ear cavity
MP	medial pallium
MS	medial septum
NA	nucleus angularis
NI	isthmal nucleus
Nis	secondary isthmal nucleus
NL	nucleus laminaris
NM	nucleus magnocellularis
NVIII	VIIIth nerve
O	operculum
OT	optic tectum
P	posterior thalamic nucleus
PA	preoptic area
PT	posterior tuberculum
RW	round window

SON	superior olivary nucleus
Stv	ventral striatum
Tec	optic tectum
Teg	tegmentum
TM	tympanic membrane
TS	torus semicircularis
VH	ventral hypothalamic nucleus
VM	ventromedial thalamic nucleus

References

Aertsen AMHJ, Vlaming MSMG, Eggermont JJ, Johannesma PIM (1986) Directional hearing in the grassfrog (*Rana temporaria* L.). II. Acoustics and modelling of the auditory periphery. Hear Res 21:17–40.

Bala ADS, Spitzer MW, Takahashi TT (2003) Prediction of auditory spatial acuity from neural images on the owl's auditory space map. Nature 424:771–774.

Beranek LL (1986) Acoustics. New York: American Institute of Physics.

Berger K (1924) Experimentelle Studien über Schallperzeption bei Reptilien. Z Vergl Physiol 1:517–540.

Blauert J (1997) Spatial Hearing. Cambridge, MA: MIT Press, pp. 372–392.

Bolt JR, Lombard RE (1985) Evolution of the amphibian tympanic ear and the origin of frogs. Biol J Linn Soc 24:83–99.

Brand A, Behrend O, Marquardt T, McAlpine D, Grothe B (2002) Precise inhibition is essential for microsecond interaural time coding. Nature 417:543–547.

Brandt C, Christensen-Dalsgaard J (2001) Responses to three-dimensional vibrations and sound stimuli in single fibers from the 8th cranial nerve of the grass frog, *Rana temporaria.* In: Elsner N, Kreuzberg GW (eds), Göttingen Neurobiology Report 2001. Stuttgart: Georg Thieme Verlag, p. 386.

Bregman A (1990) Auditory Scene Analysis. The Perceptual Organization of Sound. Cambridge, MA: The MIT Press.

Brückner S, Hyson RL (1998) Effect of GABA on the processing of interaural time differences in nucleus laminaris neurons in the chick. Eur J Neurosci 10:3438–3450.

Calford MB (1988) Constraints on the coding of sound frequency imposed by the avian interaural canal. J Comp Physiol A 162:491–502.

Carr C, Code RA (2000) The central auditory system of reptiles and birds. In: Dooling RJ, Fay RR, Popper AN (eds), Comparative Hearing: Birds and Reptiles. New York: Springer-Verlag, pp. 197–248.

Carr CE, Konishi M (1990) A circuit for detection of interaural time differences in the brain stem of the barn owl. J Neurosci 10:3227–3246.

Carr CE, Soares D, Parameshwaran S, Perney T (2001) Evolution and development of time coding systems. Curr Opin Neurobiol 11:727–733.

Christensen KR, Christensen-Dalsgaard J (1997) Directional hearing in the natterjack toad, *Bufo calamita.* In Elsner N, Wässle H (eds), Göttingen Neurobiology Report 1997. Stuttgart: Georg Thieme Verlag, p. 334.

Christensen-Dalsgaard J, Elepfandt A (1995) Biophysics of underwater hearing in the clawed frog, *Xenopus laevis.* J Comp Physiol A 176:317–324.

Christensen-Dalsgaard J, Jørgensen MB (1996) Sound and vibration sensitivity of VIIIth nerve fibers in the grass frog, *Rana temporaria.* J Comp Physiol A 179:437–445.

Christensen-Dalsgaard J, Kanneworff M (2005) Binaural interaction in the frog dorsomedullary nucleus. Brain Res Bull (in press).

Christensen-Dalsgaard J, Manley GA (2005) Directionality of the lizard ear. J. Exp Biol 208:1209–1217.

Christensen-Dalsgaard J, Narins PM (1993) Sound and vibration sensitivity in the frogs *Leptodactylus albilabris* and *Rana pipiens pipiens*. J Comp Physiol A 172:653–662.

Christensen-Dalsgaard J, Kanneworff M, Jørgensen MB (1997) Extratympanic sound sensitivity of frog auditory fibers. In: Lewis ER, Long GR, Lyon RF, Narins PM, Steele CR, Hecht-Poinar E (eds), Diversity in Auditory Mechanics, Singapore: World Scientific, pp. 64–68.

Christensen-Dalsgaard J, Jørgensen MB, Kanneworff M (1998) Base response characteristics of auditory nerve fibers in the grass frog (*Rana temporana*). Hear Res 119:155–163.

Clack JA (1993) Homologies in the fossil record: the middle ear as a test case. Acta Biotheor 41:391–409.

Clack JA (1997) The evolution of tetrapod ears and the fossil record. Brain Behav Evol 50:198–212.

Coles RB, Aitkin LM (1979) The response properties of auditory neurones in the midbrain of the domestic fowl (*Gallus gallus*) to monaural and binaural stimuli. J Comp Physiol 134:241–251.

Coles RB, Guppy A (1988) Directional hearing in the barn owl (*Tyto alba*). J Comp Physiol A 163:117–133.

Coles RB, Lewis DB, Hill KG, Hutchings ME, Gower DM (1980) Directional hearing in the Japanese quail (*Coturnix coturnix japonica*). II. Cochlear physiology. J Exp Biol 86:153–170.

Diego-Rasilla FJ, Luengo RM (2004) Heterospecific call recognition and phonotaxis in the orientation behavior of the marbled newt, *Triturus marmoratus*. Behav Ecol Sociobiol 55:556–560.

Edwards CJ, Kelley DB (2001) Auditory and lateral line inputs to the midbrain of an aquatic anuran; neuroanatomic studies in *Xenopus laevis*. J Comp Neurol 438:148–162.

Eggermont JJ (1988) Mechanisms of sound localization in anurans. In: Fritzsch B, Ryan MJ, Wilczynski W, Hetherington TE, Walkowiak W (eds), The Evolution of the Amphibian Auditory System. New York: John Wiley & Sons, pp. 307–336.

Ehret G, Keilwerth E, Kamada T (1993) The lung-eardrum pathway in three treefrog and four dendrobatid frog species: some properties of sound transmission. J Exp Biol 195: 329–343.

Elepfandt A, Eistetter I, Fleig A, Günther E, Hainich M, Hepperle S, Traub B (2000) Hearing threshold and frequency discrimination in the purely aquatic frog *Xenopus laevis* (Pipidae): measurement by means of conditioning. J Exp Biol 203:3621–3629.

Endepols H, Walkowiak W, Luksch H (2000) Chemoarchitecture of the anuran auditory midbrain. Brain Res Rev 33:179–198.

Endepols H, Feng AS, Gerhardt HC, Schul J, Walkowiak W (2003) Roles of the auditory midbrain and thalamus in selective phonotaxis in female gray treefrogs (*Hyla versicolor*). Behav Brain Res 145:63–77.

Epping WJM, Eggermont JJ (1985) Relation of binaural interaction and spectro-temporal characteristics in the auditory midbrain of the grassfrog. Hear Res 19:15–28.

Eurich C, Roth G, Schwegler H, Wiggers W (1995) Simulander: a neural network model for the orientation movement of salamanders. J Comp Physiol A 176:379–389.

Fay RR, Feng AS (1987) Mechanisms for directional hearing among nonmammalian vertebrates. In Yost WA, Gourevitch G (eds), Directional Hearing. New York: Springer-Verlag, pp. 179–213.
Feng AS (1980) Directional characteristics of the acoustic receiver of the leopard frog (*Rana pipiens*): a study of the eighth nerve auditory responses. J Acoust Soc Am 68: 1107–1114.
Feng AS (1981) Directional response characteristics of single neurons in the torus semicircularis of the leopard frog (*Rana pipiens*). J Comp Physiol 144:419–428.
Feng AS (1982) Quantitative analysis of intensity-rate and intensity-latency functions in peripheral auditory nerve fibers of northern leopard frogs (*Rana p. pipiens*). Hear Res 6:241–246.
Feng AS (1986) Afferent and efferent innervation patterns of the cochlear nucleus (dorsal medullary nucleus) of the leopard frog. Brain Res 367:183–191.
Feng AS, Capranica RR (1976) Sound localization in anurans. I. Evidence of binaural interaction in the dorsal medullary nucleus of the bullfrog (*Rana catesbeiana*). J Neurophysiol 39:871–881.
Feng AS, Capranica RR (1978) Sound localization in anurans. II. Binaural interaction in superior olivary nucleus of the green tree frog (*Hyla cinerea*). J Neurophysiol 41: 43–54.
Feng AS, Lin WY (1991) Differential innervation patterns of three divisions of frog auditory midbrain (torus semicircularis). J Comp Neurol 306:613–630.
Feng AS, Lin WY (1996) Neuronal architecture of the dorsal nucleus (cochlear nucleus) of the frog (*Rana pipiens pipiens*). J Comp Neurol 366:320–334.
Feng AS, Schellart NAM (1999) Central auditory processing in fish and amphibians. In: Fay RR, Popper AN (eds), Comparative Hearing: Fish and Amphibians. New York: Springer-Verlag, pp. 218–268.
Feng AS, Shofner (1981) Peripheral basis of sound localization in anurans. Acoustic properties of the frog's ear. Hear Res 5:201–216.
Feng AS, Gerhardt HC, Capranica RR (1976) Sound localization behavior of the green treefrog (*Hyla cinerea*) and the barking treefrog (*H. gratiosa*). J Comp Physiol 107: 241–252.
Fischer FP, Köppl C, Manley GA (1988) The basilar papilla of the barn owl *Tyto alba*: a quantitative morphological SEM analysis. Hear Res 34:87–101.
Fletcher NH (1992) Acoustic systems in biology. Oxford: Oxford University Press.
Fletcher N, Thwaites S (1979) Physical models for the analysis of acoustical systems in biology. Q Rev Biophys 12:25–65.
Foster RE, Hall WC (1978) The organization of central auditory pathways in a reptile, *Iguana iguana*. J Comp Neurol 178:783–832.
Frishkopf LS, Goldstein MH (1963) Responses to acoustic stimuli from single units in the eighth nerve of the bullfrog. Acoust Soc Am 35:1219–1228.
Gerhardt HC (1995) Phonotaxis in female frogs and toads: execution and design of experiments. In: Klump GM, Dooling RJ, Fay RR, Stebbins WC (eds), Methods in Comparative Psycho-Acoustics, Basel: Birkhäuser Verlag, pp. 209–220.
Gerhardt HC, Klump (1988) Phonotactic responses and selectivity of barking treefrogs (*Hyla gratiosa*) to chorus sounds. J Comp Physiol A 163:795–802.
Gerhardt HC, Rheinlaender J (1980) Accuracy of sound localization in a miniature dendrobatid frog. Naturwissenschaften 67:362–363.
Gleich O, Manley GA (2000) The hearing organs of birds and crocodilia. In: Dooling

RJ, Fay RR, Popper AN (eds), Comparative Hearing: Birds and Reptiles. New York: Springer-Verlag, pp. 70–138.

Gleich O, Narins PM (1988) The phase response of primary auditory afferents in a songbird (*Sturnus vulgaris* L.) Hear Res 32:81–91.

Gold JI, Knudsen EI (2000) A site of auditory experience-dependent plasticity in the neural representation of auditory space in the barn owl's inferior colliculus. J Neurosci 20:3469–3486.

Goodrich (1930) Studies on the Structure and Development of the Vertebrates, Vol. 1. New York: Dover (reprint 1958).

Gooler DM, Condon CJ, Xu J, Feng AS (1993) Sound direction influences the frequency-tuning characteristics of neurons in the frog inferior colliculus. J Neurophysiol 69: 1018–1030.

Grothe B (2003) New roles for synaptic inhibition in sound localization. Nat Rev Neurosci 4:1–11.

Heffner RS, Heffner HE (1992) Evolution of sound localization in mammals. In: Webster DB, Fay RR, Popper AN (eds), The Evolutionary Biology of Hearing. New York: Springer-Verlag, pp. 691–715.

Hetherington TE, Lindquist E (1999) Lung-based hearing in an 'earless' anuran amphibian. J Comp Physiol 184:395–401.

Hill KG, Lewis DB, Hutchings ME, Coles RB (1980) Directional hearing in the Japanese quail (*Coturnix coturnix japonica*). I. Acoustical properties of the auditory system. J Exp Biol 68:135–151.

Hoy (1992) The evolution of hearing in insects as an adaptation to predation from bats. In: Webster DB, Fay RR, Popper AN (eds), The Evolutionary Biology of Hearing. New York: Springer-Verlag, pp. 115–129.

Hyde PS, Knudsen EI (2000) Topographic projection from the optic tectum to the auditory space mapin the inferior colliculus of the barn owl. J Comp Neurol 21:8586–8593.

Hyson RL, Overholt EM, Lippe WR (1994) Cochlear microphonic measurements of interaural time differences in the chick. Hear Res 81:109–118.

Jaslow AP, Hetherington TE, Lombard RE (1988) Structure and function of the amphibian middle ear. In: Fritzsch B, Ryan MJ, Wilczynski W, Hetherington TE, Walkowiak W (eds), The Evolution of the Amphibian Auditory System. New York: John Wiley & Sons, pp. 69–91.

Jørgensen MB (1991) Comparative studies of the biophysics of directional hearing in anurans. J Comp Physiol A 169:591–598.

Jørgensen MB (1993) Vibrational patterns of the anuran eardrum. In: Elsner N, Heisenberg M (eds), Gene-Brain-Behaviour. Proceedings of the 21st Göttingen Neurobiology Conference. Stuttgart: Georg Thieme Verlag, p. 231.

Jørgensen MB, Christensen-Dalsgaard J (1997a) Directionality of auditory nerve fiber responses to pure tone stimuli in the grassfrog, *Rana temporaria*. I. Spike rate responses. J Comp Physiol A 180:493–502.

Jørgensen MB, Christensen-Dalsgaard (1997b) Directionality of auditory nerve fiber responses to pure tone stimuli in the grassfrog, *Rana temporaria*. II. Spike timing. J Comp Physiol A 180:503–511.

Jørgensen MB, Gerhardt HC (1991) Directional hearing in the gray tree frog *Hyla versicolor*: eardrum vibrations and phonotaxis. J Comp Physiol A 169:177–183.

Jørgensen MB, Kanneworff M (1998) Middle ear transmission in the grass frog, *Rana temporaria*. J Comp Physiol A 182:59–64.

Jørgensen MB, Schmitz B, Christensen-Dalsgaard J (1991) Biophysics of directional hearing in the frog *Eleutherodactylus coqui*. J Comp Physiol A 168:223–232.

Joseph AW, Hyson RL (1993) Coincidence detection by binaural neurons in the chick brain stem. J Neurophysiol 69:1197–1211.

Kaulen R, Lifschitz W, Palazzi C, Adrian H (1972) Binaural interaction in the inferior colliculus of the frog. Exp Neurol 37:469–480.

Keller CH, Hartung K, Takahashi TT (1998) Head-related transfer functions of the barn owl: measurement and neural responses. Hear Res 118:13–34.

Klump GM (1995) Studying sound localization in frogs with behavioral methods. In: Klump GM, Dooling RJ, Fay RR, Stebbins WC (eds), Methods in Comparative Psycho-Acoustics. Basel: Birkhäuser Verlag, pp. 221–233.

Klump GM (2000) Sound localization in birds. In: Dooling RJ, Fay RR, Popper AN (eds), Comparative Hearing: Birds and Reptiles. New York: Springer-Verlag, pp. 249–307.

Klump GM, Gerhardt HC (1989) Sound localization in the barking treefrog. Naturwissenschaften 76:35–37.

Klump GM, Larsen ON (1991) Azimuthal sound localization in the European starling (*Sturnus vulgaris*): I. Physical binaural cues. J Comp Physiol A 170:243–251.

Klump GM, Windt W. Cuno E (1986) The great tit's (*Parus major*) auditory resolution in azimuth. J Comp Physiol A 158:383–390.

Knudsen EI (1980) Sound localization in birds. In: Popper AN, Fay RR (eds), Comparative Studies of Hearing in Vertebrates. Berlin: Springer-Verlag, pp. 289–322.

Knudsen EI (1982) Auditory and visual maps of space in the optic tectum of the owl. J Neurosci 2:1177–1194.

Knudsen EI (2002) Instructed learning in the auditory localization pathway of the barn owl. Nature 417:322–328.

Knudsen EI, Konishi M (1978a) A neural map of auditory space in the owl. Science 200:795–797.

Knudsen EI, Konishi M (1978b) Center-surround organization of auditory receptive fields in the owl. Science 202:778–780.

Knudsen EI, Konishi M (1979) Mechanisms of sound localization in the barn owl (*Tyto alba*). J Comp Physiol 133:13–21.

Knudsen EI, Blasdel GG, Konishi M (1979) Sound localization by the barn owl (*Tyto alba*) measured with the search coil technique. J Comp Physiol 133:1–11.

Konishi M (1973) How the owl tracks its prey. Am Scientist 61:414–424.

Konishi M (2000) Study of sound localizaton by owls and its relevance to humans. Comp Biochem Physiol A 126:459–469.

Köppl C (1997) Phase locking to high frequencies in the auditory nerve and cochlear nucleus magnocellularis of the barn owl, *Tyto alba*. J Neurosci 17:3312–3321.

Köppl C, Carr CE (2003) Computational diversity in the cochlear nucleus angularis of the barn owl. J Neurophysiol 89:2313–2329.

Köppl C, Gleich O, Manley GA (1993) An auditory fovea in the barn owl cochlea. J Comp Physiol A 171:695–704.

Kubke MF, Massoglia DP, Carr CE (2002) Developmental changes underlying the formation of the specialized time coding circuits in barn owls (*Tyto alba*). J Neurosci 22:7671–7679.

Kühne R, Lewis B (1985) External and middle ears. In: King AS, McLelland J (eds), Form and Function in Birds, Vol. 3. London: Academic Press, pp. 227–271.

Larsen ON (1995) Acoustic equipment and sound field calibration. In: Klump GM, Dooling RJ, Fay RR, Stebbins WC (eds), Methods in Comparative Psycho-Acoustics. Basel: Birkhäuser Verlag, pp. 31–45.

Larsen ON, Popov AV (1995) The interaural canal does enhance directional hearing in quail (*Coturnix coturnix japonica*). In: Burrows M, Matheson T, Newland PL, Schüppe H (eds), Neural systems and behavior. Proceedings of the 4th International Conference in Neuroethology. Stuttgart:Georg Thieme Verlag, p. 313.

Larsen ON, Dooling RJ, Ryals BM (1997) Roles of intracranial air pressure in bird audition. In: Lewis ER, Long GR, Lyon RF, Narins PM, Steele CR, Hecht-Poinar E (eds), Diversity in Auditory Mechanics. Singapore: World Scientific, pp. 11–17.

Lewald J (1990) The directionality of the ear of the pigeon (*Columba livia*). J Comp Physiol A 167:533–543.

Lewald J, Dörrscheidt GJ (1998) Spatial-tuning properties of auditory neurons in the optic tectum of the pigeon. Brain Res 790:339–342.

Lewis ER, Narins PM (1999) The acoustic periphery of amphibians: anatomy and physiology. In: Fay RR, Popper AN (eds), Comparative Hearing: Fish and Amphibians. New York: Springer-Verlag, pp. 101–154.

Lombard RE, Bolt J (1979) Evolution of the tetrapod ear: an analysis and reinterpretation. Biol J Linn Soc 11:19–76.

Lombard RE, Straughan IR (1974) Functional aspects of anuran middle ear structures. J Exp Biol 61:71–93.

Luksch H, Walkowiak W (1998) Morphology and axonal projection patterns of auditory neurons in the midbrain of the painted frog, *Discoglossus pictus*. Hear Res 122:1–17.

Manley GA (1972) The middle ear of the tokay gecko. J Comp Physiol 81:239–250.

Manley GA (1981) A review of the auditory physiology of the reptiles. Prog Sens Physiol 2:49–134.

Manley GA (1990) Peripheral Hearing Mechanisms in Reptiles and Birds. New York: Springer-Verlag.

Manley GA (2000) The hearing organs of lizards. In: Dooling RJ, Fay RR, Popper AN (eds), Comparative Hearing: Birds and Reptiles. New York: Springer-Verlag, pp. 139–196.

Manley GA (2004) The lizard basilar papilla and its evolution. In: Manley GA, Popper AN, Fay RR (eds), Evolution of the Vertebrate Auditory System. New York: Springer-Verlag, pp. 200–224.

Manley GA, Clack J (2004) An outline of the evolution of vertebrate hearing organs. In: Manley GA, Popper AN, Fay RR (eds), Evolution of the Vertebrate Auditory System. New York: Springer-Verlag, pp. 1–26.

Manley GA, Köppl C (1998) Phylogenetic development of the cochlea and its innervation. Curr Opin Neurobiol 8:468–474.

Manley GA, Köppl C, Konishi M (1988) A neural map of interaural intensity difference in the brainstem of the barn owl. J Neurosci 8:2665–2677.

McCormick CA (1999) Anatomy of the central auditory pathways of fish and amphibians. In: Fay RR, Popper AN (eds), Comparative Hearing: Fish and Amphibians. New York: Springer-Verlag, pp. 155–217.

Megela-Simmons A, Moss CF, Daniel KM (1985) Behavioral audiograms of the bullfrog (*Rana catesbeiana*) and the green treefrog (*Hyla cinerea*). J Acoust Soc Am 78:1236–1244.

Melssen WJ, Epping WJM (1990) A combined sensitivity for frequency and interaural

intensity difference in neurons in the auditory midbrain of the grassfrog. Hear Res 44:35–50.
Melssen WJ, Epping WJM (1992) Selectivity for temporal characteristics of sound and interaural time difference of auditory midbrain neurons in the grassfrog: a system theoretical approach. Hear Res 60:178–198.
Melssen WJ, Epping WJM, van Stokkum IHM (1990) Sensitivity for interaural time and intensity difference of auditory midbrain neurons in the grassfrog. Hear Res 47:235–256.
Michelsen A (1994) Directional hearing in crickets and other small animals. Fortschr Zool 39:195–207.
Michelsen A (1998) Biophysics of sound localization in insects. In Hoy RR, Popper AN, Fay RR (eds), Comparative Hearing: Insects. New York: Springer-Verlag, pp. 18–62.
Michelsen A, Jørgensen M, Christensen-Dalsgaard J, Capranica RR (1986) Directional hearing of awake, unrestrained treefrogs. Naturwissenschaften 73:682–683.
Mogdans J, Knudsen EI (1994) Representation of interaural level difference in the VLVp, the first site of binaural comparison in the barn owl's auditory system. Hear Res 74: 148–164.
Moiseff A (1989) Bi-coordinate sound localization by the barn owl. JComp Physiol A 164:637–644.
Moiseff A, Konishi M (1981) Neuronal and behavioral sensitivity to binaural time differences in the owl. J Neurosci 1:40–48.
Morse PM (1948) Vibration and sound, 2nd ed (reprint 1986). New York: American Institute of Physics.
Narins PM, Ehret G, Tautz J (1988) Accessory pathway for sound transfer in a neotropical frog. Proc Natl Acad Sci USA 85:1255–1265.
Nelson BS, Stoddard PK (1998) Accuracy of auditory distance and azimuth perception by a passerine bird in natural habitat. Anim Behav 56:467–477.
Overholt EM, Rubel EW, Hyson RL (1992) A circuit for coding interaural time differences in the chick brainstem. J Neurosci 12:1698–1708.
Palmer AR, Pinder AC (1984) The directionality of the frog ear described by a mechanical model. J Theor Biol 110:205–215.
Parameshwaran S, Carr CE, Perney TM (2001) Expression of the Kv3.1 potassium channel in the avian auditory brainstem. J Neurosci 21:485–494.
Park TJ, Dooling RJ (1991) Sound localization in small birds: absolute localization in azimuth. J Comp Psychol 105:121–133.
Passmore NI, Telford SR (1981) The effect of chorus organization on mate localization in the painted reed frog (*Hyperolius marmoratus*). Behav Ecol Sociobiol 9:291–293.
Passmore NI, Capranica RR, Telford SR, Bishop PJ (1984) Phonotaxis in the painted reed frog (*Hyperolius marmoratus*). J Comp Physiol A 154:189–197.
Payne RS (1971) Acoustic location of prey by barn owls. J Exp Biol 54:535–573.
Pettigrew AG, Anson M, Chung SH (1981) Hearing in the frog: a neurophysiological study of the auditory response in the midbrain. Proc R Soc Lond B 212:433–457.
Pettigrew AG, Carlile S (1984) Auditory responses in the torus semicircularis of the cane toad, *Bufo marinus*. I. Field potential studies. Proc R Soc Lond B 222:231–242.
Pettigrew JD, Larsen ON (1990) Directional hearing in the plains-wanderer *Pedionomus torquatus*. In: Rowe M, Aitkin L (eds), Information Processing in Mammalian Auditory and Tactile Systems. New York: Alan R. Liss, pp. 179–190.

Pinder AC, Palmer AR (1983) Mechanical properties of the frog ear: vibration measurements under free- and closed-field acoustic conditions. Proc R Soc Lond B 219:371–396.

Poganiatz I, Nelken I, Wagner H (2001) Sound-localization experiments with barn owls in virtual space: influence on interaural time difference on head-turning behavior. J Assoc Res Otolaryngol 2:1–21.

Purgue A (1997) Tympanic sound radiation in the bullfrog, *Rana catesbeiana*. J Comp Physiol A 181:438–445.

Purgue A, Narins PM (2000) Mechanics of the inner ear of the bullfrog (*Rana catesbeiana*): the contact membranes and the periotic canal. J Comp Physiol A 186:481–488.

Rheinlaender J, Klump G (1988) Behavioral aspects of sound localization. In: Fritzsch B, Ryan MJ, Wilczynski W, Hetherington TE, Walkowiak W (eds), The Evolution of the Amphibian Auditory System. New York: John Wiley & Sons, pp. 297–305.

Rheinlaender J, Gerhardt HC, Yager DD, Capranica RR (1979) Accuracy of phonotaxis by the green treefrog (*Hyla cinerea*). J Comp Physiol 133:247–255.

Rheinlaender J, Walkowiak W, Gerhardt HC (1981) Directional hearing in the green treefrog: a variable mechanism? Naturwissenschaften 68:430–431.

Rice WR (1982) Acoustical location of prey by the marsh hawk: adaptation to concealed prey. Auk 99:409–413.

Rosowski JJ, Saunders JC (1980) Sound transmission through the avian interaural pathway. J Comp Physiol A 130:183–190.

Sakaluk SK, Bellwood JJ (1984) Gecko phonotaxis to cricket calling song: a case of satellite predation. Anim Behav 32:659–662.

Saunders JC, Duncan RK, Doan DE, Werner YL (2000) The middle ear of reptiles and birds. In: Dooling RJ, Fay RR, Popper AN (eds), Comparative Hearing: Birds and Reptiles. New York: Springer-Verlag, pp. 13–69.

Schmidt R (1988) Mating call phonotaxis in female American toads: lesions of central auditory system. Brain Behav Evol 32:119–128.

Schmitz B, White TD, Narins PM (1992) Directionality of phase locking in auditory nerve fibers of the leopard frog *Rana pipiens pipiens*. J Comp Physiol A 170:589–604.

Schwartz J, Gerhardt HC (1989) Spatially mediated release from auditory masking in an anuran amphibian. J Comp Physiol A 166:37–41.

Schwartz J, Gerhardt HC (1995) Directionality of the auditory system and call pattern recognition during acoustic interference in the gray tree frog, *Hyla versicolor*. Audit Neurosci 1:195–206.

Schwartzkopff J (1950) Beitrag zur Problem des Richtungshörens bei Vögeln. Z Vergl Physiol 32:319–327.

Schwartzkopff J (1952) Untersuchungen über die Arbeitsweise des Mittelohres und das Richtungshören der Singvögel unter verwendung von Cochlea-Potentialen. Z Vergl Physiol 34:46–68.

Smith CA, Konishi M, Schuff N (1985) Structure of the barn owl's (*Tyto alba*) inner ear. Hear Res 17:237–247.

Soares C, Carr CE (2001) The cytoarchitecture of the nucleus angularis in the barn owl (*Tyto alba*). J Comp Neurol 429:192–203.

Soares D, Chitwood RA, Hyson RL, Carr CE (2002) Intrinsic neuronal properties of the chick nucleus angularis. J Neurophysiol 88:152–162.

Sullivan WE, Konishi M (1984) Segregation of stimulus phase and intensity coding in the cochlear nucleus of the barn owl. J Neurosci 4:1787–1799.

Szpir MR, Sento S, Ryugo DK (1990) Central projections of cochlear nerve fibers in the alligator lizard. J Comp Neurol 295:530–547.

Takahashi T, Moiseff A, Konishi M (1984) Time and intensity cues are processed independently in the auditory system of the owl. J Neurosci 4:1781–1786.

ten Donkelaar H, Bangma GC, Barbas-Henry HA, de Boer-van Huizen R, Wolters JG (1987) The brain stem in a lizard, *Varanus exanthematicus*. Adv Anat Embryol Cell Biol 107:56–60.

van Bergeijk WA (1966) Evolution of the sense of hearing in vertebrates. Am Zool 6: 371–377.

Vlaming MSMG, Aertsen AMBJ, Epping WJM (1984) Directional hearing in the grass-frog (*Rana temporaria* L.). I. Mechanical vibrations of tympanic membrane. Hear Res 14:191–201.

Volman SF, Konishi M (1989) Spatial selectivity and binaural responses in the inferior colliculus of the great horned owl. J Neurosci 9:3083–3096.

Volman S, Konishi M (1990) Comparative physiology of sound localization in four species of owls. Brain Behav Evol 36:196–215.

Wagner H (1993) Sound-localization deficits induced by lesions in the barn owl's space map. J Neurosci 13:371–386.

Wagner H (2002) Directional hearing in the barn owl: psychophysics and neurophysiology. In: Tranebjærg L, Christensen-Dalsgaard J, Andersen T, Poulsen T (eds), Genetics and the Function of the Auditory System. Proceedings of the 19th Danavox Symposium. Copenhagen: Holmens Trykkeri, pp. 331–351.

Wagner H, Takahashi TT, Konishi M (1987) Representation of interaural time difference in the central nucleus of the barn owl's inferior colliculus. J Neurosci 7:3105–3116.

Walkowiak W (1980) The coding of auditory signals in the torus semicircularis of the fire-bellied toad and the grass frog: responses to simple stimuli and to conspecific calls. J Comp Physiol 138:131–148.

Walkowiak W, Berlinger M, Schul J, Gerhardt HC (1998) Significance of forebrain structures in acoustically guided behaviour in anurans. Eur J Morphol 37:177–181.

Wang J, Narins PM (1996) Directional masking of phase locking in the amphibian auditory nerve. J Acoust Soc Am 99:1611–1620.

Wang J, Ludwig TA, Narins PM (1996) Spatial and spectral dependence of the auditory periphery in the northern leopard frog. J Comp Physiol A 178:159–172.

Werner YL (2003) Mechanical leverage in the middle ear of the American bullfrog, *Rana catesbeiana*. Hear Res 175:54–65.

Werner YL, Montgomery LG, Safford SD, Igic PG, Saunders JC (1998) How body size affects middle-ear structure and function and auditory sensitivity in gekkonoid lizards. J Exp Biol 201:487–502.

Wever EG (1978) The Reptile Ear. Princeton: Princeton University Press.

Wever EG (1985) The Amphibian Ear. Princeton: Princeton University Press.

White TD, Schmitz B, Narins PM (1992) Directional dependence of auditory sensitivity and frequency selectivity in the leopard frog. J Acoust Soc Am 92:1953–1961.

Wightman FL, Kistler DJ, Perkins ME (1987) A new approach to the study of human sound localization. In: Yost WA, Gourevitch G (eds), Directional Hearing. New York: Springer-Verlag, pp. 26–48.

Wilczynski W (1988) Brainstem auditory pathways in anuran amphibians. In: Fritzsch

B, Ryan MJ, Wilczynski W, Hetherington TE, Walkowiak W (eds), The Evolution of the Amphibian Auditory System. New York: John Wiley & Sons, pp. 209–231.

Wilczynski W, Resler C, Capranica RR (1987) Tympanic and extratympanic sound transmission in the leopard frog. J Comp Physiol A 161:659–669.

Wilczynski W, Rand AS, Ryan MJ (2001) Evolution of calls and auditory tuning in the *Physalaemus pustulosus* species group. Brain Behav Evol 58:137–151.

Will U (1988) Organization and projections of the area octavolateralis in amphibians. In: Fritzsch B, Ryan MJ, Wilczynski W, Hetherington TE, Walkowiak W (eds), The Evolution of the Amphibian Auditory System. New York: John Wiley & Sons, pp. 185–208.

Will U, Fritzsch B (1988) The eighth nerve of amphibians. In: Fritzsch B, Ryan MJ, Wilczynski W, Hetherington TE, Walkowiak W (eds), The Evolution of the Amphibian Auditory System. New York: John Wiley & Sons, pp. 159–183.

Woodworth RS, Schlosberg H (1962) Experimental Psychology. New York: Holt, Rinehart and Winston, pp. 349–361.

Xu J, Gooler DM, Feng AS (1994) Single neurons in the frog inferior colliculus exhibit direction-dependent frequency selectivity to isointensity tone bursts. J Acoust Soc Am 95:2160–2170.

Zhang H, Xu J, Feng AS (1999) Effects of GABA-mediated inhibition on direction-dependent frequency tuning in the frog inferior colliculus. J Comp Physiol A 184: 85–98.

5

Comparative Mammalian Sound Localization

Charles H. Brown and Bradford J. May

1. Introduction

In natural environments, the approach of a competitor, a predator, a relative, a mate, or one's prey may be conveyed by subtle fluctuations within the acoustic environment. In many instances it is likely that the early detection of an intruder is conveyed not by a sound that is unusual or uncommon because of its amplitude or frequency composition, but rather by a sound that is distinctive chiefly because it occurred at an "inappropriate" location within the acoustic landscape. Here, one's ability to survive depends not on unusual sound detection capabilities, but rather on a sound localization system that permits a listener to effortlessly, yet vigilantly, track the relative positions of the sources of sounds that signal safety or danger. Moreover, the absence of a "safe" sound may be as significant to many birds and mammals as is the presence of an "unsafe" one; for an intruder's approach may be disclosed by either the production of unexpected sounds, or by the abrupt cessation of "expected" sounds that were previously sustained or ongoing in some regular pattern. Movements made unstealthily will disrupt the chorus of cicadas or the sounds of birds, or other animals, and a ripple of silence may spread across the landscape signaling that something (or someone) is nearby. The subtlest acoustic changes may be biologically the most telling. Clumsy predators are apt to go hungry, and an evolutionary premium has been placed upon the ability of most animals to quickly discern the position of a sound that does not belong (or the position of an unexpected cessation of those sounds that do belong). In the struggle for survival, the determination of the origin of a signal may be assigned a level of importance that equals or exceeds that of being able to recognize the sound, or being able to identify the perpetrator of the disturbance. It is in this biological context that the mechanisms underlying sound localization evolved, and through the course of the succession of animals on earth the sound localization abilities of many species have come to exhibit remarkable acuity and species specificity.

The position of the source is a cardinal perceptual attribute of sound. Under normal conditions, for human listeners, the source of a stimulus is instantly and effortlessly assigned a position with reference to the orientation of the listener. The localization of sound is seemingly reflexive, the perception of direction is "instantaneous," and localization does not appear to be derived by some kind of deductive cognitive process. That is, under most conditions listeners do not actively think about having to triangulate the possible origin of the sound given what they heard at their two ears. Just as a sound is perceived as having some magnitude, pitch (or noisiness), loudness, and duration, it also is perceived as having a distance dimension (it is near or far), an elevation dimension (above or below), and an azimuth dimension (left or right of the observer). Only when listeners wear earphones do sounds routinely lack a coherent or natural spatial image, and under these conditions the normal filtering characteristics of the external ear and ear canal have been bypassed, and the normal correlation between the timing and amplitude of the signals at the two ears has been violated.

Batteau et al. (1965) noted that many sounds presented through earphones are reported to have an origin somewhere inside the listener's head. They showed that sounds presented through earphones would be perceived as having a normal external position and could be accurately located in space if the signals fed to the left ear and right ears originate from microphones positioned approximately 17.5 cm apart (a normal head width) and if the microphones were fitted with replicas of human pinnae. The apparent origin of the signal is "external" to the listener when sounds are presented this way, and if the position of a sound source delivered to the microphone array is moved to the left or to the right, the perceived location of the sound source moves accordingly. If the artificial pinnae are removed from the microphones, or if the normal free-field-to-eardrum transfer functions are artificially manipulated, localization accuracy suffers (Batteau et al. 1965; Wightman and Kistler 1989a,b, 1992; Middlebrooks 1992, 1999).

It is known that the position of sound is a core dimension of auditory perception in adult human subjects, and there is good reason to believe that the same is true for human infants, and for most vertebrates. That is to say, the position of a startling sound appears to "command" most vertebrates to orient toward its site of origin. Although auditory experience may modify and adjust localization during development (Knudsen 1983, Knudsen et al. 1984), reflexive orientation to sound position is evident at or near birth in a wide variety of subjects including laughing gulls (Beer 1969, 1970), Peking ducklings (Gottlieb 1965), infant cats (Clements and Kelly 1978a), rats (Potash and Kelly 1980), guinea pigs (Clements and Kelly 1978b), and humans (Muir and Field 1979; Wetheimer 1961). The data suggest that most vertebrates, including both altricial and precocial species, are able to reflexively locate the origin of sound nearly as soon as the ear canal opens and they are able to hear.

In many organisms sound localization mechanisms may initiate and actively guide saccadic eye movements to the site of potentially important events. Animals with binocular frontal visual systems, such as most primates, have limited peripheral or hemispheric vision, and these species may be particularly dependent on a high-acuity directional hearing system to rapidly direct the eyes to the location of a disturbance (Harrison and Irving 1966). Furthermore, the more restricted the width of the horizontal binocular visual field in various mammals, the greater the acuity of their sound localization abilities (Heffner and Heffner 1985, 1992). This enhanced acuity may be critical for accurately aiming the eyes.

The perception of many events is bimodal. Speech perception, for example, is influenced by both visual information regarding tongue and lip configuration, and by the corresponding acoustic signal. When these two modalities of information are out of synchrony, or artificially separated in space, the result is disturbing to both adult and infant human subjects (Aronson and Rosenbloom 1971; Mendelson and Haith 1976). The preservation of the normal congruence between visual and auditory space is important for the development of sound location discriminations in animals (Beecher and Harrison 1971). Animals appear to be prepared to learn to direct visually guided responses toward objects positioned at the origin of sounds, and correspondingly unprepared to learn to direct responses toward objects that have been repositioned so that the contiguity of visual and auditory space has been violated (Beecher and Harrison 1971, Harrison et al. 1971; Harrison 1990, 1992). For an organism to be able to react appropriately to events occurring at different locations in space it is necessary that the visual and acoustic perceptual maps be aligned and in register. Visual deprivation early in development alters sound localization in rats (Spigelman and Bryden 1967), cats (Rauschecker and Harris 1983; Rauschecker and Kniepert 1994), ferrets *Mustela putorious* (King et al. 1988; King and Parsons 1999), and barn owls (*Tyto alba*) (Knudsen and Knudsen 1985, 1990; Knudsen and Brainard 1991; Knudsen et al. 1991). Theoretically it is possible that spatial maps are organized by visual experience. Thus, visual deprivation would lead to impairments in sound localization. Alternatively, it is possible that visual deprivation produces compensatory sharpening of directional hearing. Some recent findings are consistent with both of these contrasting perspectives, and these data are described in Section 4.5.

Although sound localization mechanisms evolved because of their significance to survival in the natural world (Masterton et al. 1969), sound localization abilities have nearly always been studied in synthetic, quiet, echo-free environments (or even with earphones), and testing has often been conducted with tones, clicks or band-limited bursts of noise. The intent of this tradition has been to assess the absolute limits of precision of directional hearing, though at the expense of exploring how well sound localization abilities function under more normal conditions. The sections that follow describe the physical cues for sound localization available to terrestrial vertebrates and the behavioral methodologies commonly used to assess the sound localization capabilities of animals, and then go on to survey the sound localization abilities of selected mammals.

2. Localization Cues

The origin of sound in space is referenced relative to the orientation of the listener, and sound position is accordingly expressed in terms of its azimuth, elevation, and distance from the listener.

2.1 Sound Source Azimuth: The Horizontal Coordinate of Sound Location

Sound localization is dependent upon the comparison of the sound waves incident at each ear in most terrestrial vertebrates. These interaural (or binaural) differences are the result of two factors: (1) the difference in distance (Δd) the sound wave must travel to reach the tympanic membrane of the two ears and (2) differences in the transfer function of the signal incident at each ear. In the case of pure tones, or very narrow bandwidth signals, differences in the transfer function at each ear are reduced to interaural level (or amplitude) differences of the waveform incident at each ear. We defer discussion of spectral cues and head-related transfer functions (HRTFs) to Section 5, and we begin the examination of directional hearing with simple sine waves. The first factor, the difference in propagation distance for the near and far ear, results in differences in the time-of-arrival and in differences in the phase of the signal at each ear. When a sound is presented from a position off to one side of a listener (not on the midline, or at 0° azimuth), corresponding points in the sound wave will necessarily be received by the "near" ear (the ear on the side of the head which is toward the source of the sound) before it reaches the "far" ear. The velocity of sound in air is nominally 343 m/s; given this velocity, for each additional cm the sound wave must travel to reach the far ear, the wave will arrive 29 μs later than it will at the near ear. Hence, interaural differences in the time-of-arrival of corresponding points in the sound wave may serve as one of the principal cues for directional hearing.

For the case in which the stimulus is a simple sustained cyclic wave, such as a pure tone, differences in the arrival time of the near- and far-ear waveforms will result in interaural differences in the phase of the wave as long as the arrival time difference is not equal to the period (or integral multiples of the period) of the signal. For example, the additional time required for the sound wave to reach the far ear may be a fraction of the period of the wave, such as one fourth of the period. In this case the corresponding interaural difference in signal phase would be one fourth of 360°, or 90°. Increments or decrements in the arrival time of near- and far-ear waveforms would result in corresponding increases or decreases in the difference in interaural phase. If the position of the source of the sound were moved so that the arrival time difference is increased from a fraction of the period to exactly match the period of the signal, the near- and far-ear waves would again be in phase. The sound wave incident at the far ear would be precisely one cycle behind the corresponding wave at the near ear. For the special cases in which the arrival time difference between the near- and far-ear waves happens to coincide with two times, three times, or other integral multiples of the period of the signal, the near- and far-ear waves will again be in register, and there will be no interaural differences in phase. Although these phase ambiguities apply to tonal stimuli, they are probably unimportant for most complex stimuli for which other cues for sound localization are available. Furthermore, even with tones, it is unlikely that the source of a sound will be located such that the arrival time differences will equal the period (or multiples of the period) of the signal; consequently, interaural differences in arrival time are usually reflected by interaural phase differences, and these differences may serve as a cue for localizing the azimuth of the source.

Interaural level differences (ILDs) are an additional cue for the perception of the azimuth of sound position. ILDs may occur when the origin of the sound is off to one side, and it is a consequence of the shape of the torso, head, pinna, and external ear canal, as well as the properties of sound diffraction, reflection, and refraction with these structures. The magnitude of sound diffraction, reflection and refraction is dependent on the relative dimensions of the wavelength of the sound wave and the size and shape of the reflective structures. In general, ILDs are most important for signals composed of wavelengths that are less than the diameter of the listener's head. Shorter wavelengths (e.g., higher-frequency signals) produce more prominent ILDs, and the characteristics of these differences are highly dependent on the specific geometry of the listener's head and pinna.

2.2 Geometrical Considerations

As a first approximation, an acoustically opaque sphere may model the head with the ears diametrically opposed (Rayleigh 1876, 1945). With this idealization, the shape and dimensions of the pinna are ignored. Furthermore, the idealized model assumes that the head is immobile (unable to scan the sound field), and that a point sound source is positioned greater than 1 m from the listener. Under these conditions, a plane may approximate the wave front. Interaural distance differences (Δd) will occur for all sound locations other than those that lie on the median plane. As depicted in Figure 5.1, for a sound source to the right of a listener at azimuth X, the additional distance that the sound must travel to reach the far ear (left ear) is given by the sum of the linear distance $r(\sin X)$ and the curvilinear distance $r(X)$. That is, the difference (Δd) in the sound path-length for the two ears is given by Eq. (5.1),

$$\Delta d = r(X + \sin X) \tag{5.1}$$

where Δd is the distance difference in cm, r is the radius of the listener's head in cm, and the sound source azimuth angle X is measured in radians.

The path length difference to the two ears is acoustically realized by the interaural difference in time-of-arrival of corresponding points in the waveforms incident at the two ears. Time-of-arrival differences (Δt) are calculated by dividing the distance difference by the velocity of sound. Given a sound velocity in air of 343 m/s, then the relationship between Δt and azimuth is provided by Eq. (5.2),

$$\Delta t = r(X + \sin X) / 3.43 \times 10^4 \tag{5.2}$$

where Δt is the temporal difference in μs, r is the radius of the observer's head in cm, and the azimuth angle X is measured in radians.

Three factors merit emphasis. First, Δt approaches a maximum value as az-

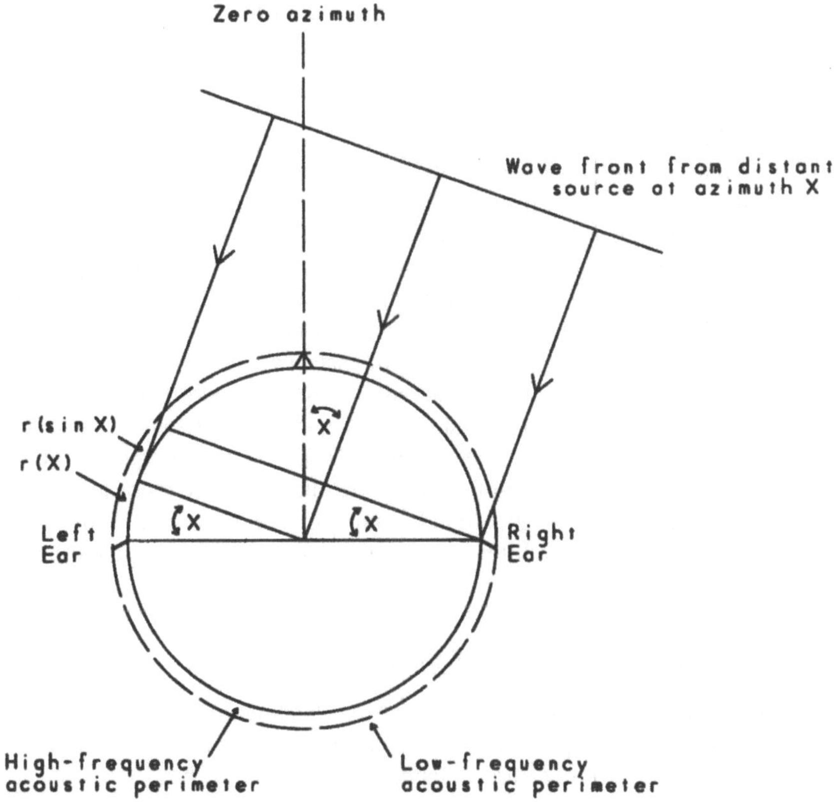

FIGURE 5.1. Geometrical considerations for interaural differences in ongoing time or phase differences as a function of frequency. The signal is presented at azimuth X from at a distant location from the listener. The interaural differences in signal phase obtained with low-frequency tones are produced by a larger effective acoustic radius of the head compared to that obtained by a high-frequency signals. High- and low-frequency signals are scaled relative to head size. A high-frequency signal is one in which the wavelength of the sound is equal to or less than two times the radius of the listener's head, while a low-frequency tone is one in which the wavelength is eight times the radius of the head or greater. In the intermediate frequency zone (defined by the interval $2r < X < 8r$) the effective radius of the head gradually changes from the two boundary conditions illustrated here. See Kuhn (1977) for a detailed treatment of these phenomena.

imuth X approaches $\pi/2$ radians or $3\pi/2$ radians (90° or 270°). Assuming r = 8.75 cm, the usual value assigned for humans, then maximum Δt = 656 μs. That is, at $\pi/2$ radians (90°) the sound wave arrives at the far ear 656 μs after it arrives at the near ear. Second, for any given azimuth X, Δt varies directly with r. As a result, listeners with large heads will experience a greater interaural time-of-arrival difference than will subjects with small heads. Consequently, if the neural processing resolution of time-of-arrival differences is approximately equal across mammalian species, then species with large heads will be able to perceive finer changes in azimuth using this cue than will species with small heads. Third, interaural time-of-arrival differences do not define a specific locus in three-dimensional space. That is, sources that differ in elevation may still have the same interaural path-length difference. Furthermore, the hemifield behind a listener is a mirror image of that in front, and locations above, behind, or below a listener may have the same interaural time-of-arrival difference as those in the front hemifield. If an organism's trunk and limbs did not serve as acoustic obstacles influencing interaural time-of-arrival differences, equal differences in time-of-arrival would be given by all points falling on a surface of a cone centered on the aural axis. In most instances, however, the limbs and torso are acoustic obstacles that impact significantly on path-length differences (and hence, interaural time-of-arrival differences) for sound sources positioned above, behind, or below a listener (Kuhn 1979). As a consequence of this complication, while equal time-of-arrival differences may be empirically measured for displacements of the sound source in the horizontal and vertical planes, equal time-of-arrival differences are unlikely to correspond to the surface of a cone.

2.3 Sound Frequency and Effective Radius

Time-of-arrival cues are produced for both the leading edge of the wavefront, and for ongoing time or phase differences in the waveforms. Interaural differences in time-of-arrival cues are influenced by frequency, head size, and azimuth. Kuhn (1977, 1987) has shown that the effective acoustic radius of the head is larger than the skull perimeter when low-frequency stimuli are presented, but equal to the skull perimeter when high-frequency sounds are presented. In general, when the wavelength of the stimulus is less than or equal to the diameter of the skull (a high-frequency signal) the effective acoustic radius approximates that of the perimeter of the skull. When the wavelength of the stimulus is greater than or equal to four times the diameter of the skull (a low-frequency signal) the effective acoustic radius expands to a larger value with a magnitude that is probably governed by the degree of prognathism, the protrusion of the jaw and nose, and by the size of the pinna. In humans the effective acoustic radius of the head for low-frequency signals is about 150% of that for high-frequency signals (Kuhn 1977), and the transition in the effective radius occurs at about 2000 Hz. In animals with pronounced prognathism of the nose and jaw, and hypertrophied pinna even greater expansion of the effective acoustic radius of the head for low-frequency signals would be expected. This phenomenon is not intuitively apparent, and it is attributed to a frequency dependency in the pattern of standing waves created around acoustic barriers (Kuhn 1977).

The functional acoustic radius for the leading edge of a sound wave, however, is equal to the skull perimeter, and it is not influenced by the relative frequency of the signal (Kuhn 1977). Thus, as depicted in Figure 5.1, low-frequency signals have an enhanced difference in interaural phase because the effective acoustic radius of the head is expanded for the fine structure of these signals.

The cues that are available for time-domain processing are influenced by the rise and fall times of the signal and the complexity of the frequency spectrum and envelope (or amplitude modulation) of the waveform. Signals, which seem to begin and end imperceptibly, have a slow rise and fall time (gradual onset and offset) and lack a crisp leading edge. Consequently, time-domain localization would likely be restricted to the comparison of interaural differences in the phase of the fine structure of the signal, or to the comparison of interaural differences in the amplitude contour, or envelope, of the waveform (Henning 1974; McFadden and Pasanen 1976). In the case of an unmodulated, slow onset and offset pure tone, time-domain processing would necessarily be restricted to an analysis of the interaural phase differences of the fine structure of the tone. However, in spectrally and temporally complex signals, the envelope will be modulated, and the envelope of these modulations will be incident at the near and far ear at correspondingly different times-of-arrival. Human subjects are able to localize signals by processing interaural differences in signal envelopes (Henning 1974; McFadden and Pasanen 1976), and these envelope cues influence sound localization in other mammals as well (Brown et al. 1980). Thus, time-domain processing of localization cues may be analyzed by processing interaural differences of the cycle-by-cycle fine structure of the signal, or by the processing of interaural differences in the time-of-arrival of the more global modulations of the envelope of complex signals (Middlebrooks and Green 1990).

2.4 Azimuth Ambiguity

Interaural differences in signal phase may provide ambiguous information regarding the position of the source. By way of example, assume that the radius of the head is 8.75 cm, and that the maximum time difference is 656 ms for the fine structure of the signals in question. In this example, as a simplification, ignore the fact the effective acoustic radius may change for signals of different frequency. This interaural time difference (ITD) ($\Delta t = 656\ \mu s$) would result in interaural phase differences of 90°, 180°, and 360° for pure tones of 380 Hz, 760 Hz, and 1520 Hz, respectively. This example illustrates two points. First, the relationship between interaural phase difference and spatial location is frequency dependent. A phase difference of 30° indicates one position for a tone of one frequency, but a different position for a tone of another frequency. Second, more than one location may produce the same difference in interaural phase when the period of the waveform is equal to or less than twice the maximum interaural difference in time-of-arrival. In this simplified example for human listeners, such location ambiguities will occur for frequencies with periods less than or equal to 1312 μs. Here, a 760-Hz stimulus will produce an 180° difference in interaural phase when the stimulus is presented either at azimuth $\pi/2$ radians or $3\pi/2$ radians (90° to the right or left). Hence, interaural phase information alone will not discriminate between these two locations. Similarly, for all frequencies greater than 760 Hz, the interaural difference in signal phase produced for a source at any given azimuth will be perfectly matched by at least one other azimuth. The possibility of ambiguity in azimuth for interaural phase differences of mid-range and high-frequency signals suggests that phase information should be utilized in sound localization only for low-frequency signals. Furthermore, the smaller the head size, the higher the frequency limit for unambiguous localization via interaural phase. A small rodent with a maximum Δt of only 100 μs will not experience ambiguous azimuths for phase differences of signals below 5000 Hz in frequency.

The perception of interaural differences in the phase of the fine structure is restricted to relatively low-frequency signals. Both physiological and behavioral observations indicate that the mammalian auditory system is unable to resolve interaural differences in signal phase for frequencies above some critical value. The critical value may differ for various species, and it is usually observed in the region of 1 kHz to 5 kHz (Klumpp and Eady 1956; Kiang et al. 1965; Rose et al. 1967; Anderson 1973; Brown et al. 1978a; Johnson 1980).

The evolutionary significance of interaural temporal processing for directional hearing is seen in physiological specializations dedicated to this function. Auditory specializations for measuring ITDs are first observed in the brainstem, where powerful endbulb synapses securely couple auditory nerve inputs to cochlear nucleus bushy cells. These calyceal endings faithfully transmit the timing of auditory inputs to binaural neurons in the medial and lateral superior olive. The ITD sensitivity of neurons in the olive and inferior colliculus have been measured with binaural beat stimuli that establish a dynamic time delay for tones or amplitude modulated noise in the two ears (Yin and Kuwada 1983; Batra et al. 1997; Ramachandran and May 2002). ITD-sensitive neurons appear to encode a specific time delay by their maximum (medial superior olive—MSO) or minimum discharge rates (lateral superior olive—LSO). A cross-species comparison of the distribution of best neural delays suggests an overrepresentation of ITD cues that fall within the biological constraints imposed by the effective radius of the head.

Although the geometrical model presented in Figure 5.1 may describe interaural differences in time-of-arrival rather accurately, the same is not true for interaural differences in signal level, or ear differences in HRTFs. In the case of ILDs, subtle variations in the shape of the skull and pinnae have a pronounced impact on the magnitude of the observed differences in interaural level. Using a Shilling artificial human head, Harris (1972) conducted measurements of ILDs with either no pinna, or large pinna or small pinna chosen to sample human pinna variations. These studies were conducted with a microphone diaphragm placed in the position of the tympanic membrane, and differences in the sound pressure level incident at each eardrum were measured as a function of the azimuth of the sound source, the frequency of the signal, and the size of the pinnae (large pinna, small pinna or no pinna). Harris's measurements, presented in Figure 5.2, show that at low frequencies (e.g., 500 Hz) ILDs were very small, while at high frequencies (e.g., 8 kHz) they were prominent. The results also

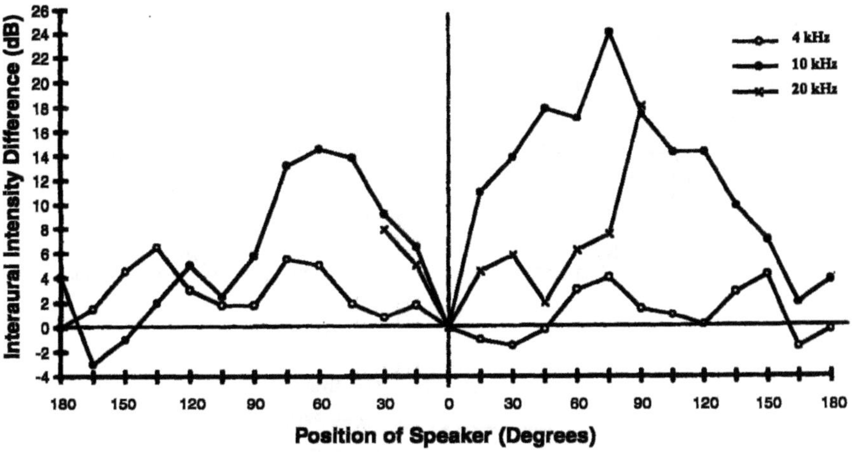

FIGURE 5.2. Sound shadows (ILDs) at representative frequencies produced by rotating a speaker around an artificial head fitted with large, small, or no pinnae. Points further from the center of the circle indicate that the signal level was more intense at the ear on that side of midline. (From Harris 1972. Reprinted with permission.)

indicate that this is an area in which mathematical models do not substitute for empirical measurements. For example, it is surprising that at some azimuth and frequency combinations, ILDs were greater for the no-pinnae condition than they were for either the large- or small-pinnae conditions.

Harrison and Downey (1970) used probe microphones placed by the tympanic membrane to measure ILDs in humans, rats, bats and squirrel monkeys. Their data showed that interaural level differences tended to increase with frequency, and they encountered very large ILDs with nonhuman subjects. Figure 5.3 displays ILDs for an individual squirrel monkey (*Saimiri sciureus*). In general, as signal frequency was increased, ILDs also increased, and at certain azimuth and frequency combinations ILDs could exceed 20 dB. However, because the magnitude of ILDs was influenced by small variations in the morphology of the head and pinnae, ILDs did not vary monotonically with changes in azimuth. It is possible that with tonal, or narrow-bandwidth signals two or more azimuths may give rise to the same overall ILDs, and sound position may then be ambiguous. Broad-bandwidth, high-frequency signals may be accurately localized via the ILD mechanism, however. Brown et al. (1978a) have argued that at each azimuth, the left and right ears will have a spectral transfer function, and the difference between the near- and far-ear functions will give rise to a binaural difference spectrum (Fig. 5.4). The properties of the binaural difference spectrum may be unique to each location, and if the stimulus were broadband, then accurate sound localization would be realized. It is possible that the binaural difference spectrum can also be used to derive sound source elevation as well (Rice et al. 1992), and this will be described in Section 5.

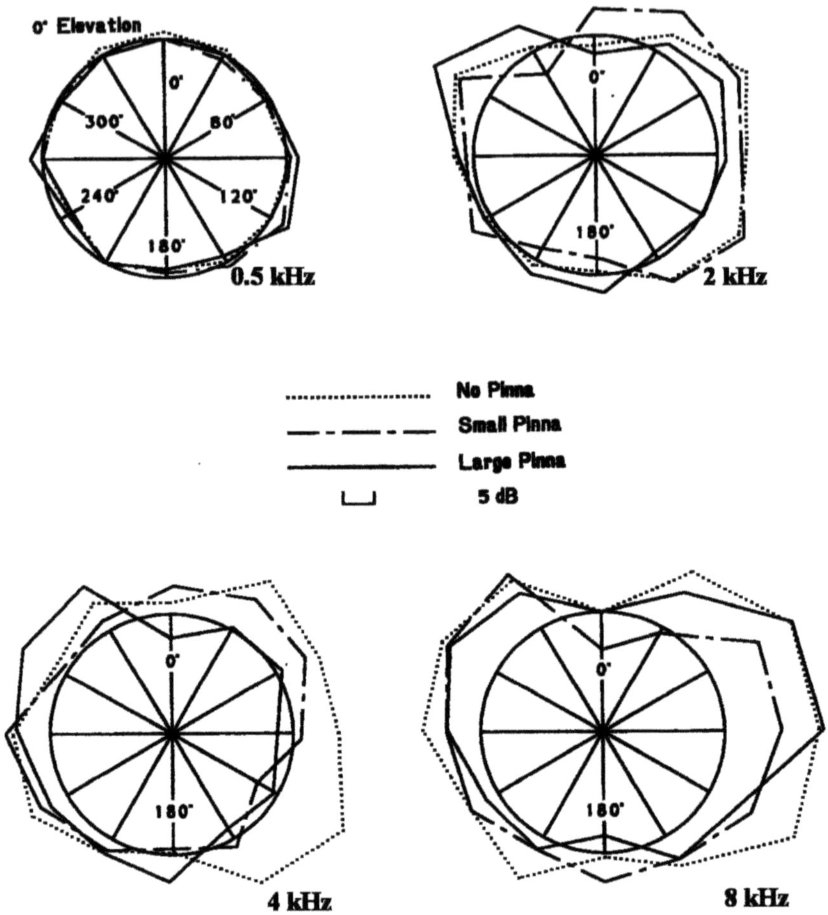

FIGURE 5.3. Interaural level differences measured in squirrel monkeys (*Saimiri sciurens*) as a function of the position of speaker azimuth (0° to 180° on either side of zero azimuth) at three tone frequencies. (From Harrison and Downey 1970. Reprinted with permission.)

In summary, the localization of sound azimuth may be dependent on the perception of interaural differences in time-of-arrival, signal level, and spectral differences. At signal frequencies above the limit for which interaural phase differences become ambiguous, ILDs or spectral differences may become a viable cue. Thus, sound localization may be governed by a multicomponent perceptual system. It is possible that some mammals may be more dependent on one mechanism, while other mammals are more dependent on the other. Species differences in the relative development of several brainstem nuclei are consistent with this possibility. Furthermore, it is likely that head size differences and

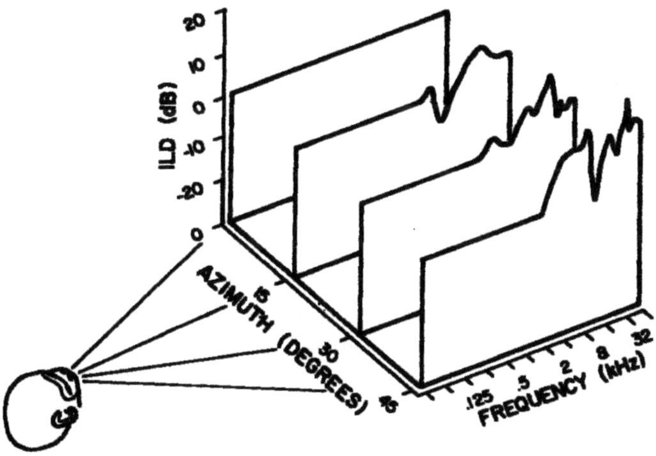

FIGURE 5.4. Hypothetical interaural sound pressure level differences as a function of the azimuth of the source and signal frequency. Negative level differences in the difference spectrum are generated when the signal level incident at the far ear exceeds that recorded at the near ear. (From Brown et al. 1978a. Reprinted with permission.)

pinna size morphology may amplify the significance of one mechanism relative to that for the other.

2.5 The Vertical Coordinate of Sound Location

The perception of source height, or elevation, may be particularly significant for the arboreal primates, marine mammals, and other nonterrestrial organisms. At mid-canopy levels in the rain forest and in marine habitats, biologically significant sounds may arise from positions above and below the listener, as well as from the right and left. If listeners were unable to move their pinnae, and if the right and left ears were acoustically symmetrical, then vertical localization would not involve binaural processing, unless, of course, listeners strategically cocked their heads (Menzel 1980). The relative level of different frequency components of a complex sound change as the location of the source is moved in elevation, and these variations in level are the property of the listener's HRTFs. Although the general properties of these transfer functions are similar for both ears, they are not bilaterally identical, and left- and right-ear asymmetries in the elevation-dependent sound transformation functions are important for vertical sound localization in cats (Musicant et al. 1990; Rice et al. 1992), humans (Shaw 1974a,b; Middlebrooks et al. 1989; Middlebrooks 1992, 1999), and barn owls *Tyto alba* (Payne 1962; Norberg 1977; Knudsen et al. 1979).

In human listeners (and probably in most terrestrial mammals), the perception of vertical position is largely dependent on the fact that the transformation function of the external ear is elevation dependent. Using high-frequency, broad-bandwidth signals, the apparent spectral content of the sound changes with elevation of the stimulus (Butler 1969; Gardner 1973; Hebrank and Wright 1974; Kuhn 1979). Because the asymmetries and convolutions of the pinna and external ear canal must be relatively large compared to the wavelength of the signal for the expression of elevation dependent differences in the external ear transformation function, this cue would require relatively high-frequency, broad-bandwidth signals, and high-frequency hearing (Shaw 1974a,b; Kuhn 1979; Wightman and Kistler 1989a,b). However, lower-frequency signals may reflect off the ground and the organism's torso in an elevation-dependent manner (Kuhn 1979, Brown et al. 1982), and it is possible that some degree of vertical localization is possible with low-frequency signals. Nearly all mammals have excellent high-frequency hearing, and this general trait in many cases may be at least as critical for vertical localization as it is for horizontal localization.

While humans have short, essentially fixed pinna, most terrestrial mammals have extended, mobile pinna; and asymmetries in pinna shape or orientation (Searle et al. 1975) may enhance vertical localization just as ear canal asymmetries enhance the perception of acoustic elevation in the barn owl, *Tyto alba* (Payne 1962; Norberg 1977; Knudsen et al. 1979). Although marine mammals have either no pinna or small pinnae, accurate vertical localization may still be possible (Renaud and Popper 1975). Much work remains to be conducted regarding marine mammal localization; it is unknown how sound is propagated around the head, torso, and ear canal of various marine mammals, and it is unknown if left–right asymmetries exist in the transformation functions for sounds presented at different elevations.

2.6 The Distance Coordinate of Sound Location

The perception of acoustic proximity (distance or depth) is very poorly understood, yet its analog has been well studied in the visual system. In visual perception, both binocular and monocular cues may provide information regarding the relative proximity of visual targets. The chief binocular cue is binocular disparity; a near object is seen from two slightly different angles by the two eyes. When the observer views a near object, more distant objects in the background will necessarily fall on different areas of the left and right retinas. A second binocular cue is convergence, the inward turn of the eyes required to maintain stereoscopic vision. This cue becomes more pronounced as the visual target is positioned progressively closer to the subject (Kaufman 1979). Because there is very little change in either of these binocular cues for two targets positioned at 10 m or 20 m, for example, it is likely that relative distance judgments for distal targets are more dependent on monocular cues than on binocular ones.

The monocular cues for distance perception (for a review see Kaufman 1979) include size constancy (retinal image size varies with changes in object distance); interposition (near objects are in front of, or partially obscure, more distant objects); linear perspective (parallel lines appear to converge at the horizon); textural perspective (the density of items per unit of retinal area increases with distance); aerial perspective (distant objects appear to lose their color saturation and appear to be tinged with blue); relative brightness (objects at greater distances from a light source have less luminance than do objects positioned closer to the source); and relative motion parallax (the apparent location of distant objects is shifted less by a change in the position of the viewer than are the perceived locations of closer objects).

In light of the richness of our appreciation of the cues underlying visual depth and distance perception, it is surprising that so little is known about either the putative perceptual cues underlying the perception of acoustic distance, or the abilities of various species to perceive differences in the proximity of acoustic events. Nevertheless, there is good reason to believe that the perception of acoustic proximity has undergone intense selection for many species. Payne (1962) showed that, in a totally darkened room, barn owls (*Tyto alba*) were able to accurately fly from an elevated perch to broadcasts of recordings of the rustling noise produced by the movements of a mouse. Because the barn owl flies headfirst, yet captures prey feet-first, it must be able to accurately estimate the acoustic azimuth, elevation and distance to be able to position its body for the strike. If the owl were unable to perceive acoustic proximity, it would risk breaking its descent either too soon or too late. Playback experiments have shown that great tits, *Parus major* (McGregor and Krebs 1984; McGregor et al. 1983), howler monkeys *Aloutta palliata* (Whitehead 1987), and gray-cheeked mangabeys *Lophocebus albigena* (Waser 1977) use acoustic cues to gauge distance and judge the possibility of incursions into one's territory by a rival individual or group. Distance perception is also important for prey capture and object avoidance in bats (Denzinger and Schnitzler 1998; Masters and Raver 2000). It is likely that the perception of acoustic distance is important for many species.

Although binocular vision is important for distance and depth perception, there is little available evidence to suggest that binaural hearing is either important, or unimportant, for the perception of acoustic proximity. It is likely that many of the monocular and binocular cues for distance perception have a rough analog in the acoustic domain.

2.6.1 Monaural Cues for Auditory Distance Perception

The prime candidates for monaural distance perception include:

1. Amplitude, sound level or auditory image constancy (the amplitude of the signal varies with distance usually in accordance with the inverse-square law (Gamble 1909; Coleman 1963). Hence, the raucous calls of the hornbill, *Bycanistes subcylindricus*, grow softer as the bird flies to a more distant part of the forest.
2. Frequency spectrum at near distances (von Békésy 1938). At distances less than 4 feet, the low-frequency components of complex signals are relatively more prominent than are mid-frequency and high-frequency components, and

as the source of the signal is moved progressively closer to the listener low-frequency components become even more prominent.

3. Frequency spectrum at far distances (Hornbostel 1923; Coleman 1963). The molecular absorption coefficient for sound in air depends on humidity, temperature, and frequency. At a temperature of 20°C, and a relative humidity of 50%, the absorption coefficient of a 10-kHz tone is about 20-fold greater than that for a 1-kHz tone (Nyborg and Mintzer 1955). Hence, high frequencies are attenuated more rapidly than are low frequencies, and at successively greater transmission distances, the frequency spectrum of complex signals shows a progressive loss of the high-frequency components (Waser and Brown 1986). This cue resembles aerial perspective in the visual domain. That is, just as more distant views are characterized by the loss of longer wavelength hues, more distant sounds are characterized by the loss of high-frequency components.
4. Reverberation. The temporal patterning of signals becomes "smeared" as the delayed reflected waves overlay the direct wave (Mershon and Bowers 1979). Hence, the ratio of direct to reflected waves can provide distance information. This phenomenon is more likely to provide usable information in forested habitats than it is in open habitats.
5. Temporal distortion. Changes in wind velocity, wind direction, or convection current flow result in changes in the duration and pitch of signals transmitted through a nonstationary medium. Signals broadcast from greater distances are probably more susceptible to disturbance by this phenomenon, but this has not been studied in detail (Brown and Gomez 1992).
6. Movement parallax. The relative location of distant sources is shifted less by a change in location of a listener than are the perceived locations of closer sources. This cue is a direct analog to relative motion parallax in the visual domain. It is probable that this cue requires rather large displacements in space for it to play a role in distance judgments for head cocking and other rotational movements of the head and neck may be insufficient to aid distance judgments in some situations (Simpson and Stanton 1973).

2.6.2 Binaural Cues for Auditory Distance Perception

Binaural cues for the perception of acoustic distance include:

1. Binaural intensity ratio. When the source of a signal is at a position other than 0° azimuth the signal may be greater in amplitude at the near ear relative to the amplitude of the signal at the far ear. This difference in sound amplitude, the binaural intensity ratio, varies as a function of head size, azimuth, signal frequency, and transmission distance (Hartley and Fry 1921; Firestone 1930).
2. Binaural differences in signal phase. In addition to the binaural intensity ratio, empirical measurements have shown that binaural differences in signal phase vary as a function of transmission distance as well as head size, azi-

muth, and signal frequency (Hartley and Fry 1921). Thus, it is possible that binaural differences in signal phase may help cue transmission distance.

3. Acoustic field width. At the front row of the auditorium the orchestra may occupy a whole hemifield, while at the rear of an auditorium, the orchestra occupies a more restricted portion of the acoustic field. Hence, the perceived distance to an acoustic source that is not a point source varies inversely with the perceived width of the acoustic field. Although this putative cue is likely binaural in origin, it resembles the monocular cue of textural perspective in the visual domain.
4. Scattered sound direction and field width. In forested habitats, sound becomes scattered by tree trunks. The greater the transmission distance, the greater the magnitude of the field width of the scattered sound, and the perceived width of the field of this scatter may influence distance judgments.

There are very little data to indicate the relative potency of the various putative monaural and binaural cues for judgments of distance, and much research remains to be done in this area. The utility of these cues for the perception of acoustic proximity depends on how reliably they change with distance. The initial cue listed above, auditory image constancy, is simply a change in signal amplitude, while all the other cues enumerated here are associated with a change in sound quality, sound distortion, or a change in sound characteristics at each ear. The only cue, which has received full examination, is auditory image constancy (e.g., amplitude constancy); however, studies of sound transmission in natural habitats have shown that amplitude may fluctuate 20 dB or more in short intervals of time (Wiley and Richards 1978; Waser and Brown 1986). Fluctuations of this magnitude may lead to errors in judgment of three or four doublings of acoustic distance (a sound presented at 25 m under unfavorable conditions may be received at a lower amplitude than the same sound broadcast at 100 m presented under more favorable conditions). Hence, sound amplitude per se is generally regarded as a poor index of transmission distance.

In all habitats, the natural environment degrades sounds, and these more complicated habitat-induced changes in sound quality may more reliably cue acoustic proximity. Brown and Waser (1988) have shown that exemplars of representative vocal classes are degraded differently by the acoustics of natural habitats. Changes in sound quality have been measured with respect to the frequency composition of the call and with respect to the temporal patterning of the signal (Brown and Waser 1988, Brown and Gomez 1992).

Figure 5.5 shows sound spectrograms of the blue monkey (*Cercopithecus mitis*) grunt utterance at the source (panel A), and three recordings of the same call after having been broadcast 100 m in savanna (panels B–D). While the signal displayed in panel B retains the overall structure of the source (panel A), the signal shown in panel C is missing the low-frequency portion of the call (the band of energy at about 500 Hz), and the signal displayed in panel D is missing the two higher-frequency components of the call (the bands of energy at about 1500 Hz, and 3000 Hz). These recordings were conducted in succes-

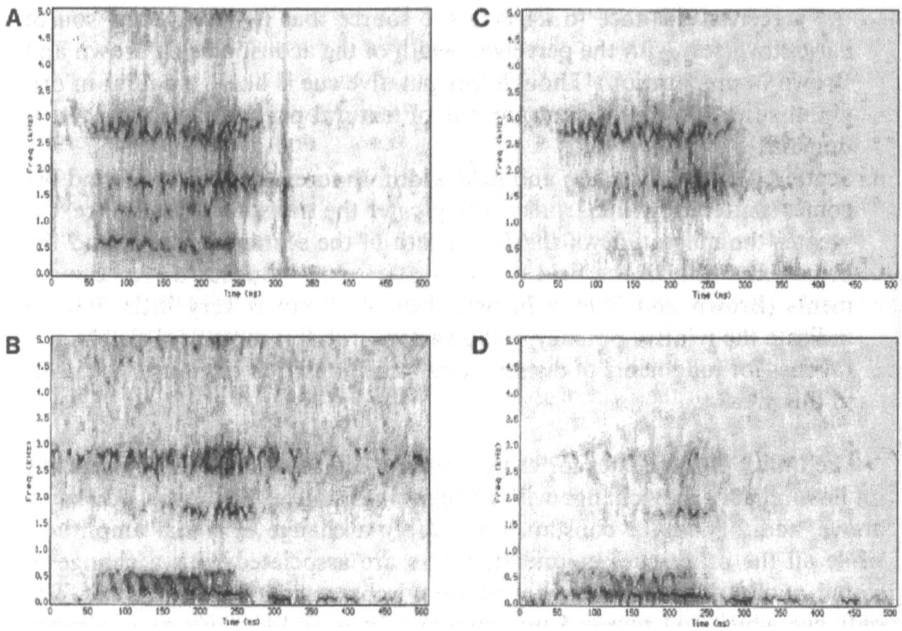

FIGURE 5.5. Sound spectrograms of a grunt call given by a blue monkey (*Cercopithecus mitis*). (**A**) The call at the source. The signal is composed of three energy bands centered at approximately 500 Hz, 1500 Hz, and 2500 Hz. (**B–D**) Spectrograms of the call recorded at a transmission distance of 100 m in the savanna habitat. (**B**) The recording was noisy but all elements of the call were present. (**C**) The 500-Hz-frequency band was absent. (**D**) The 1500-Hz and 2500-Hz-frequency bands were strongly attenuated. (From Brown and Gomez 1992. Reprinted with permission.)

sion under rather stable climatic conditions within a 2-hour interval at the same site (for a review of the factors in natural habitats that lead to different patterns of distortion see Brown and Gomez 1992). These recordings dramatize the fact that the structure of signals may be altered by the acoustics of the habitat. Environmentally induced degradation of acoustic signals occurs in probably all natural settings, and some types of distortion may be useful for estimating the distance to the source.

It is possible to adopt signal-processing techniques to quantitatively measure the magnitude of habitat-induced distortion of vocalizations (Brown and Waser 1988; Brown and Gomez 1992). The data show that some vocalizations (e.g., the blue monkey's boom) are relatively unchanged by the acoustics of the habitat, while other calls (e.g., the blue monkey's chirp or pyow) are more susceptible to degradation. The overall pattern of these degradation scores indicates that different utterances are degraded in different ways by environmental acoustics. Just as some vocalizations are ventriloquial while others are easily localized in azimuth (Brown 1982a), the present observations suggests that some vocalizations may be good for revealing acoustic proximity, while other utterances may obscure the relative proximity of the vocalizer. Presumably, the presence or absence of "distance information" in various calls is relevant to the social function of different vocalizations. Many forest monkeys emit calls that appear to mark the position of the vocalizer. These calls may be involved in regulating the spacing, distribution, and movements of individuals out of visual contact.

Recent work has shown that soundscapes, the background sounds in natural habitats, contain highly nonrandom structures (Nelken et al. 1999), and selection may have favored the evolution of sound processing strategies that exploit the coherence in soundscapes to render signals more separable from the background, and hence more audible, and potentially more locatable. Thus, relative to their audibility in white noise, signals are more audible in masking noises, which exhibit comodulation (Hall et al. 1984; Moore 1999) common to the spectro-temporal fluctuations of natural soundscapes. It is possible that variations in signal structures that influence sound localization in both azimuth and distance are related to release from masking phenomena (Hall et al. 1984; Moore 1999), and the factors that influence signal detection in natural environments may be related to directional hearing. This is a promising area for future research.

3. Sound Localization Methodology

Many animals will orient toward, and approach, the origin of some sounds. The accuracy of approach has been used to study sound localization in the gray-cheeked mangabey monkey (*Lophocebus albigena*) (Waser 1977), tree frogs (*Hyla cinera* and (*H. gratiosa*) (Feng et al. 1976), cats (Casseday and Neff 1973), and many other species. In some instances, food or some other reward has been used to maintain this behavior. In such approach procedures, the accuracy of localization is dependent on the ability of the auditory system to process a change in sensation associated with a change in the position of the source, and in the ability of the motor systems of the animal to accurately guide the subject towards the perceived location of the acoustic target. Species differences in the acuity of localization, measured by the approach procedure, may be due to differences in the precision of the perceptual system, or alternatively these apparent acuity differences may be due to variations in the accuracy of motor systems.

Orientation paradigms have also been developed to measure the acuity of localization. With these methods a head turn or body turn is used to indicate the perception of sound direction (Knudsen and Konishi 1978; Knudsen et al. 1979; Whittington et al. 1981; Brown 1982a; Perrot et al. 1987; Makous and Middlebrooks 1990; Huang and May 1996a,b; May and Huang 1996). Figure 5.6 illustrates this procedure using results from a food-reinforced orientation task. The acoustic stimuli were brief bursts of broadband noise that were presented from one of eight randomly selected locations in an anechoic room.

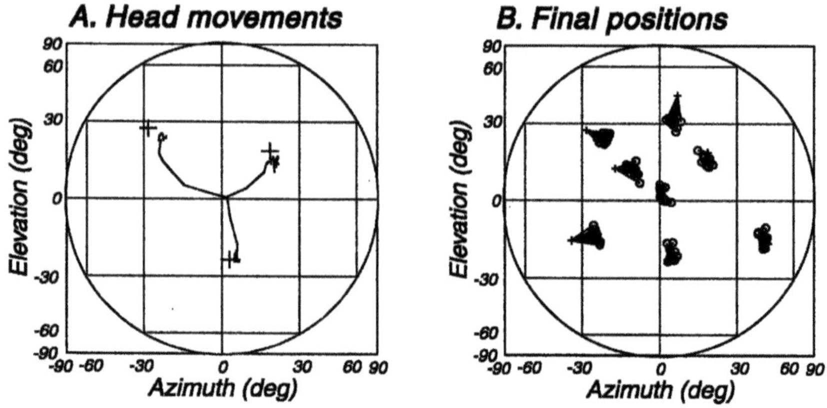

FIGURE 5.6. Sound orientation accuracy of a representative cat. The subject was required to direct its head to the source of a broad bandwidth noise burst. The sound source changed randomly across eight possible speaker locations (*plus symbols*). (**A**) The path of head movements from a fixation point (0° azimuth, 0° elevation) to a final stable position for tests with three different target locations. (**B**) The final head orientation for all tests in one session. *Lines* connect each response to the actual speaker location. (Adapted from Huang and May, 1996a.)

Figure 5.6A tracks the cat's head movements on three successive trials from a standardized fixation point (0° azimuth, 0° elevation) to the location of the speakers (plus symbols). In each case, the subject rapidly acquires the target location then holds the response for several seconds as it waits for a visual cue that signals the end of the trial. Figure 5.6B shows the cat's final stable head orientation for all trials in the testing session (open symbols).

The results shown in Figure 5.6 indicate that spectrally rich sounds evoke orientation responses that are accurate in the vertical as well as the horizontal plane (May and Huang 1996). Both orientation and approach procedures are categorized as egocentric methods (Brown and May 1990). Here localization is made not with reference to an external acoustic marker, but rather with reference to the subject's physical orientation in space. With egocentric procedures, differences in sound localization acuity may be due to limitations in the accuracy of the perceptual system, or to limitations of the motor system.

Behavioral tasks in which listeners have been trained to operate response levers to indicate the detection of a change in sound location have been used with both human (Mills 1958) and animal subjects (Brown et al. 1978a). These ear-centered, or otocentric, procedures are designed to assess the acuity of the perceptual system, and they do not require the participation of the spatial/motor system (Brown and May 1990). Hence, with these procedures listeners report the detection of a change in sound location, but they do not indicate where the sound originated relative to their own orientation.

Given the procedural variations possible between these different methodologies it is important to note that independent measurements of sound localization acuity in normal animals appear to be remarkably consistent and robust. There is a high degree of agreement in the results using both egocentric and otocentric methods within (Heffner and Heffner 1988d), and between laboratories (Brown and May 1990). Using the approach procedure under field conditions in the natural habitat, Waser (1977) showed that one of the Cercopithecoidea monkeys, the gray-cheeked mangabey (*Lophocebus albigena*), was able to localize the whoopgobble, a complex long-distance vocalization (Brown 1989), with an average error of only 6° (Fig. 5.7). Under laboratory conditions, using otocentric methods with two other species of the Cercopithecoidea monkeys (*Macaca nemestrina* and *M. mulatta*), the localization of complex vocal signals ranged from 3° to 15° depending on the specific acoustical characteristics of the utterance (Brown et al. 1978b, 1979). The mean localization error of macaque monkey broad bandwidth or frequency-modulated calls, those that are most comparable to the mangabey's gobble, is about 3°. It is remarkable that a field phonotaxis study conducted in the monkey's native habitat in Uganda and a laboratory investigation yield results that are so similar.

When comparable stimuli are used, the congruence in the data produced by different laboratories employing different methods is even more striking. Figure 5.8 shows ITD thresholds measured using earphones (Houben and Gourevitch 1979), and those calculated from free-field measurements of the acuity of localization (Brown et al. 1978a) as a function of tone frequency. Data for human subjects (Klumpp and Eady 1956) are compared with macaque monkey data. These data show that the physical characteristics of the signal have a strong impact on the accuracy of sound localization. This is true for both simple synthetic signals, such as tones, and complex natural signals, such as vocalizations.

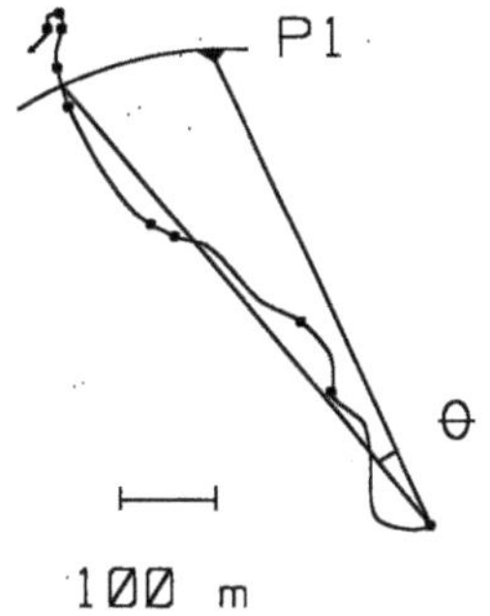

FIGURE 5.7. Sound initiated approach in gray-cheeked mangabeys (*Lophocebus albigena*) evoked by the playback of a whoopgobble vocalization. The playback was conduced in the Kibale forest in Uganda with native populations of mangabeys. P1 is the location of the broadcast loud speaker. At the time of broadcast the focal subject was located at the apex of angle θ. Mangabeys are arboreal, and only rarely descend to the ground. The track that the monkey takes is then partially governed by the location of the branches of adjacent trees in the forest. Owing to the thickness of rain forest vegetation field assistants cannot follow the path of the monkey directly, but are able to observe the movement of the monkey to identified trees in the forest (denoted by periods in the figure). The angle θ is the discrepancy between the mean direction of approach and the playback site. (From Waser 1977. Reprinted with permission.)

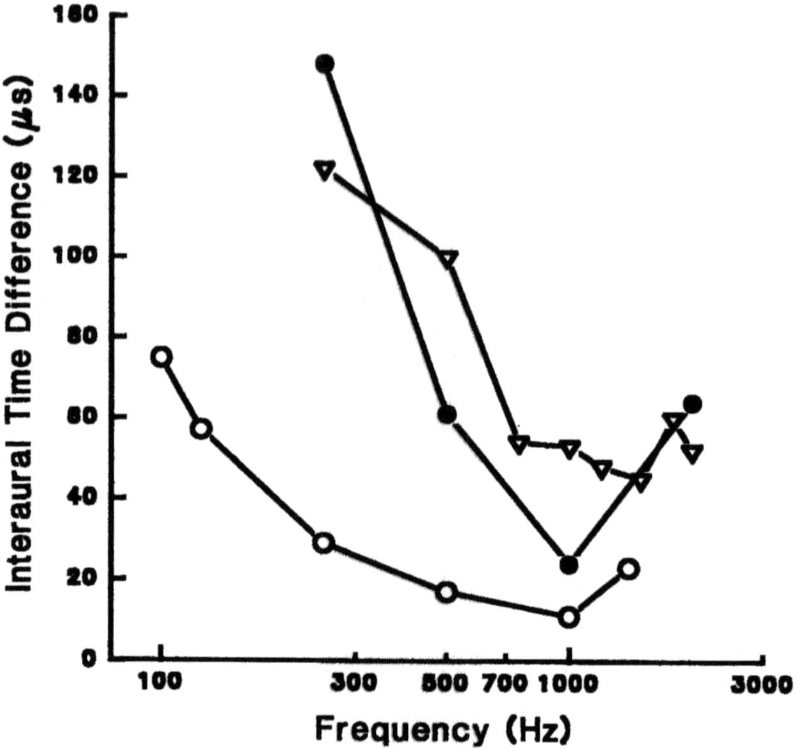

FIGURE 5.8. ITD thresholds. Thresholds, measured in microseconds, are displayed as a function of signal frequency. Macaque monkeys (*Macaca mulatta* and *M. nemestrina*) thresholds (*black squares*) are transposed from free-field localization measurements (Brown et al. 1978a); interaural time difference thresholds measured with earphones from monkeys (*open triangles*: Houben and Gourevitch 1979); and humans (*open circles*: Klumpp and Eady 1956). (From Brown et al. 1978a. Reprinted with permission.)

Furthermore, the data show that measurements of a species' acuity for sound localization are robust, and relatively independent of method. These observations indicate that it is possible to measure with high precision the acuity of sound localization that is representative of the abilities of the species, but that the data derived are dependent on the physical characteristics of the test signal.

4. The Acuity of Sound Localization

4.1 The Perception of Acoustic Azimuth

The just detectable change in the position of the sound source, the minimum audible angle (MAA), has generally been regarded as the most precise index of the acuity of localization. Figure 5.8 presents individual psychometric sound localization functions for three macaque monkeys. The test signal was a macaque coo vocalization. The psychometric functions were derived from monkeys who had been trained to hold a contact-sensitive key when sounds were pulsed repetitively from a source at 0° azimuth, directly in front of the monkey, and release contact with the key when the sound was pulsed from a source at any other azimuth. The monkey's rate of guessing (its catch-trial rate) was very low, less than 8% (this rate is displayed over the 0° point in Fig. 5.9). The monkey's ability to detect a change in the azimuth of the sound source increased with the magnitude of change in source location reaching about 100% correct by 30°. These psychometric functions conform to the class ogive shape (Cain and Marks 1971), and the 50% correct detection point (the MAA) is measured in degrees and calculated from the psychometric functions.

Investigators have tended to measure the acuity of directional hearing with biologically significant stimuli, such as vocalizations (Feng et al. 1976; Waser 1977; Brown et al. 1978a, 1979; Rheinlaender et al. 1979), or more commonly with synthetic signals that are either simple, such as pure tones (Casseday and Neff 1973; Terhune 1974; Brown et al. 1978a; Heffner and Heffner 1982), or spectrally more complex, such as clicks or noise bursts (Ravizza and Masterton 1972; Brown et al. 1980; Heffner and Heffner 1982, 1983, 1987, 1988a,b). Biologically significant signals have tended to be used with phonotaxic proce-

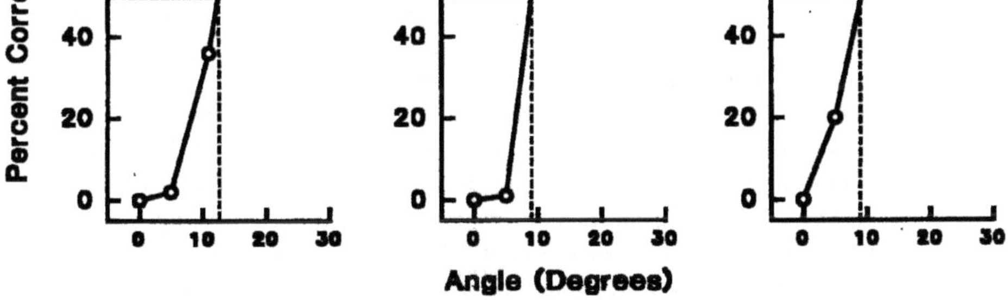

FIGURE 5.9. Psychometric functions for the localization of a macaque coo call. Functions are shown for three individual monkeys (Sidney, Miko, and Oscar). The monkey's rate of guessing (catch-trial rate) is displayed over the zero-degree point, and the monkey's percentage of correct detection for the trials presented at the four comparison locations increased with angle. The calculation of the MAA is shown by the *dashed line*. (From Brown et al. 1979. Reprinted with permission.)

dures or in studies in which the relative locatability of various natural signals was the topic of interest, while synthetic signals have tended to be used in studies that have focused on assessing the limits of the perceptual system.

4.2 Biologically Significant Signals

The different physical characteristics of various classes of complex natural stimuli, such as vocalizations, may influence the acuity of localization. In macaque monkeys, the effective bandwidth (or magnitude of frequency modulation) of the dominant frequency band of the call has a strong effect on sound localization (Fig. 5.10). Increments in the effective bandwidth of the signal enhance the accuracy of localization. MAAs for macaque coo calls span approximately a fivefold range, from about 3° to 15°. Macaque monkeys also produce a wide variety of noisy barks, grunts, and growls, and these harsh sounding, atonal, broad-bandwidth calls are all accurately localized as well (Brown et al. 1979). Complex natural signals that exhibit a broad effective bandwidth (produced either by frequency modulating a relatively tonal sound, or by generating an atonal, broad-bandwidth sound) are probably localized at the limits of resolution of the organism's perceptual system (Brown 1982b; May et al. 1986). The mate attracting calls, rallying calls, and position marking calls given by a wide variety of mammals typically exhibit a broad effective bandwidth, that likely promotes sound localization at the listener's limit of resolution.

4.3 Pure Tones

Comparative data for the localization of pure tones as a function of frequency are shown in Figure 5.11. While the human data suggest that stimulus frequency has a relatively modest effect on the localization of tones (Mills 1958), it tends to have a pronounced effect on the accuracy of localization by nonhuman mammals. At the best frequency, macaque monkeys (*Macaca mulatta* and *M. nemestrina*) (Brown et al. 1978a), harbor seals (*Phoca vitulina*) (Terhune 1974), and elephants (*Elephas maximus*) (Heffner and Heffner 1982) exhibit a resolution of about 4°, while human listeners are able to resolve about 1° (Mills 1958). At their worst frequency, human subjects are still able to resolve angles of about 3°, while most of the other mammals tested may require angles of 20° or more. Thus, human subjects tend to be more accurate at localizing the source of sounds across the frequency spectrum than are most other mammals. Testing with pure tones has almost exclusively been conducted with signals that are gated on and off slowly, and that are not modulated in amplitude so that the envelopes of the waveforms do not provide information that may influence localization. Under these conditions, human listeners localize low-frequency tones with a mechanism sensitive to interaural time differences, while the localization of high-frequency tones is governed by a mechanism sensitive to ILDs (Mills 1960). The same frequency effects have been shown to hold for monkeys (*M. mulatta* and *M.*

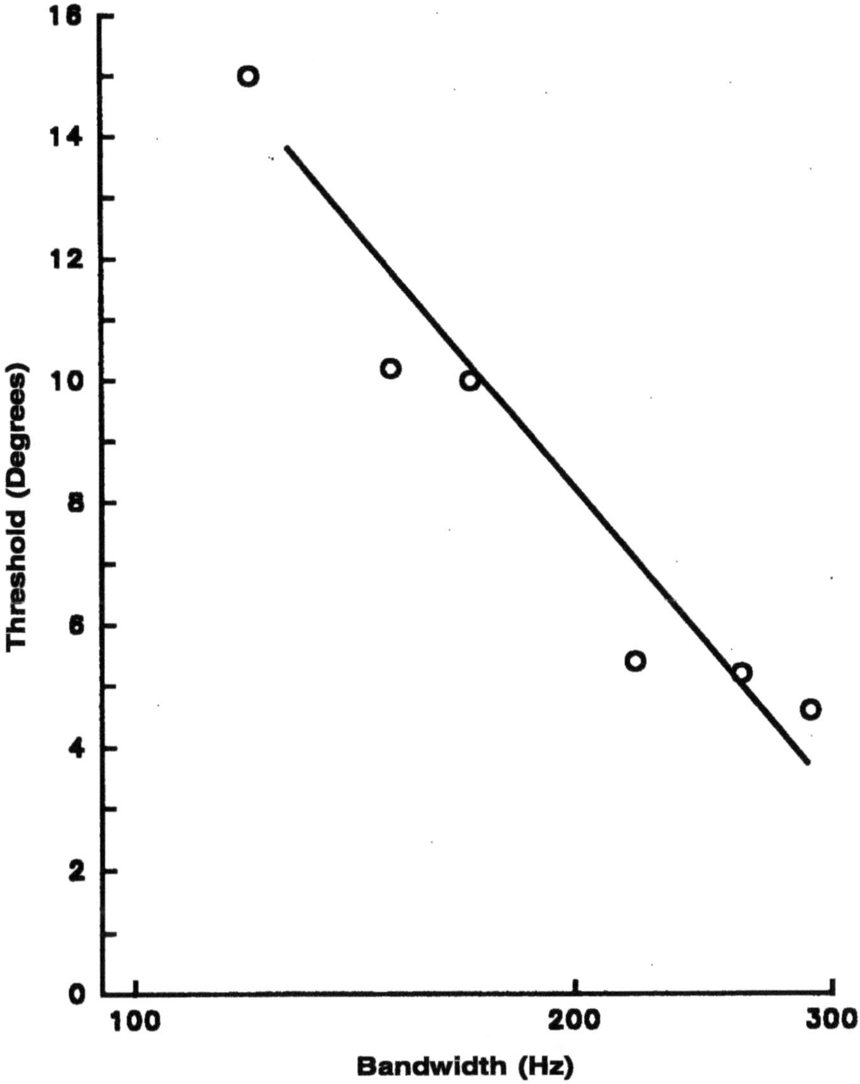

FIGURE 5.10. Macaque monkey (*Macaca mulatta* and *M. nemestrina*) localization thresholds for six coo vocalizations displayed as a function of the effective bandwidth (frequency modulated bandwidth) of the dominant band of the call. The correlation between threshold and call bandwidth was −0.59. (From Brown et al. 1978b. Reprinted with permission.)

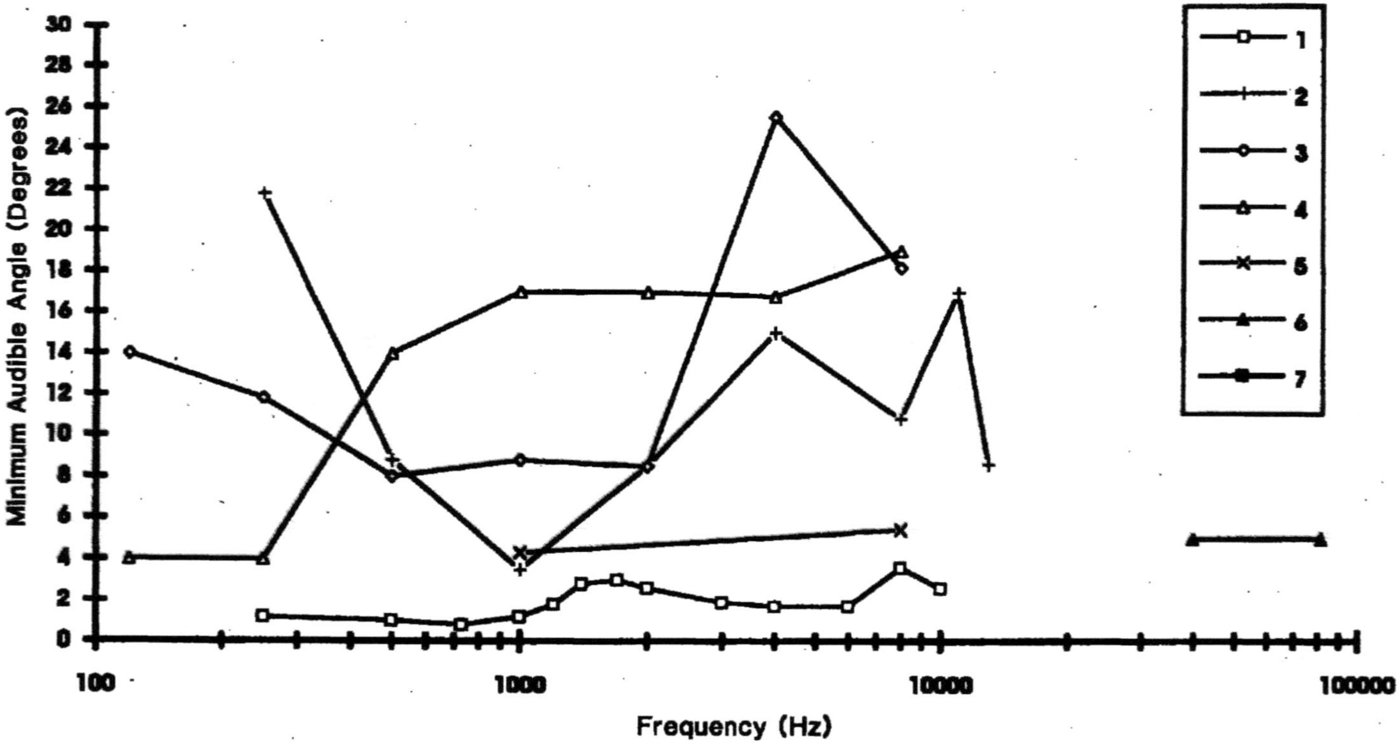

FIGURE 5.11. Sound localization acuity (MAAs) of tones as a function of signal frequency for representative mammals. 1, Human (*Homo sapiens*) (Mills 1958); 2, rhesus and pig-tailed monkeys (*Macaca mulatta* and *M. nemestrina*) (Brown et al. 1978a); 3, cat (*Felis catus*) (Casseday and Neff 1973); 4, elephant (*Elephas maximus*) (Heffner and Heffner, 1982); 5, harbor seal (*Phoca vitulina*) (Terhune 1974); 6, little brown bat (*Myotis oxygnathos*) (Ayrapet'yants and Konstantinov 1974); 7, greater horsehoe bat (*Rhinolophos ferrumequinum*) (Ayrapet'yants and Konstantinov 1974). (Fay 1988. Adapted with permission.)

nemestrina) (Brown et al. 1978a; Houben and Gourevitch 1979), and are presumed to apply for most other mammals as well.

The literature on the comparative localization of tones suggests that both mechanism for localization by human subjects are equally accurate, while in most other mammals one mechanism may be less accurate, and perhaps less significant, than the other. In this context, an extensive physiological and anatomical literature (Heffner and Masterton 1990) has shown that high-frequency localization primarily involves brainstem structures in the medial nucleus of the trapezoid body (MNTB) and LSO, while low-frequency localization primarily involves structures in the MSO. The relative development of these nuclei varies across mammals; in some species the MNTB–LSO system is undeveloped or nearly absent, while in other species the MSO system is undeveloped or nearly absent. In general, as the physical size of the mammal increases, the greater the development of the MSO system, and a concomitant reduction in the MNTB–LSO system is observed (Heffner and Masterton 1990). Thus, variations in the development of auditory structures in the ascending pathway may underlie species differences in their ability to fully utilize interaural time-of-arrival difference cues, or ILD cues. These variations may account for the observed species differences in the pure tone localization data (Fig. 5.11). However, while human subjects localize high-frequency tones well, their MNTB–LSO system is only marginally developed. Hence, although it appears that much is understood regarding the anatomical and physiological mechanisms sub serving sound localization, significant puzzles still remain.

4.4 Complex Stimuli

Comparative data for the localization of complex stimuli (e.g., vocalizations, clicks or noise bursts) are displayed in Figure 5.12. Here the MAA is plotted in reference to head size. As noted in Section 2.2, all other things being equal, both ITDs and ILDs should increase with head size. Thus, large mammals should exhibit greater sound localization acuity simply because the physical magnitude of these interaural cues increase with head size. This trend is generally observed (Fig. 5.12). However, the correlation between threshold and head size is only -0.32. Hence, some mammals are either significantly less sensitive, or more sensitive, to sound direction than would be expected by the size of their heads. Species located below the diagonal regression line shown in Figure 5.12 have better localization acuity than would be expected by their head size, while those positioned above the regression line have less acute directional hearing than would be expected. Thus, regardless of the magnitude of the physical cues available for localization, some species are particularly good localizers, while others are not.

How can these differences in the relative acuity for directional hearing be explained? Four species [gopher (Go), blind mole rat (Bm), pallid bat (Pb), dolphin (Do)] are particularly discrepant from the others tested. The gopher and blind mole rat are fosserial species, spending most of their time under-

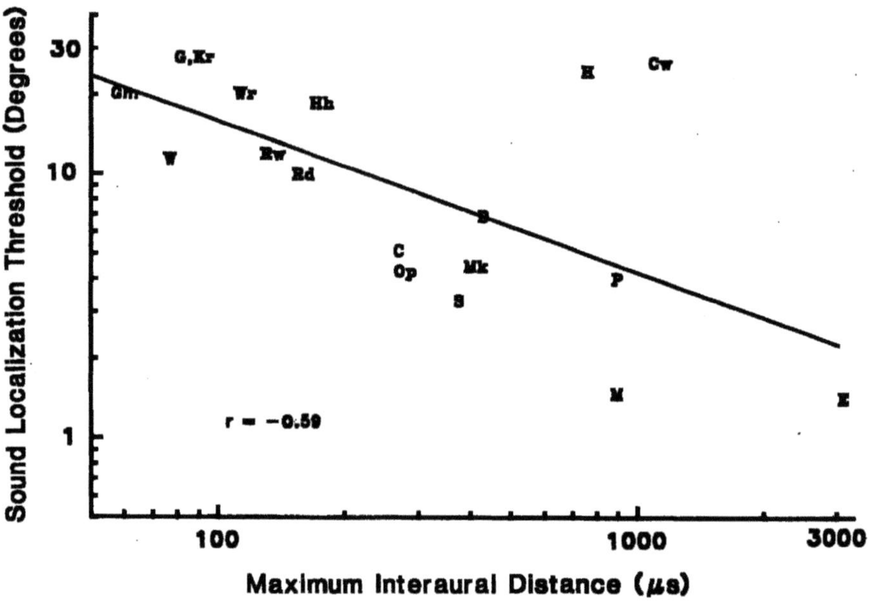

FIGURE 5.12. Sound localization threshold as a function of head size in 18 representative mammals. Acuity is displayed for a broad bandwidth sound, noise, or a click stimulus in the horizontal plane. Gm, grasshopper mouse (*Onychomys leucogaster*) (Heffner and Heffner 1988a); W, least weasel (*Mustela nivalis*) (Heffner and Heffner 1987); G, gerbil (*Meriones unguiculatus*) (Heffner and Heffner 1988c); Kr, kangaroo rat (*Dipodomys merriami*) (Heffner and Masterton 1980); Rw, wild Norway rat (*Rattus norvegicus*) (Heffner and Heffner 1985); Rd, domestic Norway rat and Wistar albino rat (*R. norvegicus*) (Kelly 1980); Wr, wood rat (*Neotoma floridiana*) (Heffner and Heffner, 1988a); Hh, hedgehog (*Paraechinus hypomelas*) (Chambers 1971); C, cat (*Felis catus*) (Heffner and Heffner 1988d); Op, opossum (*Didelphis virginiana*) (Ravizza and Masterton 1972); S, harbor seal (*Phoca vitulina*) (Terhune 1974); Mk, rhesus and pig-tailed macaque monkeys (*Macaca mulatta*) and (*M. nemestrina*) (Brown et al. 1980); D, dog (*Canis canis*) (H.E. Heffner, unpublished); H, horse (*Equus caballus*) (Heffner and Heffner 1984); M, human (*Homo sapiens*) (Heffner and Heffner 1988c); P, domestic pig (*Sus scrofa*) (Heffner and Heffner 1989); Cw, cattle (*Bos taurus*) (Heffner and Heffner 1992); E, elephant (*Elephas maximus*) (Heffner and Heffner 1982).

ground. These two species, along with the naked mole rat (Nm), have degenerate hearing characterized by poor sensitivity and poor high-frequency hearing (Heffner et al. 1987), and their sound localization acuity is also impaired. Thus, radiation into a niche in which hearing in general, and sound localization in particular, are less important biologically may result in a comparative reduction of these sensory capacities. The other two highly atypical species, the dolphin (*Tursiops truncatus*) and pallid bat (*Antrozous pallidus*), are echolocators, and selection for some forms of echolocation may also heighten sound localization acuity. If these four species are removed from the correlation, the association between head size and localization acuity increases to −0.57. The corresponding correlation has improved substantially, but much of the variance in the association between these two variables has not been accounted for. It has been argued that the relationship between vision and sound localization may be an important factor in explaining some of this variance.

In a classic paper, Harrison and Irving (1966) argued that accurate sound localization abilities are particularly important for redirecting the site of gaze for species with high-acuity tunnel vision. That is, the horizontal width of the field of high-acuity vision tends to be much narrower in animals with high-acuity binocular visual systems (such as primates) compared to animals with nonoverlapping hemispheric visual systems (such as rabbits). In most mammals, ganglion cell density varies across the vertical and horizontal coordinates of the retina, and regions of high ganglion cell density are associated with high acuity vision. Heffner and Heffner (1988c) have defined the region of best vision as that portion of the retina in which the ganglion cell density is at least 75% of the maximum observed for that species. Using this approach they have shown that mammals with comparatively narrow fields of best vision have better localization acuity compared to those with broader fields of best vision. The relationship between sound localization and best visual field width breaks down for burrowing fossorial mammals which have radiated into the subterranean habitat, and which in turn exhibit a comparative blunting of both the visual and acoustic senses. Apparently, just as acute vision is superfluous in the absence of light, acute hearing and sound localization is of little value in the absence of a free field. That is, subterranean tunnels may channel sound similarly to a waveguide, and the resulting absence of a free field may change the direction of selection for acoustic processing. In general, these observations support the notion that for many species of mammals, one key function of directional hearing systems is to acoustically guide the orientation of the visual system.

4.5 Plasticity and Sound Localization Acuity

Anecdotal reports have long suggested that some blind humans appear to develop unusually keen auditory abilities. Specifically, the perception of acoustic space in some blind individuals has appeared to significantly exceed the abilities of subjects with normal sight. These reports raise the possibility that perceptual compensation may result when visual processing centers have been reassigned to acoustic processing following the onset of blindness. Thus, it is possible that the loss of use of one sensory modality may lead to a reorganization of the cortex to favor the processing of the remaining viable sensory modalities. Recent physiological studies have obtained results consistent with the idea that early blindness may result in cross-modal reorganization of the cortex, and this reorganization may produce compensatory effects for sound localization (Kujala

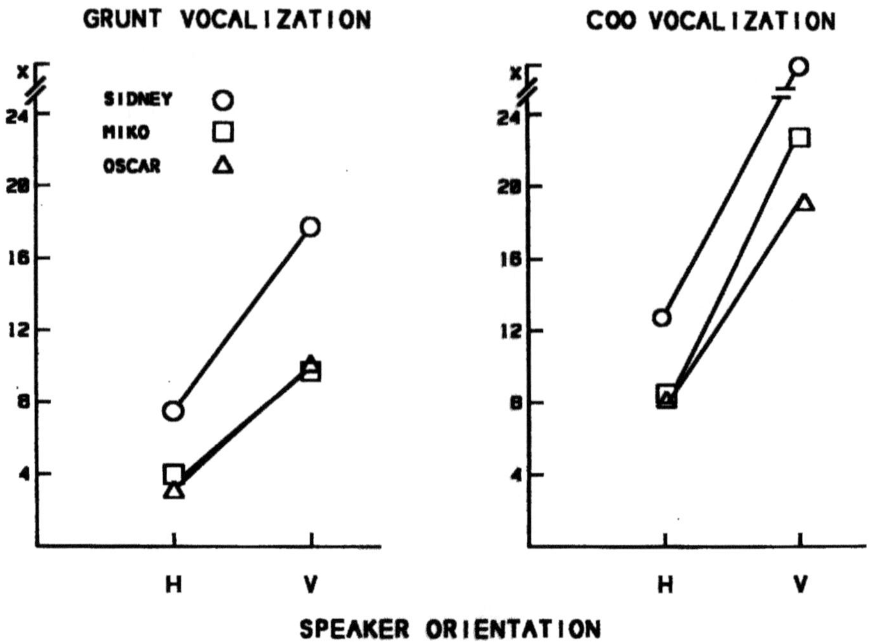

FIGURE 5.13. Horizontal (H) and vertical (V) minimum audible angles for a macaque grunt vocalization, and a macaque coo vocalization for three macaque monkeys (Sidney, Miko, and Oscar). An X indicates that the performance level of the subject never exceeded chance. (From Brown et al. 1982. Reprinted with permission.)

et al. 1992, 1995, 2000). Behavioral data consistent with this hypothesis have been reported for both cats and ferrets. When testing was conducted at a reference azimuth of 0°, MAAs were 16° and 15°, respectively, for ferrets with normal vision compared to those deprived of binocular vision at infancy (King and Parsons 1999). However, when testing was conducted at a reference azimuth of 45°, MAAs for the normal and visually deprived groups were 34° and 16° respectively (King and Parsons 1999). Thus, visual deprivation resulted in an improvement in the acuity of spatial hearing for stimuli located at lateral azimuths, but not at midline locations. King and Parsons (1999) also compared minimal audible angles for subjects blinded at adulthood, and they obtained a similar pattern of results. A complementary pattern of findings has also been reported for cats (Rauschecker and Kniepert 1994). Visually deprived cats showed enhanced sound localization abilities for signals broadcast from lateral and rear positions relative to normally sighted controls. This trend was strong for cats deprived of vision in infancy, and only approached (but did not achieve) statistical significance in adult deprived cats. Recent studies with humans have found that visually impaired, but not totally blind, subjects localize sounds with less accuracy than sighted controls (Lessard et al. 1998). However, 50% of the subjects who were totally blind were superior to sighted controls in a monaural localization task (Lessard et al. 1998). Further evidence suggests, that like cats and ferrets, blindness in humans may have a more pronounced effect for the localization of sounds presented from peripheral locations relative to localization near the midline (Roder et al. 1999). In concert, these findings support the concept of compensatory plasticity; however, the etiology and severity of blindness, as well as its age at onset, may influence its significance for directional hearing.

Although the literature on plasticity has implicated changes in cortical structures, it is also possible that plasticity is expressed by changes at subcortical sites. For example, physiological studies have shown that metabotropic receptors in the dorsal cochlear nucleus (DCN) are capable of modulating synaptic transmission in a manner that resembles neural plasticity in the cerebellum (Molitor and Manis 1997; Devor 2000). Furthermore, the DCN is implicated in spectral processing (Spirou and Young 1991; Nelken and Young 1994), and the resolution of variations in spectral content may be particularly important for the resolution of front/back confusions and the localization of lateral azimuths, spatial regions particularly susceptible to the effects of blindness.

4.5.1 The Perception of Acoustic Elevation

The literature is much more limited concerning the accuracy of perception of acoustic elevation. In arboreal living species, or in marine mammals, the determination of acoustic elevation may be as significant as the determination of azimuth. Vertical and horizontal minimum audible angles for primate vocalizations are shown for macaque monkeys (*M. mulatta* and *M. nemestrina*) in Figure 5.13. The test vocalizations were a macaque coo call, and a macaque grunt call. The grunt, which is broader in bandwidth, was localized more accurately than the coo. The median vertical localization thresholds were approximately 9° and 20°, respectively. For these same signals, the acuity of vertical localization was approximately two to three times less accurate than was localization in the horizontal plane. High-frequency hearing and high-frequency broadband stimuli are important for accurate vertical localization. If the signal contains sufficient high-frequency information, macaque monkeys may detect vertical displacements of only 3 to 4° (Fig. 5.14). This observation corresponds with the expectations based on the cues for perception of elevation discussed in Section 2.5. However, as shown in Figure 5.13, it is likely that the perception of sound azimuth is more accurate than is the perception of elevation for most signals.

Table 5.1 presents the acuity of vertical localization for representative mammals for the best signals tested. With a vertical acuity of 23°, the chinchilla (*Chinchilla laniger*) (Heffner et al. 1995) was the least acute mammal tested, while the bottlenose dolphin (*Tursiops truncatus*) (Renaud and Popper 1975) at 2° was the most precise vertical localizer. However, the literature is too sparse to permit much exploration of the role of pinna shape or size, visual field size,

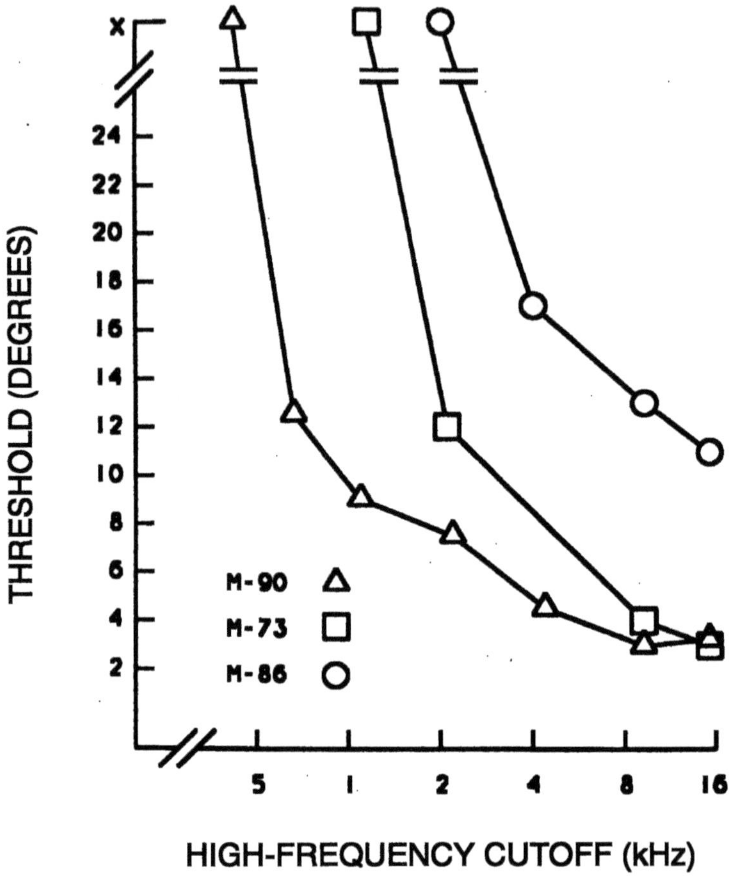

FIGURE 5.14. Vertical minimum audible angles for band-limited noise as a function of the high-frequency cutoff of the noise band for three macaque monkeys (M-90, M-73, and M-86). The low-frequency limit of the noise was 125 Hz. An X indicates that the performance level of the subject did not exceed chance. (From Brown et al. 1982. Reprinted with permission.)

or brainstem anatomical correlates with vertical acuity. Hopefully, investigators in the future will direct further attention to the problem of the perception of elevation.

5. Spectral Cues for Sound Localization

Behavioral assessments of the perception of sound source elevation by human listeners have contributed greatly to our current understanding of the role of spectral information in directional hearing. Although the basic principles of

TABLE 5.1. Vertical localization acuity in representative mammals.

Group	Species	Acuity	Source
Rodentia	Chinchilla	23°	Heffner et al. (1995)
Marsupialia	Opossum	13°	Ravizza and Masterton (1972)
Carnivora	Cat	4°	Martin and Webster (1987)
Primate	Rhesus/pig-tailed monkey	3°	Brown et al. (1982)
	Human	3°	Wettschurek (1973)
Cetactea	Dolphin	2°	Renaud and Popper (1975)

The data summarized in this table are rounded to the nearest integer, and are for the best signal tested. In some instances the test signal was a pure tone; in most cases, however, the best test signal was a band of noise, a click, or a species-specific vocalization.

these processes have been known for decades (Hebrank and Wright 1974; Butler and Belendiuk 1977; Watkins 1978), the maturation of digital signal processing techniques has resulted in significant recent advancements for psychoacoustic and physiological research in this area of the hearing sciences. Now, the salient directional features of human HRTFs are known in sufficient detail to allow the simulation of realistic auditory environments with earphones and other closed-field acoustic systems (Wightman and Kistler 1989b; Carlile and Pralong 1994; Pralong 1996; Kulkarni and Colburn 1998). In the future, these so-called virtual sound fields are likely to become a routine dimension of audio devices ranging from home entertainment centers to assistive aids for the hearing impaired.

In the laboratory, HRTF-based sounds provide an important functional context for exploring how spatial information is derived from the spectrotemporal properties of complex acoustic stimuli at processing levels ranging from the auditory nerve to cortex (Young et al. 1992; Imig et al. 1997; May and Huang 1997; Delgutte et al. 1999; Xu et al. 1999), just as ITD and ILD testing procedures have led to a better understanding of the binaural auditory system (Moore 1991). Much of our current knowledge regarding the auditory processing of spectral cues for sound localization has been gained from electrophysiological studies of the domestic cat.

Functional interpretations of the neural response patterns linked to directional hearing have been made possible by a long history of psychoacoustical studies in cats. The natural sound localization abilities of the cat have been described over a variety of stimulus conditions (Casseday and Neff 1973; Martin and Webster 1987; Heffner and Heffner 1988d; Populin and Yin 1998a), and the information processing roles of the major ascending auditory pathways have been confirmed by evaluating the behavioral deficits that follow surgical lesioning procedures (Moore et al. 1974; Casseday and Neff 1975; Neff and Casseday 1977; May 2000). This work suggests that the biological necessity for accurate sound localization has exerted a profound influence on the information processing pathways of the auditory system. Anatomical specializations for processing ILD and ITD cues are obvious in the striking binaural innervation patterns of the superior olive. Selectivity for the spectral features of complex sounds is created by the frequency-dependent convergence of inhibitory inputs within the auditory brainstem (Spirou et al. 1993, Imig et al. 2000). These neural networks are more difficult to distinguish anatomically but no less important in the auditory behaviors of the cat (Sutherland et al. 1998, May 2000).

5.1 The HRTF of the Cat

The filtering properties of the cat's head and pinna are known in detail and provide biologically relevant stimulus parameters for evaluating the neural and perceptual basis of directional hearing (Musicant et al. 1990; Rice et al. 1992). Representative HRTFs of the cat are shown in Figure 5.15 using measurements from the study of Rice et al. Each function is for a different source location and describes the gain of sound energy that propagates to the eardrum relative to the free-field amplitude spectrum of the stimulus. The data were recorded

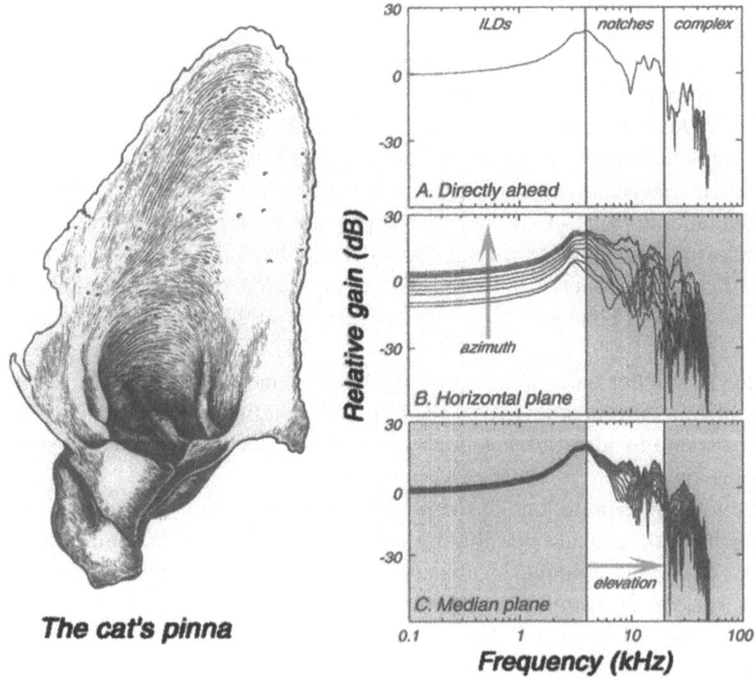

FIGURE 5.15. HRTFs of the cat. (**A**) Filtering effects of the pinna fall into three frequency domains. This example was measured with the sound directly in front of the subject (0° AZ, 0° EL). (**B**) Low frequencies convey ILDs as sounds move in the horizontal plane around the head. (**C**) Mid-frequencies exhibit a prominent notch that varies in frequency with changes in sound source elevation. High-frequency spectral cues are complex and show a less orderly relationship to the direction of a sound source. (From Rice et al. 1992. Reprinted with permission.)

by surgically implanting a probe microphone in the ear canal of an anesthetized cat. The transfer function in Figure 5.15A was measured with the sound source directly in front of the subject's head (0° AZ, 0° EL). As proposed by Rice and colleagues, three major directional properties of the HRTF are evident in this example. At low frequencies (<5 kHz), the function displays a broad amplification that rises to an energy peak around 4 to 6 kHz. At mid frequencies (5 to 20 kHz), the HRTF exhibits a single prominent energy minimum, or spectral notch. At high frequencies (>20 kHz), a complex pattern of peaks and notches is observed as the overall gain of the transfer function falls to low sound pressure levels.

Figure 5.15B summarizes how the HRTF changes with the azimuth of a sound source in the horizontal plane. The transfer functions that are superimposed in this example were obtained by placing the sound source at 11 locations in the frontal sound field (±75° in increments of 15°). In comparison to the HRTF in Figure 5.15A, the more lateralized HRTFs display a low-frequency gain that is either attenuated by the "sound shadow" of the head when the source is in the far field opposite the microphone, or amplified by the pinna when the source is in the near field. This directionally dependent change in gain is the acoustic basis of the ILD (Shaw 1974a,b; Kuhn 1987; Martin and Webster 1989). By contrast, Figure 5.15C shows how the low-frequency filtering properties of the HRTF are virtually unaffected as the sound source passes through 17 locations in the median plane (−30° to 90° in increments of 7.5°). These findings suggest that in the cat binaural processes related to the ILD provide a poor representation of elevation.

The spectral notch at the mid-frequencies of the HRTF changes in frequency as the sound source moves in the horizontal or median plane. This potential localization cue supplements ILD information at lateralized spatial locations in Figure 5.15B, but appears to have singular importance for signaling elevation changes in Figure 5.15C. Neurophysiological studies of the central auditory system have identified neural populations in the DCN and inferior colliculus (IC) that are selectively responsive to HRTF-based spectral notches (Young et al. 1992; Imig et al. 2000). The highly nonlinear spectral integration properties of these neurons may represent a pathway for encoding sound localization cues in the spectral domain (Nelken et al. 1997; Spirou et al. 1999), just as the binaural pathways of the MSO and LSO are specialized for processing interaural time and level differences.

The directional cues conveyed by the high-frequency filtering properties of the HRTF are complex and highly variable. Current analyses have not revealed global directional relationships in the spectral characteristics of high-frequency peaks and notches, but elevated frontal locations do show more high-pass filtering effects. An interesting perceptual phenomenon that may arise from this property of the HRTF is the observation that high-frequency tones or noise bands are often heard as elevated sources regardless of their actual location (Pratt 1930; Butler and Belendiuk 1977; Blauert 1997). These systematic errors in narrow-band localization can be explained by matching the proximal stimulus spectrum in the subject's ear to the directional properties of the HRTF (Middlebrooks 1992; van Schaik et al. 1999). For example, a narrow band of noise with a center frequency of 12 kHz will be attributed to locations where an individual's HRTF selectively passes those frequency components. Confusions of this nature are most apparent in vertical localization because the perception of horizontal location is enhanced by binaural directional information, like the ILD cues described in Figure 5.15B.

Three-dimensional virtual sound locations can be reproduced in a closed-field by adding false HRTF-filtering effects to the source spectrum (Wightman and Kistler 1989b; Pralong 1996; Langendijk and Bronkhorst 2000). Given this interesting perceptual effect, how can the auditory system derive HRTF effects without knowing the characteristics of the original source spectrum? This signal processing problem is avoided under normal listening conditions because the spectral shapes of natural sounds tend to have locally constant slopes that are capable of revealing the sharp peaks and notches of the HRTF (Zakarauskas and Cynader 1993). It is also true that the listener gets simultaneous "looks" at the spectrum from the different directional perspectives of the two ears. If the sound is sustained, the HRTF will change with movements of the source or the listener's head to reveal the underlying spectrum (Wightman and Kistler 1999). Animals with mobile pinna, like cats, can also translate the HRTF by moving the ears independently of the head (Young et al. 1996; Populin and Yin 1998b), a behavior that adds the dimension of proprioceptive feedback to spectral processing networks in the auditory brainstem (Kanold and Young 2001). Nevertheless, optimal localization of sound source elevation is observed for familiar sounds (McGregor et al. 1985; Blauert 1997) and the filtering effects of the listener's own ears (Wenzel et al. 1993; Hofman et al. 1998; Middlebrooks 1999).

5.2 Spectral Cues for the Discrimination of Changes in Sound Source Direction

The most common procedure for characterizing directional hearing in nonhuman animals is the MAA task in which the subject indicates the detection of a change in location by responding on a lever (May et al. 1995; Huang and May 1996b) or suppressing an ongoing behavior to avoid electrical shocks (Martin and Webster 1987; Heffner and Heffner 1988d). These methods have the advantage of relatively short training periods and produce psychometric data that allow easy quantification of directional acuity (Mills 1958).

Figure 5.16 shows average psychometric functions that were obtained by testing three cats with the MAA task (Huang and May 1996b). The cats were required to hold down on a response lever when bursts of noise were presented from a reference speaker (0° AZ, 0° EL), and to release the lever when the sound source shifted to another speaker. The comparison speakers were arranged in the median plane (Fig. 5.16A) or the horizontal plane (Fig. 5.16B). The percentage of correct lever releases is plotted in relation to the magnitude of the

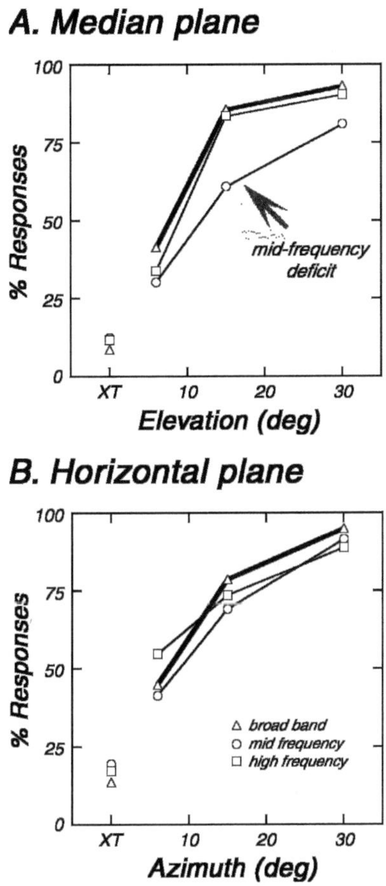

FIGURE 5.16. Effects of stimulus frequency on directional acuity in the median plane (**A**) and horizontal plane (**B**). Psychometric functions show the percentage of correct responses for directional changes relative to a reference speaker at 0° AZ, 0° EL. The percentage of incorrect responses for catch trials (XTs) are plotted as individual symbols to the left of the psychometric functions. The individual functions reflect the average responses of three cats to broadband (> 5 kHz), mid frequency (5 to 20 kHz), or high-pass noise (> 20 kHz). (Adapted from Huang and May 1996b.)

directional change between the reference and comparison speaker. Responses to catch trials (*XTs*) are indicated to the left of the psychometric functions. Catch trials were conducted just like MAA tests, but no speaker change was presented. The subject's responses to catch trials are presumed to reflect the probability of false-positive responses that result from guessing. The MAA is defined as the change in location (elevation or azimuth) that corresponds with the signal detection criterion of $d' = 1$ (based on the probabilities of correct responses to MAA trials and error responses to catch trials).

The psychometric functions in Figure 5.16 reveal the effects of frequency on directional acuity. Three frequency ranges were selected to evaluate the perceptual significance of the HRTF directional filtering effects that are summarized in Figure 5.15. Broadband noise contained spectral cues above 5 kHz. Mid-frequency noise was limited to the systematic spectral notch region from 5 to 20 kHz. High-frequency noise conveyed only the complex spectral cues above 20 kHz.

The behavioral results in Figure 5.16 indicate that best directional acuity was achieved with broadband noise, and this agrees with the results of Martin and Webster (1987). Since this stimulus condition included both mid-frequency and high-frequency spectral cues, the relative contribution of the individual spectral domains can be assessed by comparing response accuracy under broadband and the band-limited testing conditions. In general, no difference was observed in directional acuity between broadband and high-frequency noise. The subjects failed to detect changes in location more often when tests were conducted with mid-frequency noise. This deficit was most evident for sound sources in the median plane.

Computational models based on the auditory nerve encoding of spectral cues for sound localization offer an interpretation of the behavioral results in Figure 5.16 (May and Huang 1997). These models demonstrate a sensitive representation of directional change among neurons that respond best to the high-frequency components of HRTF-shaped noise. Even small changes in sound source location are capable of producing large and pervasive changes in the high-frequency HRTF, as shown in Figure 5.15. These spectral variations are reflected in the discharge rates of auditory neurons. The neural response is not inherently directional, but it is a sufficient cue for accurate performance of the MAA task where the subject is only required to respond to acoustic differences that are correlated with directional changes. Singular mid-frequency notches provide less effective information because they are more localized in frequency and smaller in magnitude than the multiple high-frequency notches. This mid-frequency deficit is less apparent for sound sources in the horizontal plane because spectral cues are augmented with binaural directional information.

5.3 Spectral Cues for the Perception of an Absolute Directional Identity

An alternative behavioral method for measuring sound localization accuracy involves training the subject to point toward the direction of a sound or approach the source. This procedure is necessary for studies in which the perceived location of the sound is an important parameter of the experiment. For example, the investigator may be interested in the systematic errors that are induced by modifying the source spectrum of the localization stimulus. After such manipulations, the subject might grossly misinterpret the actual location of the modified stimulus but still respond correctly to a change from one speaker location to another in the MAA task.

Sound-directed orientation behaviors of the head (Thompson and Masterton 1978) or eyes (Populin and Yin 1998a) have been used to characterize the perception of directional identity in cats. Unlike an approach procedure, which is constrained by source locations arrayed along the floor of the testing arena (Casseday and Neff 1973), head-orientation tasks can measure the simultaneous localization of stimulus azimuth and elevation. Orientation is a natural reflexive behavior that can be used as a short-term response metric for unexpected sounds in naïve subjects (Sutherland et al. 1998), or it can be shaped into a food-reinforced operant paradigm that is capable of sustaining long-term psychophysical analyses of the acoustic cues for directional hearing (May and Huang 1996).

Performance in a food-reinforced orientation task is summarized by the behavioral results shown in Figure 5.17 (Huang and May 1996a). This cat earned food rewards by accurately orienting its head toward randomly selected sound sources in an anechoic chamber. An electromagnetic sensor that was worn during the testing session tracked head movements. Trials in the orientation task were conducted with discrete presentations of brief noise bursts (40 ms), so the subject could not influence localization accuracy by moving its head or ears

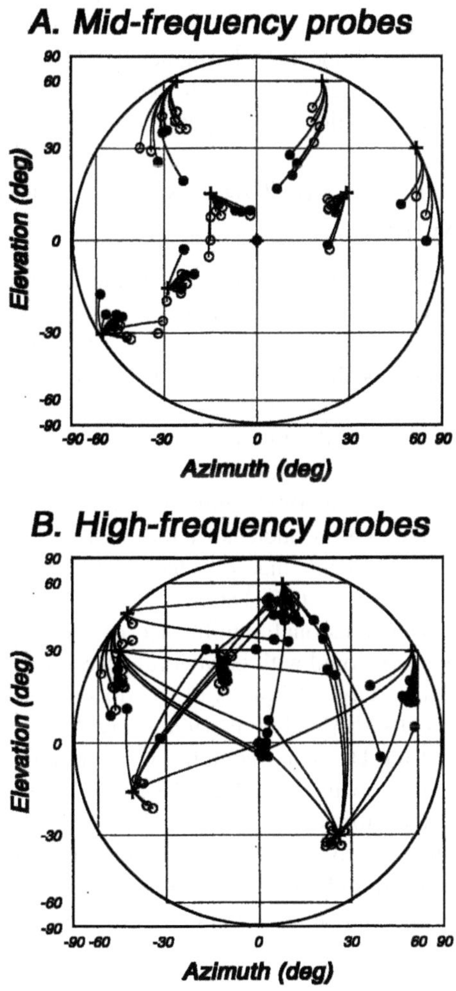

FIGURE 5.17. Effects of stimulus frequency on sound orientation behavior. Broadband stimuli were presented on 85% of the tests in these sessions to assess baseline accuracy (*open symbols*). The remaining probe trials evaluated the reliability of directional cues that were conveyed by restricted frequency regions of the head-related transfer function (*filled symbols*). (**A**) Orientation responses for 5 to 20 kHz band-pass noise versus broadband noise. (**B**) Orientation responses for 20 kHz high-pass noise versus broadband noise. Results for each condition are based on performance during one session. Additional plotting conventions are described in Figure 5.6. (Adapted from Huang and May 1966b.)

after stimulus onset. Most of the trials were conducted with bursts of broadband noise containing the full complement of HRTF-based localization cues. Orientation responses to these spectrally rich stimuli (open symbols in gray) were in good agreement with actual sound source locations (plus symbols).

The effect of frequency on the perception of sound source direction was evaluated by probing the cat's orientation behavior with bursts of mid-frequency noise (Fig. 5.17A) and high-frequency noise (Fig. 5.17B). Probe stimuli were intermingled at random intervals among the more frequent broadband noise bursts. Only slight changes in orientation accuracy were noted when the probes maintained mid-frequency cues. The reliability of mid-frequency spectral information in the perception of sound source elevation is attributed to auditory processing of directional notches in the cat's HRTF (Fig. 5.15C) (May and Huang 1997). High-frequency probes resulted in head movements that were inaccurate and highly variable, particularly with respect to the vertical coordinate of the source. These results suggest that cats do not utilize the complex filtering effects of the high-frequency HRTF for directional hearing even though the spectral cues in this high-frequency region should provide an excellent source of information for the perception of source direction (Fig. 5.16).

5.4 Spectral Processing Pathways in the Central Auditory System

Just as there are pathways in the central nervous system to enhance binaural directional hearing, behavioral and electrophysiological studies are beginning to reveal neural specializations for the auditory processing of spectral cues for sound localization. In the DCN, ascending inputs from the auditory nerve combine with a complex local inhibitory circuitry and descending projections from throughout the brain to create a notch-sensitive projection neuron that is also capable of integrating information about the orientation of the moveable pinna (Young et al. 1992; Imig et al. 2000). The target neurons for these projections in the central nucleus of the inferior colliculus (ICC) show spatially selective receptive fields that are sensitive to HRTF-filtering effects (Ramachandran et al. 1999; Davis 2002; Davis et al. 2003).

The functional significance of the putative spectral processing pathway has been explored with behavioral procedures by evaluating the auditory deficits that follow surgical lesions of the dorsal acoustic strial fibers that link the DCN to ICC (Sutherland et al. 1998; May 2000). As shown in Figure 5.18, these fibers exit the DCN and combine with the intermediate acoustic strial fibers from the ventral cochlear nucleus (VCN). The lesion was made by transecting the striae at the most dorsal limit of the nucleus. Previous studies have shown that the surgical procedure has little effect on hearing sensitivity because this more generalized auditory information ascends from the VCN to the binaural brainstem nuclei and inferior colliculus by way of the trapezoid body (Masterton et al. 1994).

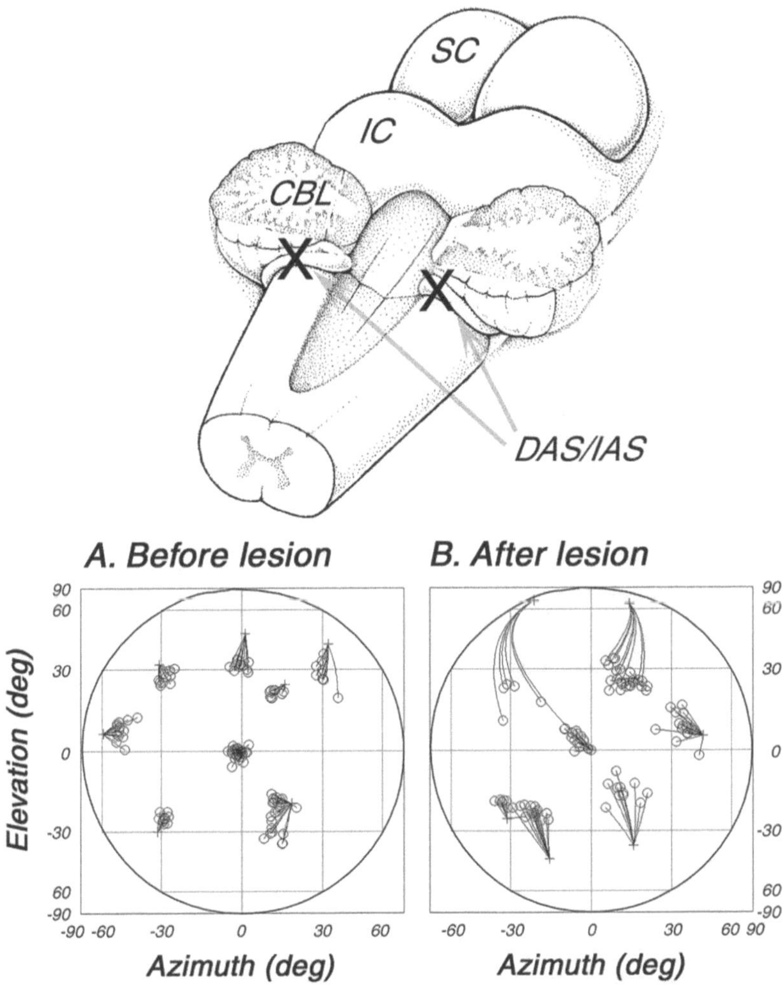

FIGURE 5.18. Effects of dorsal cochlear nucleus lesions on sound orientation behavior. The anatomical drawing illustrates the placement of surgical lesions (X). Orientation responses observed during one session before the lesion (**A**) and during another session with the same cat after the output pathways of the DCN were transected (**B**). Plotting conventions are described in Figure 5.6. CBL, cerebellum; DAS/IAS, dorsal and intermediate acoustic striae; IC, inferior colliculus; SC, superior colliculus. (Adapted from May 2000.)

The orientation plots in Figure 5.18 compare the head pointing behaviors of one cat before and after a bilateral lesion of DCN projections (May 2000). These tests were conducted with bandpass noise to restrict the domain of HRTF-based spectral information to the mid-frequency notches that exist at 5 to 20 kHz. As predicted by the results of the probe testing procedure in Figure 5.17A, the cat exhibited excellent orientation accuracy prior to the lesion (Fig. 5.18A). Large errors were noted after the lesion disrupted the spectral processing pathways of the DCN and ICC (Fig. 5.18B).

An analysis of the patterns of errors in the lesioned cat indicates that the localization deficits were statistically significant only in terms of response elevation. Regardless of the actual location of the sound source, the subject's orientation responses were seldom directed at elevations beyond ±30°. These systematic underestimations could exceed 60° for extreme source locations. It is likely that the subject maintained accurate azimuthal localization after the DCN lesion by relying on nonspectral directional information; for example, ILD and ITD cues that were processed in the intact binaural pathways of the auditory brainstem nuclei.

Cats with DCN lesions also have been studied with MAA procedures to confirm the specificity of the orientation deficits in Figure 18 (May 2000). These experiments are summarized by the psychometric functions in Figure 5.19.

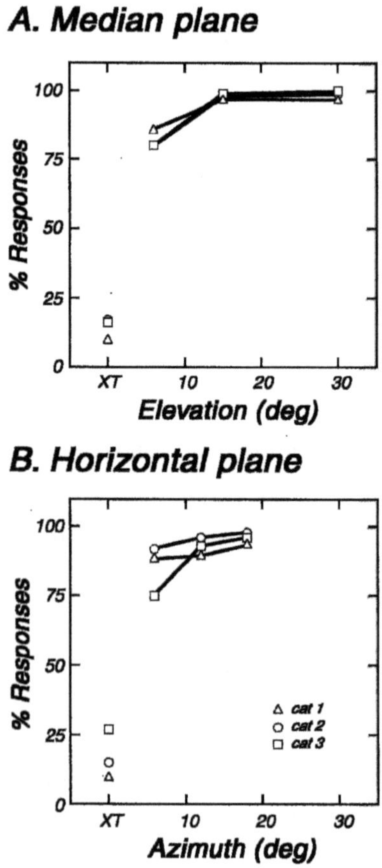

FIGURE 5.19. Effects of DCN lesions on spatial acuity. Psychometric functions summarize the detection scores of three cats for directional changes in the median plane (**A**) and horizontal plane (**B**). Responses to catch trials (XTs) are indicated by the symbols to the left of the functions. The physical dimensions of the speakers prevented testing are angular separations that were less than 6°. (Adapted from May 2000.)

Each function represents the post-lesion performance of one cat for directional changes in the median plane (A) and horizontal plane (B). The test stimuli were mid-frequency noise bursts. Although all of the subjects exhibited orientation deficits after the DCN lesion, none of the subjects showed signs of impaired spatial discrimination. These results confirm that DCN lesions do not lead to general hearing deficits and further support the idea that spatial acuity and directional identification are based on different auditory cues and processing pathways.

6. The Perception of Acoustic Proximity

The perception of acoustic distance, or acoustic proximity, has received very little formal study. Brown (1994) measured the minimal perceptible change in acoustic distance for human listeners in a forest habitat at a reference distance of 50 m. Using the speech utterance "hey" and a 1-kHz tone for the stimuli, it was determined that subjects would use changes in loudness, or sound amplitude, if the magnitude of the stimulus at its source was held constant as distance was varied. However, if signal amplitude was adjusted to compensate for changes in distance (and if random amplitude fluctuations were introduced), subjects were able to perceive changes in acoustic proximity only for the spectrally complex speech stimulus. This fact indicates that human listeners used changes in sound quality as described in Section 2.6 to detect changes in acoustic distance. Figure 5.20 shows that human listeners could perceive a 10% change in acoustic distance when the source level was fixed for both the tone and speech stimulus. This finding shows that loudness, or auditory image constancy, is an important cue for the perception of changes in acoustic proximity when it is available for processing (the amplitude of the signal is fixed). The detection of a 10% change in acoustic distance in a forested site compares closely with distance-discrimination thresholds of about 6% for reference distances of 6 to 49 m on an open athletic field (Strybel and Perrott 1984), and with distance-discrimination thresholds of about 6% for reference distances of 1 to 2 m in an anechoic room (Ashmead et al. 1990). The scattering of sound in the forested habitat will change the rate of sound attenuation with respect to distance relative to that in open environments (Waser and Brown 1986). Sound propagation is complicated because the elevation of the source and receiver, and the frequency of the signal have strong effects. Nevertheless, signals in the speech range, at the elevation of the human head, tend to be propagated better in forested than in open habitats (Waser and Brown 1986; Brown et al. 1995). That is, in forested compared to open habitats, a greater change in propagation distance will be required to produce a unit change in the level of the signal, and these acoustic influences likely account for the difference in the thresholds reported in open field and anechoic environments compared to that observed in forested environments.

Under most natural situations, sound amplitude is not the only available cue

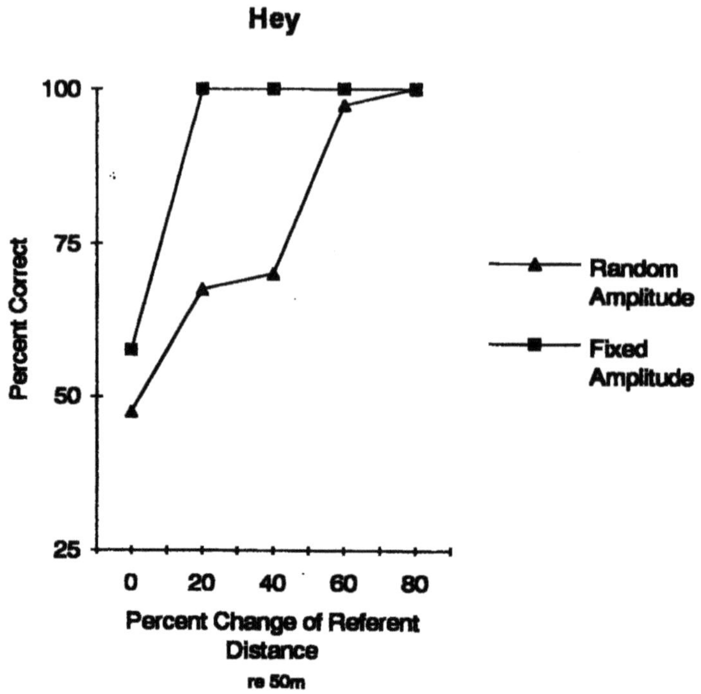
Hey
100
75
50
25
Percent Correct
0
20
40
60
80
Percent Change of Referent
Distance
re 50m
Random
Amplitude
Fixed
Amplitude

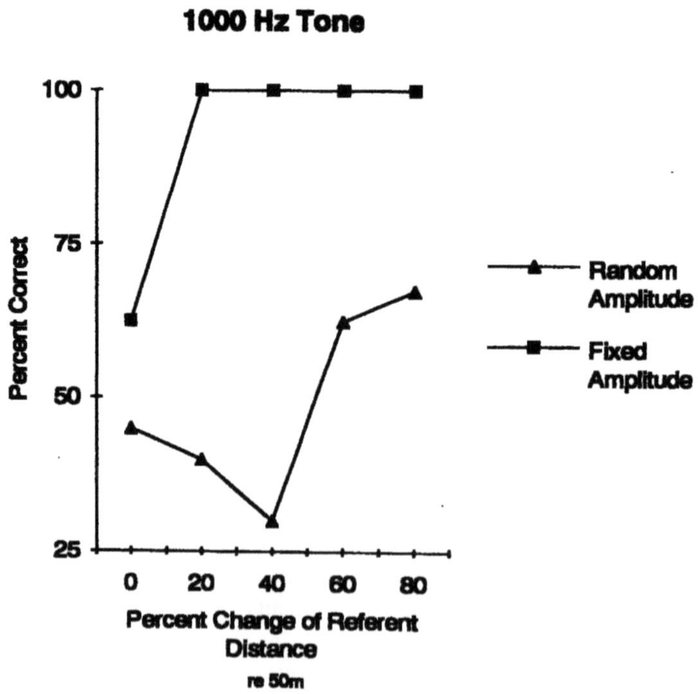
1000 Hz Tone
100
75
50
25
Percent Correct
0
20
40
60
80
Percent Change of Referent
Distance
re 50m
Random
Amplitude
Fixed
Amplitude

for the perception of a change in acoustic proximity. When the amplitude of the signal is adjusted to compensate for changes in transmission distance, and when the amplitude of the signal varies randomly trial-to-trial, loudness, or auditory image constancy, is no longer a viable cue. Nevertheless, human subjects are still able to perceive changes in acoustic proximity when tested with a complex speech stimulus. The data in Figure 5.20 show that subjects could perceive a change of 44% of the reference distance under these conditions. In an anechoic room, Ashmead et al. (1990) reported that human listeners could detect changes in distance of about 16% at reference distances of 1 to 2 m when the amplitude of the test and reference stimuli were equated. It is likely that spectral changes and reverberation were the most prominent cues underlying the perception of changes of distance.

The ability to perceive changes in sound quality associated with changes in acoustic distance has been measured in blue monkeys (*Cercopithecus mitis*). Figure 5.21 shows that blue monkeys can detect a change in proximity of 54% for the pyow vocalization broadcast in their natural habitat. This finding suggests that reflection of the wave front by tree trunks and other surfaces and frequency-specific attenuation may change or distort acoustic signals in a manner that provides a reliable and perceptually useful index of acoustic distance. It is conceivable that organisms residing in various habitats may have developed signals that are particularly well suited to permit listeners to ascertain the distance to the vocalizer. Furthermore, it is possible that some calls possess an acoustic structure that makes it possible to detect small changes in the proximity of the vocalizer, while other calls may tend to obscure the available distance cues.

7. Conclusion

Mammals have a sense of the azimuth, elevation, and distance of the source of acoustic events. However, the resolution of sound position is not equal in all three coordinates. The available data suggests that for most mammals the acuity of resolution of sound source azimuth is greater than that for elevation, and the acuity of resolution for sound source elevation is greater than that for distance. Hence, the minimal audible change in acoustic locus for azimuth, elevation, and distance may be described by the surface of an ellipsoid, a three-dimensional figure oriented such that the width is less than the height, which in turn, is less

FIGURE 5.20. The minimum perceptible change in proximity for human listeners. The test signals were the word "hey" (*top*) and a 1-kHz tone (*bottom*). The reference distance was 50 m. Testing was conducted in a forested habitat. The *triangles* indicate detection when the intensity of the signal is held constant; the *squares* indicate detection when the level of the signal is randomized and adjusted to compensate for changes in loudness with distance (From Brown 1994. Reprinted with permission.)

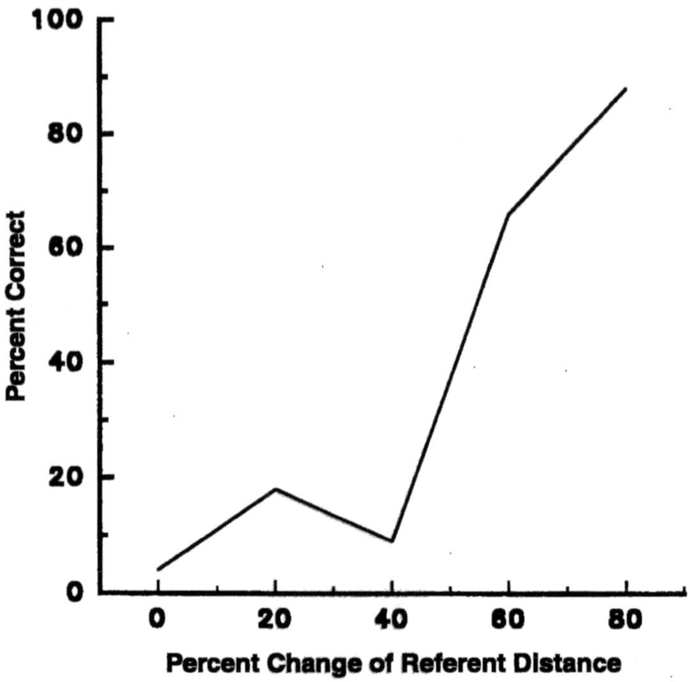

FIGURE 5.21. The minimum perceptible change in proximity in blue monkeys (*Cercopithecus mitis*). The test signal was the pyow vocalization. The reference distance was 50 m. The signal was broadcast and rerecorded at transmission distances of 5, 10, 20, 30, 40, and 50 m. Broadcasts were conducted at an elevation of 7 m in Kibale forest in western Uganda. The amplitude of the test signal was randomized between trials, and adjusted to compensate for changes in loudness with distance. (From Brown 1994. Reprinted with permission.)

than the length. A theoretical three-dimensional minimal perceptible change in locus ellipsoid is illustrated in Figure 5.22.

All three coordinates of sound source localization are important biologically. However, because the cues that underlie the perception of azimuth, elevation, and distance are so dissimilar, it is possible that subjects may experience abnormalities or disorders that impair perception in one dimension, yet leave relatively intact perception in the other two dimensions. Furthermore, it is possible that the ecology and life history of different species have led to enhanced sensitivity for localization in one coordinate relative to that in another. Terrestrial species may have been selected to maximized acuity for source azimuth, while marine organisms and arboreal species may have been selected for enhanced acuity for source elevation, and forest-living species may have been selected for greater acuity for source distance. Researchers have generated a wealth of studies of the comparative perception of sound source azimuth, and have only begun

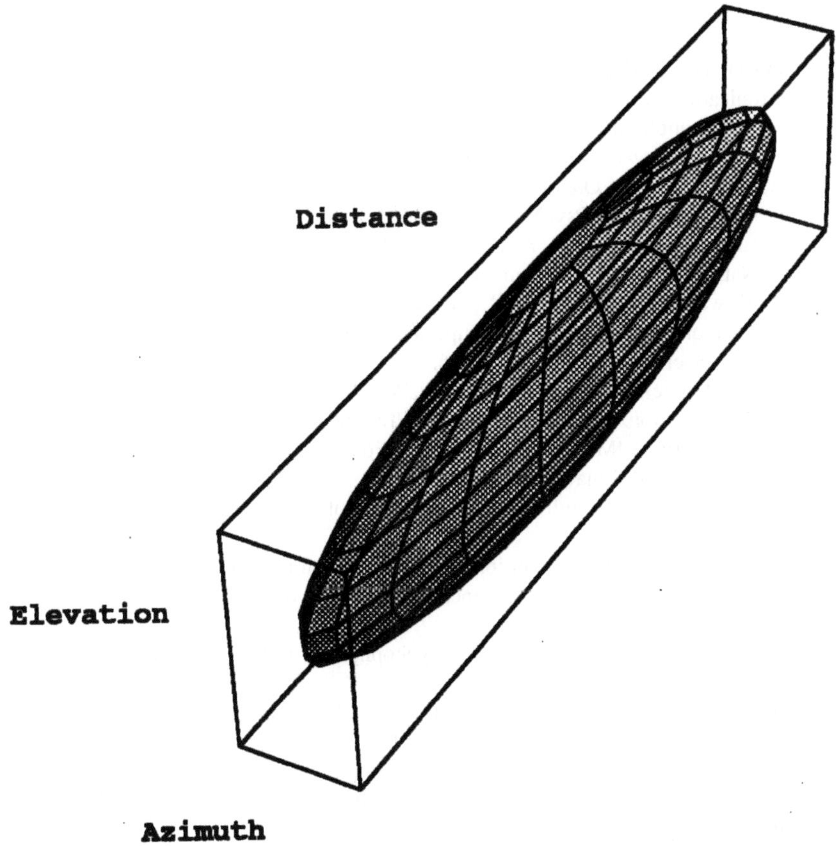

FIGURE 5.22. A theoretical volume describing the minimum perceptible change in locus of a broad-bandwidth sound. The reference locus is the center of the ellipsoid and the just perceptible change in locus in any direction is given by the distance from the center to any point on the surface of the volume. In the ellipsoid drawn here the resolution for changes in azimuth are two times that for changes in elevation, and eight times that for changes in distance. The actual dimensions of the volume describing the minimally perceptible change in space would be influenced by the acoustics of the habitat (test environment) and the temporal and spectral complexity of the test signal. (From Brown 1994. Reprinted with permission.)

to study the localization of source elevation or distance. The methodology for good comparative studies of auditory perception are well established, and we encourage researchers to focus greater attention on the elevation and distance coordinates of sound source position.

References

Anderson DJ (1973) Quantitative model for the effects of stimulus frequency upon the synchronization of auditory nerve discharges. J Acoust Soc Am 54:361–364.

Aronson E, Rosenbloom S (1971) Space perception in early infancy: perception within a common auditory-visual space. Science 172:1161–1163.

Ashmead DH, LeRoy D, Odom RD (1990) Perception of the relative distances of nearby sound sources. Percept Psychophys 47:326–331.

Ayrapet'yants ES, Konstantinov AI (1974) Echolocation in nature. An English translation of the National Technical Information Service, JPRS 63326-1-2.

Batra R, Kuwada S, Fitzpatrick DC (1997) Sensitivity of temporal disparities of low- and high-frequency neurons in the superior olivary complex. I. Heterogeneity of responses. J Neurophysiol 78:1222–1236.

Batteau DW, Plante RL, Spencer RH, Lyle WE (1965) Localization of sound: Part 5. Auditory perception. U.S. Navy Ordnance Test Station Report, TP 3109, part 5.

Beecher MD Harrison JM (1971) Rapid acquisition of an auditor location discrimination in rats. J Exp Anal Behav 16:193–199.

Beer CG (1969) Laughing gull chicks: recognition of their parent's voices Science 166: 1030–1032.

Beer CG (1970) Individual recognition of voice in social behavior of birds. In: Lehrman DS, Hinde RA, Shaw E (eds), Advances in the Study of Behavior, Vol. 3. New York: Academic Press, pp. 27–74.

Blauert J (1997) Spatial Hearing, The Psychophysics of Human Sound Localization. Cambridge, MA: MIT Press.

Brown CH (1982a) Ventroloquial and locatable vocalizations in birds. Z Tierpsychol 59: 338–350.

Brown CH (1982b) Auditor localization and primate vocal behavior. In: Snowdon CT, Brown CH, Petersen MR (eds), Primate Communication. Cambridge: Cambridge University Press, pp. 144–164.

Brown CH (1989) The active space of blue monkey and grey-cheeked mangabey vocalizations. Anim Behav 37:1023–1034.

Brown CH (1994) Sound localization. In: Fay RR, Popper AN (eds), Comparative Hearing: Mammals. New York: Springer-Verlag, pp. 57–96.

Brown CH, Gomez R (1992) Functional design features in primate vocal signals: the acoustic habitat and sound distortion. In: Nishida T, McGrew WC, Marler P, Pickford M, de Waal FMB (eds), Topics in Primatology, Vol. 1: Human Origins. Tokyo: University of Tokyo Press, pp. 177–198.

Brown CH, May BJ (1990) Sound Localization and binaural processes. In: Berkley MA, Stebbins WC (eds), Comparative Perception, Vol. I: Basic Mechanisms. New York: John Wiley & Sons, pp. 247–284.

Brown CH, Waser PM (1988) Environmental influences on the structure of primate vocalizations. In: Todt D, Geodeking P, Symme D (eds), Primate Vocal Communication Berlin: Springe-Verlag, pp. 51–66.

Brown CH, Beecher MD, Moody DB, Stebbins WC (1978a) Localization of pure tones in Old World monkeys. J Acoust Soc Am 63:1484–1492.

Brown CH, Beecher MD, Moody DB, Stebbins WC (1978b) Localization of primate calls by Old World monkeys. Science 201:753–754.

Brown CH, Beecher MD, Moody DB, Stebbins WC (1979) Locatability of vocal signals in Old World monkeys: design features for the communication of position. J Comp Physiol Psychol 93:806–819.

Brown CH, Beecher MD, Moody DB, Stebbins WC (1980) Localization of noise bands by Old World monkeys. J Acoust Soc Am 68:12–132.

Brown CH, Schessler T, Moody DB, Stebbins WC (1982) Vertical and horizontal sound localization in primates. J Acoust Soc Am 72:1804–1811.

Brown CH, Gomez R, Waser PM (1995) Old World monkey vocalizations: adaptation to the local habitat? Anim Behav 50:945–961.

Butler RA (1969) Monaural and binaural localization of noise bursts vertically in the medial sagittal plane. J Audit Res 9:230–235.

Butler RA, Belendiuk K (1977) Spectral cues utilized in the localization of sound in the median sagittal plane. J Acoust Soc Am 61:1264–1269.

Cain WS, Marks LE (1971) Stimulus and Sensation. Boston: Little, Brown.

Carlile S, Pralong D (1994) The location-dependent nature of perceptually salient features of the human head-related transfer functions. J Acoust Soc Am 95:3445–3459.

Casseday JH, Neff WD (1973) Localization of pure tones. J Acoust Soc Am 54:365–372.

Casseday JH, Neff WD (1975) Auditory localization: role of auditory pathways in brain stem of the cat. J Neurophysiol 38:842–858.

Chambers RE (1971) Sound localization in the hedgehog (*Paraechinus hypomelas*). Unpublished master's thesis, Florida State University, Tallahassee.

Clements M, Kelly JB (1978a) Directional responses by kittens to an auditory stimulus. Dev Psychobiol 11:505–511.

Clements M, Kelly JB (1978b) Auditor spatial responses of young guinea pigs (*Cavi porcellus*) during and after ear blocking. J Comp Physiol Psychol 92:34–44.

Coleman PD (1963) An analysis of cues to auditory depth perception in free space. Psychol Bull 60:302–315.

Davis KA (2002) Evidence of a functionally segregated pathway from dorsal cochlear nucleus to inferior colliculus. J Neurophysiol 87:1824–1835.

Davis K, Ramachandran R, May BJ (2003) Auditory processing of spectral cues for sound localization in the inferior colliculus. JARO 4:148–163.

Delgutte B, Joris PX, Litovsky RY, Yin TC (1999) Receptive fields and binaural interactions for virtual-space stimuli in the cat inferior colliculus. J Neurophysiol 81: 2833–2851.

Denzinger A, Schnitzler, HU (1998) Echo SPL, training experience, and experimental procedure influence the ranging performance in the big brown bat, *Eptescus fuscus*. J Comp Physiol 183:213–224.

Devor A (2000) Is the cerebellum like cerebellar-like structures? Brain Res Brain Res Rev 34:149–156.

Fay RR (1988) Hearing in Vertebrates: A Psychophysics Databook. Winnetka, IL: Hill-Fay Associates.

Feng AS, Gerhardt HC, Capranica RR (1976) Sound localization behaviour of the green tree frog (*Hyla cinerea*) and the barking tree frog (*H. gratiosa*). J Comp Physiol 107: 241–252.

Firestone FA (1930) The phase differences and amplitude ratio at the ears due to a source of pure tones. J Acoust Soc Am 2:260–270.

Gamble EA (1909) Intensity as a criterion in estimating the distance of sounds. Psychol Rev 16:416–426.

Gardner MB (1973) Some monaural and binaural facets of median plane localization. J Acoust Soc Am 54:1489–1495.
Gottlieb G (1965) Imprinting in relation to parental and specie identification by avian neonates. J Comp Physiol Psychol 59:345–356.
Hall JW, Haggard MP, Fernandes MA (1984) Detection in noise by spectro-temporal pattern analysis. J Acoust Soc Am 76:50–56.
Harris JD (1972) A florilegium of experiments on directional hearing. Acta Oto-Laryngol Suppl 298:1–26.
Harrison JM (1990) Simultaneous auditory discriminations. J Exp Anal Behav 54:45–51.
Harrison JM (1992) Avoiding conflicts between the natural behavior of the animal and the demands of discrimination experiments. J Acoust Soc Am 92:1331–1345.
Harrison JM, Downey P (1970) Intensity changes at the ear as a function of azimuth of a tone: a comparative study. J Acoust Soc Am 56:1509–1518.
Harrison JM, Irving R (1966) Visual and nonvisual auditory systems in mammals. Science 154:738–743.
Harrison JM, Downey P, Segal M, Howe M (1971) Control of responding by location of auditory stimuli: rapid acquisition in monkey and rat. J Exp Anal Behav 15:379–386.
Hartley RVL, Fry TC (1921) The binaural localization of pure tones. Physiol Rev 18: 431–442.
Hebrank J, Wright D (1974) Spectral cues used in the localization of sound sources in the median plane. J Acoust Soc Am 56:1829–1834.
Heffner HE, Heffner RS (1984) Sound localization in large mammals: localization of complex sounds by horses. Behav Neurosci 98:541–555.
Heffner HE, Heffner RS (1985) Sound localization in wild Norway rats (*Rattus norvegicus*). Hear Res 19:151–155.
Heffner RS, Heffner HE (1982) Hearing in the elephant (*Elephas maximus*): absolute sensitivity, frequency discrimination, and sound localization. J Comp Physiol Psychol 96:926–944.
Heffner RS, Heffner HE (1983) Hearing in large mammals: horses (*Equus caballus*) and cattle (*Bos taurus*). Behav Neurosci 97:299–309.
Heffner RS, Heffner HE (1985) Auditory localization and visual fields in mammals. Neurosci Abstr 11:547.
Heffner RS, Heffner HE (1987) Localization of noise, use of binaural cues, and a description of the superior olivar complex in the smallest carnivore, the least weasel (*Mustel nivalis*). Behav Neurosci 101:701–708.
Heffner RS, Heffner HE (1988a) Sound localizatrion in a predatory rodent, the northern grasshopper mouse (*Onychomys leucogaster*). J Comp Psychol 102:66–71.
Heffner RS, Heffner HE (1988b) Sound localization and the use of binaural cues in the gerbil (*Meriones unguculatus*). Behav Neurosci 102:422–428.
Heffner RS, Heffner HE (1988c) The relation between vision and sound localization acuity in mammals. Soc Neurosci Abstr 14:1096.
Heffner RS, Heffner HE (1988d) Sound localization acuity in the cat: effect of signal azimuth, signal duration, and test procedure. Hear Res 36:221–232.
Heffner RS, Heffner HE (1989) Sound localization, use of binaural cues, and the superior olivary complex in pigs. Brain Behav Evol 33:248–258.
Heffner RS, Heffner HE (1992) Visual factors in sound localization in mannuals. J Comp Neurol 317:219–232.

Heffner HE, Masterton RB (1980) Hearing in glires: domestic rabbit, cotton rat, feral house mouse, and kangaroo rat. J Acoust Soc Am 68:1584–1599.
Heffner RS, Masterton RB (1990) Sound localization: brainstem mechanisms. In: Berkley MA, Stebbins WC (eds), Comparative Perception, Vol. I: Basic Mechanisms. New York: John Wiley & Sons, pp. 285–314.
Heffner RS, Richard MM, Heffner HE (1987) Hearing and the auditory brainstem in a fossorial mammal, the pocket gopher. Neurosci Abstr 13:546.
Heffner RS, Heffner HE, Koay G (1995) Sound localization in chinchillas. II. Front/back and vertical localization. Hear Res 88:190–198.
Henning GB (1974) Detectability of interaural delay in high-frequency complex waveforms. J Acoust Soc Am 55:84–90.
Hofman PM, Van Riswick JG, Van Opstal AJ (1998) Relearning sound localization with new ears. Nat Neurosci 1:417–421.
Hornbostel EM (1923) Beobachtungen über ein- und zweiohrigs Hören. Psychol Forsch 4:64–114.
Houben D, Gourevitch G (1979) Auditory lateralization in monkeys: an examination of two cues serving directional hearing. J Acoust Soc Am 66:1057–1063.
Huang AY, May BJ (1996a) Sound orientation behavior in cats. II. Mid-frequency spectral cues for sound localization. J Acoust Soc Am 100:1070–1080.
Huang AY, May BJ (1996b) Spectral cues for sound localization in cats: effects of frequency domain on minimum audible angles in the median and horizontal planes. J Acoust Soc Am 100:2341–2348.
Imig TJ, Poirier P, Irons WA, Samson FK (1997) Monaural spectral contrast mechanism for neural sensitivity to sound direction in the medial geniculate body of the cat. J Neurophysiol 78:2754–2771.
Imig TJ, Bibikov NG, Poirier P, Samson FK (2000) Directionality derived from pinna-cue spectral notches in cat dorsal cochlear nucleus. J Neurophysiol 83:907–925.
Johnson DH (1980) The relationship between spike rate and synchrony in responses of auditory nerve fibers to single tones. J Acoust Soc Am 68:1115–1122.
Kanold PO, Young ED (2001) Proprioceptive information from the pinna provides somatosensory input to cat dorsal cochlear nucleus. J Neurosci 21:7848–7858.
Kaufman L (1979) Perception: The World Transformed. New York: Oxford University Press.
Kelly JB (1980) Effects of auditory cortical lesions on sound localization in the rat. J Neurophysiol 44:1161–1174.
Kiang NYS, Watanabe T, Thomas EC, Clark LF (1965) Discharge Patterns of Single Fibers in the Cat's Auditory Nerve. Cambridge, MA: MIT Press.
King AJ, Parsons CH (1999) Improved auditory spatial acuity in visually deprived ferrets. Eur J Neurosci 11:3945–3956.
King AJ, Hutchings ME, Moore DR, Blakemore C (1988) Developmental plasticity in the visual and auditory representations in the mammalian superior colliculus. Nature 332:73–76.
Klumpp RG, Eady HR (1956) Some measurements of interaural time difference thresholds. J Acoust Soc Am 28:859–860.
Knudsen EI (1983) Early auditory experience aligns the auditory map of space in the optic tectum of the barn owl. Science 222:939–942.
Knudsen EI, Brainard MS (1991) Visual instruction of the neural map of auditory space in the developing optic tectum. Science 253:85–87.

Knudsen EI, Knudsen PF (1985) Vision guides the adjustment of auditory localization in young barn owls. Science 230:545–548.

Knudsen EI, Knudsen PF (1990) Sensitive and critical periods for visual calibration of sound localization by barn owls. J Neurosci 10:222–232.

Knudsen EI, Konishi M (1978) A neural map of auditory space in the owl. Science 200: 795–797.

Knudsen EI, Blasdel GG, Konishi M (1979) Sound localization by the barn owl measured with the search coil technique. J Comp Physiol 133:1–11.

Knudsen EI, Knudsen, PF, Esterly SD (1984) A critical period for the recovery of sound localization accuracy following monaural occlusion in the barn owl. J Neurosci 4: 1012–1020.

Knudsen EI, Esterly SD, du Lac S (1991) Stretched and upside-down maps of auditory space in the optic tectum of blind-reared owls: acoustic basis and behavioral-coorelates. J Neurosci 11:1727–1747.

Kuhn GF (1977) Model for the interaural time difference in the azimuthal plane. J Acoust Soc Am 62:157–167.

Kuhn GF (1979) The effect of the human torso, head and pinna on the azimuthal directivity and on the median plane vertical directivity. J Acoust Soc Am 65:(S1), S8(A).

Kuhn GF (1987) Physical acoustics and measurements pertaining to directional hearing. In: Yost W, Gourevitch G (eds), Directional Hearing. New York: Academic Press, pp. 3–25.

Kujala T, Alho K, Paavilainen P, Summala H, Näätänen R (1992) Neural plasticity in processing of sound location by the early blind: an event-related potential study. Electroencephalogr Clin Neurophysiol 84:469–472.

Kujala T, Alho K, Kekoni J, Hamalainen H, Reinikainen K, Salonen O, Standertskjold-Nordenstam CG, Näätänen R (1995) Auditory and somatosensory event-related brain potentials in early blind humans. Exp Brain Res 104:519–526.

Kujala T, Alho K, Näätänen R (2000) Cross-modal reorganization of human cortical functions. Trends Neurosci 23:115–120.

Kulkarni A, Colburn HS (1998) Role of spectral detail in sound-source localization. Nature 396:747–749.

Langendijk EH, Bronkhorst AW (2000) Fidelity of three-dimensional-sound reproduction using a virtual auditory display. J Acoust Soc Am 107:528–537.

Lessard N, Paré M, Lepore F, Lassonde M (1998) Early-blind humans subjects localize sound sources better than sighted subjects. Nature 395:278–280.

Makous JC, Middlebrooks JC (1990) Two-dimensional sound localization by human listeners. J Acoust Soc Am 87:2188–2200.

Martin RL, Webster WR (1987) The auditory spatial acuity of the domestic cat in the interaural horizontal and median vertical planes. Hear Res 30:239–252.

Martin RL, Webster WR (1989) Interaural sound pressure level differences associated with sound-source locations in the frontal hemifield of the domestic cat. Hear Res 38: 289–302.

Masters WM, Raver KA (2000) Range discrimination by big brown bats (*Eptesicus fuscus*) using altered model echoes: implications for signal processing. J Acoust Soc Am 107:625–637.

Masterton B, Heffner H, Ravizza R (1969) The evolution of human hearing. J Acoust Soc Am 45:966–985.

Masterton RB, Granger EM, Glendenning KK (1994) Role of acoustic striae in hearing: mechanism for enhancement of sound detection in cats. Hear Res 73:209–222.

May BJ (2000) Role of the dorsal cochlear nucleus in the sound localization behavior of cats. Hear Res 148:74–87.
May BJ, Huang AY (1996) Sound orientation behavior in cats. I. Localization of broadband noise. J Acoust Soc Am 100:1059–1069.
May BJ, Huang AY (1997) Spectral cues for sound localization in cats: a model for discharge rate representations in the auditory nerve. J Acoust Soc Am 101:2705–2719.
May B, Moody DB, Stebbins WC, Norat MA (1986) Sound localization of frequency-modulated sinusoids by Old World monkeys. J Acoust Soc Am 80:776–782.
May BJ, Huang AY, Aleszczyk CM, Hienz RD (1995) Design and conduct of sensory experiments for domestic cats. In: Klump G, Dooling RJ, Fay RR, Stebbins WC (eds), Methods in Comparative Psychoacoustics. Basel: Birkhäuser, pp. 95–108.
McFadden D, Pasanen EG (1976) Lateralization at high frequencies based on interaural time differences. J Acoust Soc Am 59:634–639.
McGregor PK, Krebs JR (1984) Sound degradation as a distance cue in great tit (*Parus major*) song. Behav Ecol Sociobiol 16:49–56.
McGregor PK, Krebs JR, Ratcliffe LM (1983) The reaction of great tits (*Parus major*) to playback of degraded and undegraded songs: the effect of familiarity with the stimulus song type. Auk 100:898–906.
McGregor P, Horn AG, Todd MA (1985) Are familiar sounds ranged more accurately? Percept Mot Skills 61:1082.
Mendelson MJ, Haith MM (1976) The relation between audition and vision in the human newborn. Monographs of the Society for Research in Child Development, 41, Serial No. 167.
Menzel CR (1980) Head cocking and visual perception in primates. Anim Behav 28: 151–159.
Mershon DH, Bowers JN (1979) Absolute and relative cues for the auditory perception of egocentric distance. Perception 8:311–322.
Middlebrooksb JC (1992) Narrow-band sound localization related to external ear acoustics. J Acoust Soc Am 92:2607–2624.
Middlebrooks JC (1999) Individual differences in external-ear transfer functions reduced by scaling in frequency. J Acoust Soc Am 106:1480–1492.
Middlebrooks JC, Green DM (1990) Directional dependence of interaural envelope delays. J Acoust Soc Am 87:2149–2162.
Middlebrooks JC, Makous JC, Green DM (1989) Directional sensitivity of sound-pressure levels in the human ear canal. J Acoust Soc Am 86:89–108.
Mills AW (1958) On the minimum audible angle. J Acoust Soc Am 30:237–246.
Mills AW (1960) Lateralization of high frequency tones. J Acoust Soc Am 32:132–134.
Molitor SC, Manus PB (1997) Evidence for functional metabotrophic glutamate receptors in the dorsal cochlear nucleus. J Neurophysiol 77:1889–1905.
Moore BCJ (1999) Modulation minimizes masking. Nature 397:108–109.
Moore CN, Casseday JH, Neff WD (1974) Sound localization: the role of the commissural pathways of the auditory system of the cat. Brain Res 82:13–26.
Moore DR (1991) Anatomy and physiology of binaural hearing. Audiology 30:125–134.
Muir D, Field J (1979) Newborn infants orient to sounds. Child Dev 50:431–436.
Musicant AD, Chan JCK, Hind JE (1990) Direction-dependent spectral properties of cat external ear: new data and cross-species comparisons. J Acoust Soc Am 87:757–781.
Neff WD, Casseday JH (1977) Effects of unilateral ablation of auditory cortex on monaural cat's ability to localize sound. J Neurophysiol 40:44–52.
Nelken I, Young ED (1994) Two separate inhibitory mechanisms shape the responses of

dorsal cochlear nucleus type IV units to narrowband and wideband stimuli. J Neurophysiol 71:2446–2462.

Nelken I, Kim PJ, Young ED (1997) Linear and nonlinear spectral integration in type IV neurons of the dorsal cochlear nucleus. II. Predicting responses with the use of nonlinear models. J Neurophysiol 78:800–811.

Nelken I, Rotman Y, Bar Yosef O (1999) Responses of auditory-cortex neurons to structural features of natural sounds. Nature 397:154–157.

Norberg RA (1977) Occurrence and independent evolution of bilateral ear asymmetry in owls and implications on owl taxonomy. Philos Trans R Soc Lond B 282:375–408.

Nyborg W, Mintzer D (1955) Review of sound propagation in the lower atmosphere. U.S. Air Force WADA Tech. Rept 54–602.

Payne RS (1962) How the barn owl locates its prey by hearing. Living Bird 1:151–159.

Perrott DR, Ambarsoom H, Tucker J (1987) Changes in head position as a measure of auditory localization performance: auditory psychomotor coordination under monaural and binaural listening conditions. J Acoust Soc Am 85:2669–2672.

Populin LC, Yin TC (1998a) Behavioral studies of sound localization in the cat. J Neurosci 18:2147–2160.

Populin LC, Yin TC (1998b) Pinna movements of the cat during sound localization. J Neurosci 18:4233–4243.

Potash M, Kelly J (1980) Development of directional responses to sounds in the infant rat (*Rattus norvegicus*). J Comp Physiol Psychol 94:864–877.

Pralong D (1996) The role of individualized headphone calibration for the generation of high fidelity virtual auditory space. J Acoust Soc Am 100:3785–3793.

Pratt CC (1930) The spatial character of high and low tones. J Exp Psychol 13:278–285.

Ramachandran R, May BJ (2002) Functional segregation of ITD sensitivity in the inferior colliculus of decerebrate cats. J Neurophysiol 88:2251–2261.

Ramachandran R, Davis KA, May BJ (1999) Single-unit responses in the inferior colliculus of decerebrate cats. I. Classification based on frequency response maps. J Neurophysiol 82:152–163.

Rauschecker JP, Harris LR (1983) Auditory compensation of the effects of visual deprivation in the cat's superior colliculus. Exp Brian Res 50:69–83.

Rauschecker JP, Kniepert U (1994) Auditory localization behaviour in visually deprived cats. Eur J Neurosci 6:149–160.

Ravizza RJ, Masterton RB (1972) Contribution of neocortex to sound localization in opossum (*Didelphis virginiana*). J Neurophysiol 35:344–356.

Rayleigh JWS (1876) Our perception of the direction of a sound source. Nature (Lond) 14:32–33.

Rayleigh JWS (1945) The Theory of Sound, 2nd ed. New York: Dover.

Renaud DL, Popper AN (1975) Sound localization by the bottlenose porpoise *Tursiops truncatus*. J Exp Biol 63:569–585.

Rheinlaender J, Gerhardt HC, Yager DD, Capranica RR (1979) Accuracy of phonotaxis by the green treefrog (*Hyla cinerea*). J Comp Physiol 133:247–255.

Rice JJ, May BJ, Spirou GA, Young ED (1992) Pinna-based spectral cues for sound localization in cat. Hear Res 58:132–152.

Roder B, Teder-Salejarvi W, Sterr A, Rosler F., Hillyard SA, and Neville HJ (1999) Improved auditory spatial tuning in blind humans. Nature 400:162–166.

Rose JE, Brugge JF, Anderson DJ, Hind JE (1967) Phase-locked responses to low-frequency tones in single auditory nerve fibers of the squirrel monkey. J Neurophysiol 30:769–793.

Searle CL, Braida LD, Cuddy DR, Davis MF (1975) Binaural pinna disparity: another auditory localization cue. J Acoust Soc Am 57:448–455.

Shaw EA (1974a) Transformation of sound pressure level from the free field to the eardrum in the horizontal plane. J Acoust Soc Am 56:1848–1861.

Shaw EAG (1974b) The external ear. In: Keidel WD, Neff WD (eds), Handbook of Sensory Physiology, Vol. V/I. Berlin: Springer-Verlag, pp. 455–490.

Simpson WE, Stanton LD (1973) Head movement does not facilitate perception of the distance of a source of sound. Am J Psychol 86:151–159.

Spigelman MN, Bryden MP (1967) Effects of early and late blindness on auditory spatial learning in the rat. Neuropsychologia 5:267–274.

Spirou GA, Young ED (1991) Organization of dorsal cochlear nucleau type IV unit response maps and their relationship to activation by bandlimited noise. J Neurophysiol 66:1750–1768.

Spirou GA, May BJ, Wright DD, Ryugo DK (1993) Frequency organization of the dorsal cochlear nucleus in cats. J Comp Neurol 329:36–52.

Spirou GA, Davis KA, Nelken I, Young ED (1999) Spectral integration by type II interneurons in dorsal cochlear nucleus. J Neurophysiol 82:648–663.

Strybel TZ, Perrott DR (1984) Discrimination of relative distance in the auditory modality: the success and failure of the loudness discrimination hypothesis. J Acoust Soc Am 76:318–320.

Sutherland DP, Masterton RB, Glendenning KK (1998) Role of acoustic striae in hearing: reflexive responses to elevated sound-sources. Behav Brain Res 97:1–12.

Terhune JM (1974) Directional hearing of the harbor seal in air and water. J Acoust Soc Am 56:1862–1865.

Thompson GC, Masterton RB (1978) Brain stem auditory pathways involved in reflexive head orientation to sound. J Neurophysiol 41:1183–1202.

van Schaik A, Jin C, Carlile S (1999) Human localisation of band-pass filtered noise. Int J Neural Syst 9:441–446.

von Békésy GV (1938) Über die Entstehung der Entfernungsempfin dung beim Hören. Akust Z 3:21–31. [Available in English in Weaver EG (ed), Experiments in Hearing. New York: John Wiley, 1960, pp. 301–313.]

Waser PM (1977) Sound localization by monkeys: a field experiment. Behav Ecol Sociobiol 2:427–431.

Waser PM, Brown CH (1986) Habitat acoustics and primate communication. Am J Primatol 10:135–154.

Watkins AJ (1978) Psychoacoustical aspects of synthesized vertical locale cues. J Acoust Soc Am 63:1152–1165.

Wenzel EM, Arruda M, Kistler DJ, Wightman FL (1993) Localization using nonindividualized head-related transfer functions. J Acoust Soc Am 94:111–23.

Wetheimer M (1961) Psychomotor coordination of auditory and visual space at birth. Science 134:1692.

Wettschurek RG (1973) Die absoluten Unterschiedswellen der Richtungswahrnehmung in der Medianebene beim naturlichen Hören, sowie beim Hören über ein Kunstkopf-Übertragungssystem. Acoustica 28:197–208.

Whitehead JM (1987) Vocally mediated reciprocity between neighboring groups of mantled howling monkeys, Aloutta palliata palliata. Anim Behav 35:1615–1627.

Whittington DA, Hepp-Reymond MC, Flood W (1981) Eye and head movements to auditory targets. Exp Brain Res 41:358–363.

Wightman FL, Kistler DJ (1989a) Headphone simulation of free-field listening. I: Stimilus synthesis. J Acoust Soc Am 85:858–867.

Wightman FL, Kistler DJ (1989b) Headphone simulation of free-field listening. II: Psychophysical validation. J Acoust Soc Am 85:868–878.

Wightman FL, Kistler DJ (1992) The dominant role of low-frequency interaural time differences in sound localization. J Acoust Soc Am 91:1648–1661.

Wightman FL, Kistler DJ (1999) Resolution of front-back ambiguity in spatial hearing by listener and source movement. J Acoust Soc Am 105:2841–2853.

Wiley RH, Richards DG (1978) Physical constraints on acoustic communication in the atmosphere: implications for the evolution of animal vocalization. Behav Ecol Sociobiol 3:6–94.

Xu L, Furukawa S, Middlebrooks JC (1999) Auditory cortical responses in the cat to sounds that produce spatial illusions. Nature 399:688–691.

Yin TC, Kuwada S (1983) Binaural interaction in low-frequency neurons in inferior colliculus of the cat. II. Effects of changing rate and direction of interaural phase. J Neurophysiol 50:1000–10019.

Young ED, Spirou GA, Rice JJ, Voigt HF (1992) Neural organization and responses to complex stimuli in the dorsal cochlear nucleus. Philos Trans R Soc Lond B Biol Sci 336:407–413.

Young ED, Rice JJ, Tong SC (1996) Effects of pinna position on head-related transfer functions in the cat. J Acoust Soc Am 99:3064–3076.

Zakarauskas P, Cynader MS (1993) A computational theory of spectral cue localization. J Acoust Soc Am 94:1323–1331.

6

Development of the Auditory Centers Responsible for Sound Localization

M. Fabiana Kubke and Catherine E. Carr

1. Introduction

Knudsen and Konishi (1978) found space-specific auditory neurons in the midbrain of barn owls, providing the first animal model for the neuronal computation of sensory maps. Over the following years, comparable findings were obtained in mammals. These two vertebrate taxa have provided us with an increasing wealth of information as to how sensory maps are computed in the nervous system, and how sound localization circuits develop.

The organization of the auditory system in vertebrates is as varied as the vertebrates themselves, making it difficult to typify an auditory system that would apply to all vertebrate classes. Despite this variation, some commonalities exist, the majority of which apply to most classes (Webster et al. 1992a). The formation of the auditory system results from two fundamental processes. A carefully choreographed sequence of events is initiated during embryonic development and followed by subsequent modifications and refinement of neuronal connections via experience-mediated plasticity. Early developmental processes are responsible for the establishment of the basic neuronal pathways that constitute the foundation of the adult auditory system characteristic of each species. The early developmental template is then reorganized by experience-dependent mechanisms. Thus, both embryonic development and experience-dependent modification need to be considered if auditory function in the adult is to be understood.

This chapter concentrates on the description of the structure and development of sound localization pathways, primarily in birds and mammals, trying to highlight the similarities and differences observed between these two groups. The developmental program that leads to the establishment of the basic sound localization circuit and the experience-dependent modifications to this basic plan that give rise to the adult processing of sound location are summarized. Although these processes have been separated in this chapter for didactic purposes, they are deeply interdependent and overlap both spatially and temporally, making it difficult at times to classify any one event within a discrete category.

Thus, the inclusion of a process into one or another category, which may appear arbitrary at times, has been done trying to preserve the linearity within experimental approaches.

2. Development of the Peripheral Auditory System

Auditory stimuli reach the peripheral sensory papilla, which consists of an array of sensory hair cells. Three features of the sound stimuli are extracted to compute the spatial auditory map that is used for sound localization: frequency spectrum, temporal parameters associated with phase of the sound stimulus, and sound intensity information. Frequency is mapped along the papilla such that a hair cell will respond maximally to a given frequency based on its topographic position within the cochlea. In addition, hair cells show specializations for frequency-specific responses such as spatial gradients in the length of their stereocilia, and variation in the dynamics of ion channels.

The formation of the cellular elements of the ear and the first-order auditory (and vestibular) neurons appears to be governed by cell lineage relationships and several transcription factors and signaling molecules have been implicated in the different phases of this morphogenetic program (Baker and Bronner-Fraser 2001; Fekete and Wu 2002). Ganglion cells arise from the same ectodermal invagination that forms the inner ear, and the neural crest contributes to the formation of the glial and Schwann sheath cells in the cochleovestibular ganglion (D'Amico-Martel and Noden 1983; Fritzsch et al. 2002). Prior to and during invagination of the otic cup, ganglionic neurons begin their delamination process and migrate a short distance into the underlying mesenchyme, where they differentiate to form the cochleovestibular ganglion (Rubel and Fritzsch 2002). Proliferation of ganglion neurons in chickens proceeds in an apical to basal gradient, whereas in the mouse it occurs in a basal to apical gradient (Rubel 1978).

The formation of connections between hair cells and ganglion cells is the basis of the cochleotopic (tonotopic) map that is maintained throughout the auditory pathway. The distal processes of ganglion cells provide the afferent innervation to the ear by projecting back to the original site of delamination (Rubel and Fritzsch 2002). The distal processes of ganglion neurons invade the papilla starting at the basal end in both birds and mammals, and some of the early fiber outgrowth may happen even before hair cells become postmitotic (Rubel and Fritzsch 2002). The initial contact is followed by the appearance of membrane thickenings and synaptic vesicles that will form the synaptic apparatus. The precision of the adult innervation results from the synaptic pruning of inappropriate connections that occurs over the first week postnatal and is accompanied of significant loss of ganglion cells (Pujol et al. 1998). Neurotrophins play a major role in the establishment and maintenance of inner ear innervation and synapses are morphologically differentiated prior to the onset of function (Rubel and Fritzsch 2002).

3. Anatomical Development of the Central Auditory System

The neurons of the cochlear ganglion transmit acoustic information from the inner ear to the brain, maintaining the cochleotopic organization that translates into a precise tonotopic map. Primary afferent innervation and synaptic activity are essential for normal development of neurons in the cochlear nucleus, affecting cell survival and morphological maturation (Rubel and Fritzsch 2002).

3.1 Organization of the Central Auditory System

The brainstem auditory pathways of birds and mammals are schematized in Figure 6.1. In all vertebrates, the auditory nerve projects to the hindbrain cochlear nuclei. There are three subdivisions of the cochlear nuclei in mammals, namely the anterior ventral cochlear nucleus (AVCN) and posterior ventral cochlear nucleus (PVCN) and a separate dorsal cochlear nucleus (DCN). The principal cell type in AVCN is the bushy cell, which receives innervation from type I spiral ganglion cells in the form of endbulbs of Held, and whose morphology varies depending on its position within the nucleus. The PVCN also contains several cell types, including multipolar and octopus cells (Cant and Benson 2003). The DCN, a cerebellar-like structure, is found in marsupials and eutherian mammals but is absent in monotremes (Bell 2002; Cant and Benson 2003). Primary afferents terminate in the deeper cell layers of DCN, preserving tonotopic organization. In the cochlear nuclei of mammals, frequencies are mapped such that cells with low best frequency (BF) are found in the ventrolateral region, and cells with high BF are found in the dorsomedial region of the nuclei. Axons from the AVCN provide the primary innervation to the superior olivary complex (SOC). Interaural time differences (ITDs) arising from the spatial separation of the ears are processed by neurons in the medial superior olive (MSO). These neurons are bilaterally innervated by axons from spherical bushy cells and project, in turn, to the ispilateral nuclei of the lateral lemniscus (NLL). Bushy cells also make calyceal synapses on contralateral medial nucleus of the trapezoid body (MNTB) neurons, which, in turn, make boutonlike terminations on the principal neurons of the ispilateral lateral superior olive (LSO), where interaural intensity information is processed. Both olivary and cochlear nuclei project to different nuclei of the lateral lemniscus and to the central nucleus of the inferior colliculus (IC) (Fig. 6.1A).

Birds have two separate cochlear nuclei (Fig. 6.1B), nucleus magnocellularis (NM) and nucleus angularis (NA), similar to the mammalian AVCN and PVCN, respectively, but appear to lack a homolog of DCN (Ariëns Kappers et al. 1936). In birds, NM is the most caudal of the nuclei associated with the cochlear system. More rostrally, NM contains large rounded cells with short dendrites similar to the bushy cells of AVCN, which, like their mammalian counterparts, receive innervation in the form of endbulbs of Held. This merges into a caudolateral region characterized by stellate cells that instead receive innervation in the form of terminal boutons (Jhaveri and Morest 1982; Köppl, 1994). NM

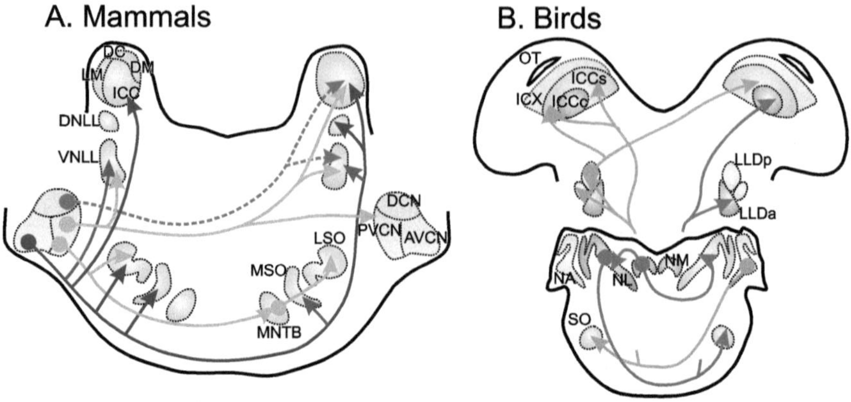

FIGURE 6.1. Schematic showing the connections in the mammalian and avian auditory brainstem. (**A**) Mammals: The auditory nerve bifurcates to give rise to an ascending and a descending branch. The ascending branch innervates the AVCN, and the descending branch innervates first the PVCN and then the DCN. The projections of the cochlear nuclei are denoted as different lines (AVCN, *dark lines*; PVCN, *light lines*; and DCN, *dotted lines*). The cochlear nuclei send ascending projections to the olivary and periolivary nuclei, which include the MNTB, MSO, and LSO. The IC in mammals can be subdivided into a central nucleus, an external cortex, and a dorsal cortex. The central nucleus can be divided into a pars centralis, pars medialis, and pars lateralis, each of which can be distinguished by the packing density of their neurons. Stellate cells from VCN and fusiform and giant cells from DCN project to the contralateral central nucleus of the IC, giving rise to banded inputs. The central nucleus receives bilateral input from LSO and a mostly ispilateral input from MSO, also forming banded, tonotopically organized projections. It also receives bilateral projections from DNLL in a banded organization, and a more diffuse projection from VNLL. In the ferret superior colliculus deep layers an auditory space map is found that depends on signals provided by the superficial visual layers. Auditory inputs originate in the nucleus of the brachium of the IC (nBIC). Projections from superior colliculus to nBIC may be involved in visual–auditory calibration during the formation of the space map. (Adapted from Romand and Avan 1997 and Cant and Benson 2003. (**B**) Birds: In barn owls, separation into time and sound level pathways (dark lines and light lines, respectively) begins with the cochlear nuclei. VIIIth nerve afferents divide to innervate both the level-coding NA and the time-coding nucleus magnocellularis NM. NM projects bilaterally to NL, which in turn projects to SO, LLDa, and to the core region of ICCc. The SO projects back to NA, NM and NL (projections are not drawn). In birds, the IC is subdivided into two principal subnuclei, the lateral ICX and the more medial ICC. The ICC can be further subdivided into the ICCc, corresponding to the NL recipient zone, and the ICCls and ICCms. The ICC is tonotopically organized with cells tuned to low-BF more superficial and high-BF cells deeper. In the IC, time and sound level pathways combine in the central nucleus, then project to the external nucleus, which contains a map of auditory space. (From Kubke et al. 1999. Journal of Comparative Neurology © 1999.)

projects bilaterally to the nucleus laminaris (NL), where the ITDs are first computed and transformed into a place map. The tonotopic representation in NM and NL is such that cells with higher BF are found in more rostromedial regions of the nucleus, and cells with progressively lower BF are found progressively towards more caudolateral regions of the nucleus (Kubke and Carr 2000). Projections of NA and NL may overlap in the superior olive (SO), which provides descending inhibitory input to NA, NM, and NL (Carr et al. 1989; Yang et al. 1999). Both NL and NA in addition project to the auditory midbrain, sending collaterals to the lemniscal nuclei.

Auditory information reaches the auditory telencephalon of birds via two different ascending pathways. One of them originates from LLI and at least in barn owls, to a lesser extent, from the dorsal part of the anterior portion of the dorsal nucleus of the lateral lemniscus (LLDa) to terminate in the telencephalic nucleus basalis (Wild et al. 2001). The other ascending projection originates in the lemniscal nuclei and IC, and projects to the telencephalon via the auditory thalamus (Wild 1987). In mammals, the cochlear nuclei and the SOC make bilateral projections to the IC, which, in turn, projects to the telencephalon via the auditory thalamus (Karten 1967). Mammalian lemniscal nuclei, like those of birds, also show direct projections to auditory thalamus (Webster et al. 1992b).

3.2 Development of the Auditory Hindbrain

In general terms, the first neurons to be generated in the auditory system are the peripheral cochleovestibular ganglion neurons, followed by neurons in the central nervous system. Cell birth occurs in different nuclei in partially overlapping time spans, with neurons in the IC being the last to be generated. Axon outgrowth and synaptogenesis generally occur at the time that the individual nuclei are first identifiable.

3.2.1 Rhombomeric Organization

The hindbrain of vertebrates shows a series of segmental units known as rhombomeres, where restricted gene expression patterns may contribute to the final determination of progenitors within the neuroepithelium (Keynes and Lumsden 1990; Krumlauf et al. 1993). The rhombomeric origin of the hindbrain auditory nuclei of chickens has been studied by two groups of researchers who showed that that the cochlear nuclei and NL not only originate from different sets of rhombomeres, but that their precursors also occupy restricted positions within each of them (Marín and Puelles 1995; Cambronero and Puelles 2000; Cramer et al. 2000). In chickens, NA originates from rhombomeres 3 to 6 (r3–r6), NM from segmental units r5/6–r8, and NL from r5–r6 (Marín and Puelles 1995; Cambronero and Puelles 2000; Cramer et al. 2000). Marín and Puelles's data indicate that each rhombomere contributes to most (if not all) of the frequency axis, but whether cells with the same best frequency, but originating from different rhombomeres, have similar or different physiological properties has yet to be determined (Marín and Puelles 1995). In other vertebrates, information regarding early embryonic relationships relies on gene expression patterns including gain and loss of function studies. In the auditory system, the most severe mutants are those that affect the identity of rhombomeres r5 and r6, since they affect otocyst development, and r4, which provides the neural crest population involved in the innervation of the ear. The conspicuous effects of these mutations on the peripheral auditory system have diminished attention to their role in the differentiation of the auditory hindbrain.

The selection of progenitor pools in the auditory brainstem is probably regulated by genetic cascades similar to those that govern the differentiation of peripheral structures (Rubel and Fritzsch 2002). Expression of basic helix–loop–helix (bHLH) family of genes is seen in the neuroectoderm associated with proliferative zones (Ferreiro et al. 1992; Jasoni et al. 1994). As is the case in the peripheral structures, neurogenesis may be regulated by lateral inhibitory mechanisms involving the Delta–Notch signaling pathways (Panin and Irvine 1998). It has been proposed that signaling through Notch may maintain neurogenesis by retaining an uncommitted population of cells within the neuroepithelium (Myat et al. 1996). Future studies may show how specific combinations of genes affect specific cell identities within the ascending auditory pathway. For example, the DCN fails to differentiate in the *BETA2/NeuroD* null mouse (Liu et al. 2000).

3.2.2 Cell Birth

Shortly after neural tube closure, cells in the neuroepithelium that will give rise to the auditory nuclei begin to undergo their terminal mitosis. Neurons are generated in the ventricular zone medial and adjacent to the rhombic lip over different yet partially overlapping periods of time (Harkmark 1954; Cant 1998).

In mammals, the earliest neurons to be born are those of the SOC and NLL, with each nucleus being born in a specific order (Cant 1998). The sequence of development appears to reflect the final morphology of cells, with large cells being born earlier than smaller cells (Altman and Bayer 1980; Martin and Rickets 1981; Cant 1998). In the cochlear nuclei as well, the order in which cells are born is correlated with their final cell fate. Pyramidal cells of the DCN are born first, followed by outer layer cells, while granule cells are the last to undergo terminal mitosis (Altman and Bayer 1980; Martin and Rickets 1981; Cant 1998). Thus, in the mammalian auditory nuclei, cell birth date appears to be related to the final fate of the progenitor cell, suggesting that there may be a coupling between the exit of the cell cycle and the commitment to a particular cell type (Rakic 1974; McConnell and Kaznowski 1991; Cant 1998).

Neurogenesis in birds has been studied by two groups (Rubel et al. 1976; Kubke et al. 2004). NM neurons in chicken and barn owl (*Tyto alba*) go through their final cell division first, followed by NL cells. Cell birth in the barn owl NA is delayed with respect to these two nuclei. Studies with [^{3}H]thymidine have shown that the period of neurogenesis in barn owls is longer than that of chickens (Kubke et al. 2004), although the extent of the lengthening of neurogenesis may be underestimated since the onset of neurogenesis in chickens has not been precisely defined. The homogeneous cell populations of the avian NM and NL make them unsuitable to determine the relationship between cell birth and cell fate, but it will be interesting to know if the different cell types in NA are born on different days (Hausler et al. 1999; Soares and Carr 2001). Cell birth in the auditory midbrain occurs after that of the auditory hindbrain (see Section 3.3) (Kubke et al. 2004).

Many aspects of the development of the auditory nervous system proceed along the tonotopic axis but this does not appear to be the case for cell birth (see Section 5). Although neurogenesis is not homogeneous within each auditory nucleus, the origin of these gradients is not yet entirely clear. In birds, cell birth correlates well with the position of the progenitor cells within specific rhombomere boundaries, with cells from more caudal rhombomeres being born first (Puelles and Martinez-de-la-Torre 1987; Marín and Puelles 1995; Kubke et al. 2004). Rhombomeric origin may be an important developmental determinant, as strengthened by Marín and Puelles's suggestion that projection patterns within the ascending circuit respect these segmental boundaries (Marín and Puelles 1995). For example, these authors proposed that NM neurons originating from more caudal rhombomeres (and thus born earlier) project to NL cells that originate from more caudal rhombomeres themselves. Since each rhombomere in chickens appears to contribute to an almost complete frequency range, this specificity of connections may reflect an early organization within the auditory hindbrain yet to be uncovered.

Whether rhombomeric position plays a role in the sequence of neurogenesis and projection patterns in mammals remains to be established. However, several observations suggest that this may be the case. In the LSO, MSO, and MNTB, cell birth occurs in a medial to lateral gradient dorsally and a lateral to medial gradient ventrally (Altman and Bayer 1980). Although it was proposed that this sequence of cell birth might be related to the tonotopic axis, this does not appear to be so. Instead, this sequence may be related to their final projection patterns, possibly reflecting some early positional information within the neuroectoderm similar to that seen in birds (Altman and Bayer 1980; Kudo et al. 1996, 2000). In addition, the differences in the length of neurogenesis may depend upon the number of rhombomeres involved, since neurons in more rostral rhombomeres commence their final mitotic divisions at a later age.

3.2.3 Naturally Occurring Cell Death

Final cell number within a neuronal structure results from the balance between the number of cells that are generated during neurogenesis and the subsequent naturally occurring cell death. Cell death is an integral process in the development of the central auditory system and has been examined in birds.

In barn owls, cell death in NM and NL leads to a 32% reduction in the number or NM cells and about a 39% reduction in the number of NL cells (Kubke et al. 2004). In chickens, several values for cell death for NM and NL have been reported (Rubel et al. 1976; Parks 1979; Solum et al. 1997). A more recent study using stereological counting methods found 43% cell death in NM and 52% cell death in NL (Wadhwa et al. 1997), values that were further confirmed using the terminal deoxynucleotidyl transferase-mediated dUTP nick-end labeling (TUNEL) method (Alladi 2002). At least some of these differences could reside in differences in the strains of chickens or in the methods used (Edmonds et al. 1999). What factors determine the extent of cell death in this system remains to be established. Otocyst removal in the embryonic chicken results in a reduction in the volumes of NA and NM as well as a reduction in the number of NM cells by cell death, indicating that the final number of NM cells is affected by afferent input (Levi-Montalcini 1949; Parks 1979).

3.2.4 Regulation of Cell Number

It is possible that some of the differences observed with respect to cell birth and death are responsible for the differences in the relative size of the adult auditory nuclei. For example, the MSO is small in mammals with poor low-frequency hearing (Webster 1992) and the ability to discriminate between binaural phase differences has been correlated with its size (Masterton et al. 1975). Avian auditory specialists show an increase in the size of the auditory nuclei beyond that which would be expected by the increase in the size of the brainstem alone (hyperplasia) (Kubke et al. 2004). It has been proposed that this hyperplasia may be responsible for the increases in the total number of cells within the auditory circuit, and that increased cell number is associated with the emergence of auditory specialization and increased computational ability. Further investigations to determine the variation of cell birth and cell death in individual species are necessary to understand fully the role that these developmental programs play in the regulation of the size of auditory structures. Both neurogenetic factors and trophic interactions with the periphery may contribute to the final relative size of auditory structures. These regulatory mechanisms may act in concert to create auditory structures whose size conforms to the individual niche of each species.

3.2.5 Cell Migration and Morphogenesis

The early morphogenesis of the cochlear nuclei and NL in barn owls follows a sequence that is strikingly similar to that described in chickens (Harkmark 1954; Book and Morest 1990; Kubke et al. 2002). The cochlear neuron precursors migrate into a dorsolateral region of the developing hindbrain where they will later coalesce to form the cochlear nuclei and NL. Initially there is an accumulation of cells at or near the rhombic lip in the immediate vicinity of the neuroepithelium, which constitutes the anlage of the cochlear nuclei (Fig. 6.2). Shortly after terminal mitosis, NM neuroblasts begin to migrate and to extend

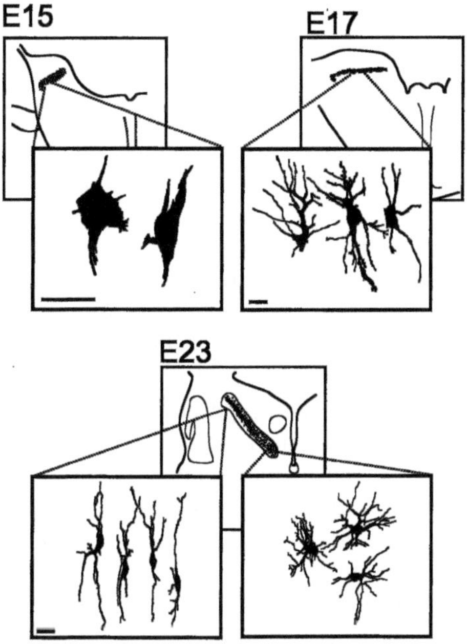

FIGURE 6.2. Schematic drawings of left hemisections through the hindbrain illustrating the early morphogenesis of the auditory hindbrain (dorsal toward the top). At E15 NL can be recognized in the more rostral regions of the hindbrain and Golgi material revealed the presence of neuroblastlike cells in the rostral ends of the cochlear anlage. By E17, the more rostral regions of the column have already differentiated into NL and assumed the typical laminar organization that is characteristic of basal birds. NL cells that form part of this laminar structure exhibit dendritic polarization with most of their dendrites projecting within the dorsoventral axis. At E23, cells in higher-frequency regions can be seen to have migrated away from the center of the lamina filling the NL neuropil, whereas lower-frequency regions maintain the laminar organization. At this stage, two distinct cell morphologies are observed. In the regions where the cells are confined to the middle of NL in a laminar structure, NL cells are clearly polarized with two dendritic tufts projecting dorsally and ventrally (*left panel*). In contrast, and on the same section, where NL has undergone the loss of the laminar organization, NL cells show dendrites distributed around their cell bodies and no obvious polarization (*right panel*). The morphology of the bitufted cells is reminiscent of the shape of the cells in NL in chickens, emus, and crocodilians, whereas the morphology of the cells after the loss of the laminar organization resembles that of the adult barn owl NL. Scale bar = 20 μm. (From Kubke et al. 2002. Copyright 2002 by the Society for Neuroscience.)

their axons from the primitive leading process (Book and Morest 1990). The cochlear and laminaris nuclei later begin to separate until they can begin to be identified between E5.5 and 6.5 in the chicken (barn owl E13 to 14) (Rubel et al. 1976; Kubke et al. 2002), although the migration of NA neurons is slightly delayed (Ariëns Kappers et al. 1936; Book and Morest 1990). The development of the cochlear nuclei in mammals is very similar to that described for birds (Cant 1998).

Differences in the auditory brainstem of different species are not limited to cell number. The morphological organization of the circuit also shows modifications that appear to reflect the auditory niche that different species occupy. Two solutions have been found throughout evolution to produce this circuit, and they are illustrated in Figure 6.3. A third solution, which does not require a map of delay, relies instead on left–right comparisons (McAlpine and Grothe 2003). Thus, the morphogenesis of the auditory nuclei is of great importance since it lays down the bauplan of the auditory pathway, determining the substrates used for sound localization. In addition, the specificity of connections and cellular architecture are crucial for accurate processing of binaural auditory signals (Agmon-Snir et al. 1998).

ITDs are computed as interaural phase disparities in NL of birds and the MSO of low-frequency hearing mammals in a circuit that conforms to the Jeffress model (Fig. 6.3) (Jeffress 1948; Carr and Konishi 1990; Overholt et al. 1992; Grothe 2000). Two elements, delay lines and coincidence detectors, are used such that a coincidence detector will respond maximally when the interaural time difference is equal and opposite to the retardation imposed by the delay line. This requires that axons be suitably aligned with respect to the cellular elements in order to produce an appropriate map of azimuth. Developmental and comparative studies indicate that the organization of NL in chickens represents the plesiomorphic condition, where NL forms a crescent shaped mass of cells medial and ventral to NM (Fig. 6.3) (Ariëns Kappers et al. 1936). In the apomorphic circuit characteristic of barn owls, cells are found instead dispersed in a 1-mm thick neuropil. In both chicken and barn owls, NL initially appears as a flat, compact layer of bitufted cells resembling the organization seen in the adult chicken (Fig. 6.2) (Kubke et al. 2002). Later in development, barn owl NL neurons lose both their bitufted dendrite morphology and their compact layer organization to assume the adult pattern characteristic of the adult barn owl by E26 (Kubke et al. 2002). Thus, the acquisition of the apomorphic pattern can be traced to this specific period of development, when barn owl organization diverges from the plesiomorphic organization of basal birds by adding a secondary morphogenetic phase to its development. Interestingly, while this modification of cellular architecture observed in barn owls is seen also in several bird species with high-frequency hearing, a similar reorganization does not appear to emerge in mammals. Instead, the MSO of mammals with no low-frequency hearing may process other aspects of the auditory stimulus (Grothe 2000).

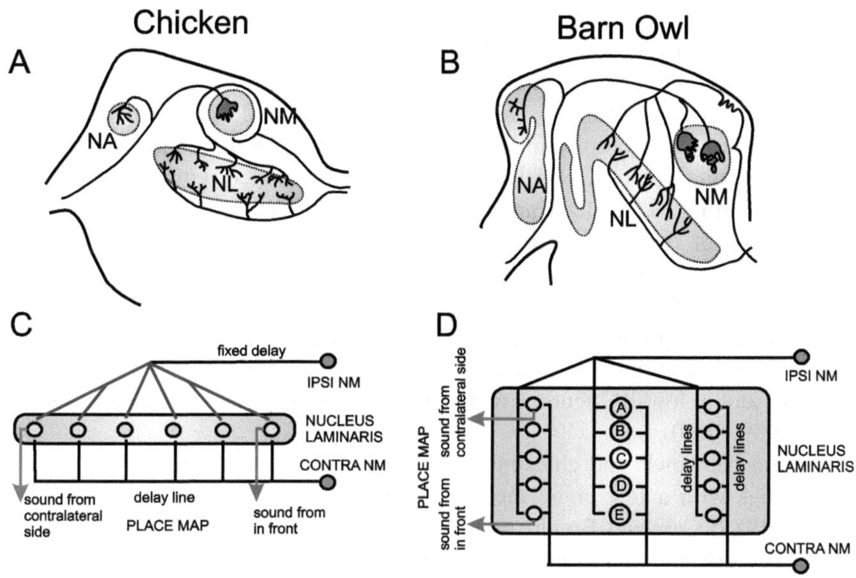

FIGURE 6.3. Schematic showing the Jeffress model for sound localization as applied to chicken and barn owl. (**A**) Schematic cross section through a chicken brainstem showing the organization of the projections from NM to NL. This organization conforms to a modified Jeffress model (**C**) with delay lines formed by the contralateral NM axons that run ventral to NL. This results in a space map oriented in the mediolateral dimension with cells in more lateral positions responding maximally to sounds originating from far contralateral space, and cells in a more medial position responding maximally to sounds originating from the front. (**B**) Schematic cross section through a barn owl brainstem showing the organization of the projections from NM to NL. The organization of the delay lines conforms to the Jeffress model (**D**). NM axons enter NL and traverse it making contact with NL neurons along their way. This results in multiple maps of ITD (**D**) with space mapped in a dorsoventral dimension. Neurons located in the dorsal edge of NL respond maximally to sounds originating from far contralateral space, and neurons located in more ventral position respond maximally to sounds originating from the front. (Reprinted from Kubke and Carr 2000. Copyright 2000, with permission from Elsevier.)

3.2.6 Morphological Maturation

The initial morphogenetic events described so far lay down the bauplan of the mature auditory system. After the basic connectivity pattern has been established (see Section 3.4), auditory neurons undergo significant morphological and biochemical changes that result in the mature forms typical of the adult auditory system. While some of this maturation is initiated during embryonic development, most of these modifications take place during early life.

Two major morphological changes take place during this time: changes in cell size and sculpting of dendritic architecture. The initial phase, characterized by the loss of migratory processes, overlaps and is followed by the formation of primary dendrites (Book and Morest 1990). The first neurons to undergo dendritic differentiation are those in the cochlear nuclei. In both mammals and birds, dendritic maturation is characterized by a prolific dendritic growth, which is followed by a period of elimination of branches. Differentiation of dendrites of fusiform and giant DCN cells in hamsters is correlated with the appearance of afferent input (Schweitzer and Cant 1984, 1985a,b), and in the rat MSO, the stabilization of dendritic branches occurs at around the time of ear opening (Rogowski and Feng 1981). In the gerbil LSO, high-BF neurons become spatially constrained during the third postnatal week, while low-BF neurons retain a broader arborization into adulthood (Sanes et al. 1992). This results in high-BF neurons exhibiting more restricted fields of dendritic arbors than low-BF neurons, and in low-BF neurons responding to a larger number of octaves (Sanes et al. 1989, 1990).

In both barn owls and chickens, adult NM neurons have large round or oval cell bodies with a few or no medium length dendrites (Fig. 6.4) (Jhaveri and Morest 1982; Carr and Boudreau 1993). NM neurons undergo a similar series of morphological transformations during development in both species. NM dendrites change in length and number over development, these changes being dependent upon the cell's position along the frequency axis. NM cells in chicken embryos are initially multipolar and spindle shaped, bearing some somatic processes, and extend an abundance of long dendrites of varying length by E11 (Jhaveri and Morest 1982; Parks and Jackson 1984). In chickens, the number of NM neurons with dendrites increases from E17 to P4 (Conlee and Parks 1983). Mature barn owl NM neurons also show a decrease in the number of dendrites from low- to high-BF regions of the nucleus. This adult pattern begins to form when cells in the rostral high-BF regions show a decrease in both dendritic number and length, beginning about 2 weeks after hatching.

The development of dendrites in NL cells in chicken embryo has been studied in detail (Smith 1981). Early on, NL neurons show the typical bitufted organization of dendrites, although the number of primary dendrites is small and there is no morphological difference between cells occupying different positions along the tonotopic axis. Between E10 and E13 NL cells grow a number of fine dendritic and filopodial processes that progresses along the tonotopic axis. The great proliferation of dendritic processes ends between E14 and E16, when there is a pruning of dendritic processes that also progresses along the tonotopic axis. The combination of these two processes results in the establishment of the gradient of dendritic length along the tonotopic axis, which is characteristic of chicken NL (Smith and Rubel 1979). The dendrites seen in the post-hatch chicken are similar to those seen at E19 (Smith 1981).

At present, it is not known whether similar developmental changes in dendritic length take place in barn owls. By E28, barn owl NL neurons have large number of short dendrites (Fig. 6.4) (Carr and Boudreau 1996). The number of dendrites decreases in the week after hatching and does not change significantly

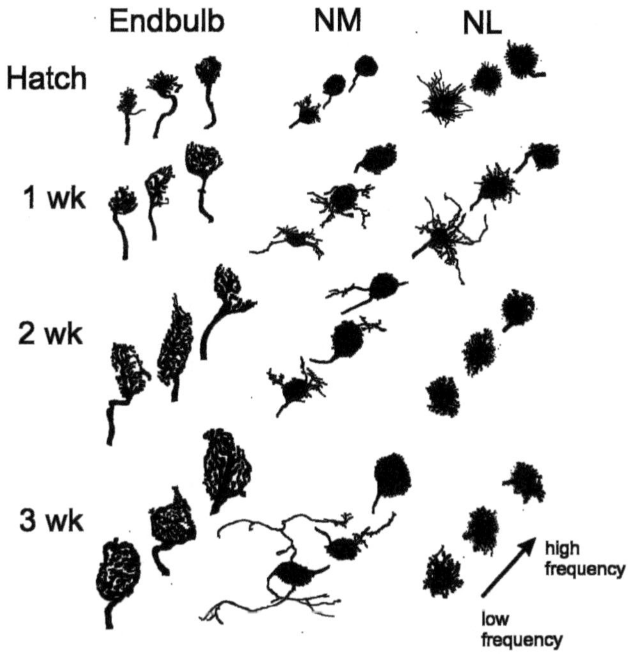

FIGURE 6.4. Camera lucida reconstructions of endbulbs of Held, and NM and NL neurons during barn owl development, showing a maturational gradient that follows the future tonotopic axis. Endbulbs are first seen as flattenings of the auditory axonal terminal, that become enlarged and fenestrated as development proceeds, to acquire an adultlike morphology by the time that the head has reached its adult size (3 weeks). The development of NM and NL cells is characterized by a prolific formation of immature dendrites that is followed by pruning. Adult NM and NL cells have many short dendritic processes distributed throughout the soma. (From Carr and Boudreau 1996. Journal of Comparative Neurology © 1996.)

thereafter, suggesting that the dendritic number is not affected by the maturation of the NM-NL circuit. It is not surprising that the development of NL neurons in barn owls differs from that in chickens, since the final architecture of this nucleus in these two species is substantially different (Fig. 6.3). Barn owl NL neurons show the dorsoventral polarity typical of chickens only early in embryogenesis (Carr et al. 1998; Kubke et al. 2002). They then lose this polarity and take on a different form, characterized by numerous short dendrites distributed around the soma (Figs. 6.2 and 6.4) (Carr et al. 1998). Furthermore, chicken NL dendrites are longer and more highly branched and continue to decrease in number and extent of branching until P25 (Smith 1981), whereas the number of barn owl NL dendrites decreases until about P6 but does not show further changes (Carr and Boudreau 1996). Thus, in the cochlear nuclei and NL of birds, the adult cell morphology is essentially present around hatching, and only secondary modifications take place during the early life of the bird.

Despite the differences seen in the overall timing of development in different species, the sequence of structural differentiation observed in the growth of dendrites in cells in the cochlear nuclei is very similar in all species. There are however, species-specific differences that may be related to specific auditory specializations. For example, barn owls show a differential loss of dendrites in the high-BF regions of NM such that by adulthood, the majority of cells in the high-BF region have virtually no dendrites. Barn owls have exceptional high-frequency hearing and have extended its range by 4 to 8 kHz compared to chickens (Konishi 1973a). Thus, the loss of dendrites in high-BF cells occurs in a region that has no counterpart in chickens, and may be a specialization for phase locking at high frequencies.

3.3 Development of the Auditory Midbrain

Despite the wealth of knowledge regarding the development of hindbrain structures, less is known about the early development of the midbrain. Formation of the mesencephalon depends on organizing activity originated at the isthmus and several gene expression patterns have been described in the developing midbrain that may play a role in its compartmentalization (Bally-Cuif and Wassef 1995; Agarwala and Ragsdale 2002). Cells in the auditory midbrain are derived from the neuroepithelium of the posterior recess of the cerebral aqueduct and are among the last to be born (Palmgren 1921; Cant 1998; Kubke et al. 2004). The cells of the IC appear to migrate in an outside in pattern such that those born the earliest lie more laterally, rostrally and ventrally in the adult IC, whereas those born last lie medially, caudally, and dorsally. Like in the cochlear nuclei, larger cells are born earlier than smaller cells. During the early postnatal period cells increase in size and dendrites mature. Dendritogenesis in the central nucleus precedes that in the dorsal cortex. Like cell birth, differentiation also appears to occur in a gradient. For example, cells in the deeper layers differentiate earlier than those of more superficial layers.

In birds, the IC develops from a dorsocaudal part of the mesencephalic vesicle, which bends and protrudes inward and rostralward, building a massive ventricular bulge (Palmgren 1921). In barn owls, when the IC can be first identified in Nissl material it resembles the tectal layer with which it is continuous, but the future core of the central nucleus of the IC (ICCc) cannot be readily differentiated from surrounding regions of the IC primordium (Kubke et al. 1999; Nieder et al. 2003). Maturation in the IC of birds has been examined in relation to synaptogenesis, and is discussed in the next section.

3.4 Internuclear Connections and Synaptogenesis

The developmental process by which internuclear connections are laid down is of fundamental importance, since it is responsible for creating appropriate cellular interaction and maintaining the tonotopic maps that are crucial for proper sound localization. In both birds and mammals, the initial projections appear to be formed before the animal responds to airborne sound and, at a macroscopic level, the precision of connectivity is present from the onset of innervation.

3.4.1 Synaptogenesis in the Cochlear Nuclei

Cochlear nerve fibers arising from ganglion cells enter the brain early in development, before the hair cells or other structures of the cochlea are mature. Some central processes can be seen even before delamination is complete (Cant 1998; Rubel and Fritzsch 2002; Molea and Rubel 2003). The vestibular and cochlear components of the VIIIth nerve are first seen in the hindbrain shortly after the cochlear neurons have undergone their terminal mitosis, and the growth of afferent fibers corresponds to the period of neurogenesis of the central targets (Knowlton 1967; Book and Morest 1990; Rubel and Fritzsch 2002). Despite the absence of a well-defined target (see Section 3.2.5) the cochlear and vestibular components of the posterior branch of the afferent nerve are segregated (Kubke and Carr 2000; Sato and Momose-Sato 2003). However, terminals from chicken auditory ganglion do not penetrate NM until after the cochleovestibular anlage is recognizable (Molea and Rubel 2003). Axons first invade the middle region of NM, close to the area of entry of the VIIIth nerve, then proceed to innervate the caudal region of NM, and by E7.5 all but the more rostromedial NM is innervated. The afferent axons exhibit restricted terminal fields that reflect the tonotopic bands of the mature NM. Innervation in NA proceeds in a tonotopic-specific sequence (Molea and Rubel 2003). A similar topography is reflected in the timing of innervation in mammals, where neurons from the basal turn of the cochlea reach their targets before those from the apical end (Rubel and Fritzsch 2002). Topographic order appears to be established in the cochlear nuclei some time before the cochlea has matured, and the initial arborization of afferent fibers is restricted to a narrow plane (Leake et al. 2002). Although topographical projections are evident before the onset of hearing they are less precise than in the adult, indicating the presence of experience-dependent refinement of synaptic organization.

Shortly after entering the brainstem, the VIIIth nerve bifurcates, with each branch projecting to a different region of the cochlear nuclei. Each branch must follow different pathfinding cues to reach its appropriate target, establishing synapses with a variety of cell types. At least some of these cues must be provided by their central targets. For example, the invasion of auditory fibers in both DCN in mammals and NM in barn owls is delayed with respect to that of other cochlear nuclei. Growth cones of the auditory nerve are seen in the outside boundaries of these nuclei, reminiscent of the waiting periods described in other developing systems. This delay cannot represent an immature state of the primary afferent, since different branches of the same neuron have been able to initiate synaptogenesis with other second-order auditory neurons. For example, although DCN neurons have migrated to their final position, afferent fibers do not penetrate it until after P1 and the adult pattern of innervation is not achieved until P10, although VCN has already received afferent innervation by birth (Schweitzer and Cant 1984; Cant 1998). Similarly, in barn owls, auditory nerve terminals are seen within NA well before the same neurons penetrate NM (Kubke, unpublished observations). Innervation of NM by afferent fibers coincides with the time at which NM establishes synapses with its target NL. The disparity of synaptogenic timing within the two branches of the same neuron remains puzzling, but it is possible that it requires signals from its target that are dependent on their maturational state. For example, barn owl NM neurons may not become competent to receive synapses until they are able to make the appropriate connections with NL.

Different branches of the auditory nerve also differ in the morphological development of their synaptic terminal. Each branch makes two morphological types of synaptic endings: endbulbs and boutons. Endbulb morphology has become associated with fast temporal coding whereas neurons in which temporal parameters do not play such a crucial role receive boutonlike terminations. Both NM and AVCN receive auditory terminals in the form of large calyceal endbulbs of Held (Ryugo and Parks 2003). The maturation of the endbulb morphology is a lengthy process that occurs over several days and at least some aspects of the morphological maturation may be regulated by activity (Rubel and Fritzsch 2002; Ryugo and Parks 2003).

Endbulb development has been studied in birds and mammals (Rubel and Fritzsch 2002). In birds, an initial small ending forms many filamentous processes that progressively envelope the somas of their target neurons to become large and fenestrated (Fig. 6.4) (Jhaveri and Morest 1982; Rubel and Parks 1988; Carr and Boudreau 1996). In both barn owls and chickens, endbulb terminals are formed a few days before the embryo responds to sound and the transformation of the plexus to the endbulb occurs around the same time as the cell loses its embryonic dendrites (Jhaveri and Morest 1982; Carr and Boudreau 1996). In barn owls, endbulb maturation continues after hatching, and their development follows a similar pattern to that described for the cat auditory nerve.

In mammals, the initial contact between afferent fibers and cochlear neurons consists of small and immature endings, which gradually increase in size and complexity (Ryugo and Fekete 1982; Limb and Ryugo 2000). In cats, endbulbs arise between P7 and P10 as long flat swellings on the spherical cells of the AVCN that will later adopt the typical calyceal morphology. By P10, the afferents make contacts with the soma and the number of somatic contacts increases as they acquire their mature form and by P12, the endbulb has become closely apposed to the cell body (Cant 1998). Mammals have provided good evidence of the role of activity in this process of morphological maturation. Afferents can be divided into high and low spontaneous rate fibers and the endbulbs of high spontaneous rate fibers are larger and less numerous than those of lower spontaneous rate fibers (Ryugo and Sento 1991; Ryugo et al. 1996). That activity plays a role in the maturation of the endbulbs is further supported by evidence from animals with hearing deficits. Deaf white cats that show complete hearing loss have profound structural abnormalities in the structure of their endbulbs including a decrease in the terminal branching, a reduction in synaptic density and an enlargement of synaptic size, whereas individuals with partial hearing exhibit morphologies that are intermediate between those of hearing and completely deaf cats (Ryugo et al. 1997, 1998). In the deaf *shaker2* mutant mouse endbulbs also show a loss of morphological complexity, including a reduction in endbulb arborizations, which is comparable to that seen in deaf white cats (Limb and Ryugo 2000). It has yet to be determined whether activity has a direct effect on the morphology of the terminal, or whether it requires feedback mechanisms from the postsynaptic cell. The latter possibility could explain how two branches of a single axon are capable of forming and maintaining synaptic structures as different as endbulbs and boutons.

3.4.2 Projections from the Cochlear Nuclei

Although the initial pattern of connections have been established by the time of birth, secondary modifications, characterized by increases in size and number of synapses, continue over the first few weeks of age. This process overlaps with the morphological and biochemical maturation of the target neurons. By the time of onset of physiological hearing, all connections have an adult-like appearance.

Processes from the cochlear nuclei to the SOC in mammals can be seen at their target cells when they complete their migration. Axons initially exhibit numerous thin processes that later mature into broad branches (Cant 1998). Immature synapses and growth cones are found in the SOC of the ferret as early as E28, and between E15 and E17 in the rat, followed by the formation of collateral branches in the ascending projections from the cochlear nuclei between E18 and P5 (Kandler and Friauf 1993). In ferrets, rats, and gerbils, the cochlear nuclei have made appropriate connections by the time of birth but the process of synaptic maturation and establishment of synaptic number is still underway. There is a high degree of tonotopic specificity from the onset of synaptogenesis and a general lack of aberrant connections (Kil et al. 1995). By the time of birth in the gerbil, cochlear nuclei axons are already seen making contact with lateral and medial dendrites of ispilateral and contralateral MSO, respectively (Kil et al. 1995). Calyces of Held onto neurons in the MNTB mature over the first two postnatal weeks, following a sequence similar to that described for the calyces in AVCN. Projections from the cochlear nuclei and SOC to IC are also present at birth in the rat (Friauf and Kandler 1990). Stellate cells from the ventral cochler nucleus (VCN) and fusiform and giant cells from the DCN project to the contralateral central nucleus of the IC giving rise to banded inputs (Cant and Benson 2003). The central nucleus also receives bilateral input from the LSO and a mostly ispilateral input from the MSO, as well as bilateral projections from the dorsal nucleus of the lateral lemniscus (DNLL) forming banded, tonotopically organized projections (Shneiderman and Henkel 1987; Oliver et al. 1997). It also receives a more diffuse projection from the VNLL (Kudo 1981; Shneiderman et al. 1988; Bajo et al. 1993). The fibers from the DNLL extend towards their targets as early as E15, and have crossed the midline by E19. By the time of birth, these fibers have already invaded the ventromedial high-BF layers of IC (Gabriele et al. 2000b). By P4 the earliest indication of input segregation into afferent rich and afferent sparse bands encompassing the entire tonotopic axis is seen, although adultlike patches are not seen until P12 (Gabriele et al. 2000a). Unilateral cochlear ablation disrupts these patterns (Doubell et al. 2000). Dendrites in cat IC become reorganized during early postnatal life (Meininger and Baudrimont 1981).

In birds, axons from the cochlear nuclei and NL reach their targets before the nuclei have completed their differentiation within the auditory anlage. In chickens, NM axons terminate on dorsal dendrites of ispilateral NL and ventral dendrites of contralateral NL in a discrete tonotopic and symmetrical order (Fig. 6.2) (Parks and Rubel 1975; Young and Rubel 1983). Neurons that will form NM migrate into their final position between E5.5 and E7.5 and have extended their contralateral axon through the dorsal cochlear commissure as early as E6 (Book and Morest 1990). By E6.5, the axons of NM cells can be seen in the ventral NL but dorsal dendrites of NL are innervated somewhat later. The dorsal and ventral terminal fields become symmetrical between E9 and E17 (Young and Rubel 1986).

Neurons from the NL project to the IC before their final differentiation and the initial projections of NL onto the LLD and ICCc appear to be tonotopically precise (Kubke et al. 1999). During barn owl embryonic development, the connections between the optic tectum and the IC that mediate the calibration of visual and auditory maps are also laid down (Luksch et al. 2000; Nieder et al. 2003). External nucleus of the IC (ICX) axons grow into the optic tectum (OT) from E16 on and at the day of hatching (E32), the projection displays a dorsoventral topography comparable to that in the adult barn owl. The reciprocal projection from the optic tectum to the IC forms by E18 with at least part of this projection formed by the neurons of the tectal stratum griseum. In both mammals and birds, the development of projections appears to maintain the topographic organization of tonotopic bands, and no obvious aberrant connections are seen during development (Kandler and Friauf 1993; Kil et al. 1995; Kubke et al. 1999).

3.4.3 Organization of the Patterns of Connections Required for ITD Coding

The organization of the patterns of connections between NM and NL are of crucial importance since this arrangement underlies the ability to code ITD. In particular, two features of the organization of NL (and MSO) appear to be crucial. The bitufted morphology of NL neurons and the changes in length of dendritic tufts along the tonotopic axis play a major role in coincidence detection. In addition, ipsi- and contralateral inputs are segregated, such that ventral dendrites are innervated by contralateral NM axons, while dorsal dendrites receive only inputs from the ispilateral NM. Although this organization is set up during embryonic development, afferent input plays a role in the maturation of this circuit. Otocyst removal does not prevent the formation of the dendritic length gradient, but the final length of the dendritic tufts is profoundly affected (Smith et al. 1983). Aberrant connections to inappropriate dendritic neuropil can also be seen after otocyst removal. However, the developmental patterns that determine the precise territorial domains of ipsi- and contralateral inputs remained elusive until new evidence from Cramer et al. (2002). Members of the Eph/ephrin signaling pathway are expressed in the developing auditory system (Cramer et al. 2002; Person et al. 2004). Dorsal dendrites in chicken NL express EphA4 and ephrin B2 from E9 to E11 within a gradient along the tonotopic axis, with higher levels of expression in higher BF regions (Person et al. 2004). Disruption of the patterns of EphA4 expression results in an increase of the number of NM axons that invade their inappropriate dendritic neuropil (Cramer et al. 2004). Thus, Ephs and ephrins may operate in the developing NM–NL circuit to signal the innervation territories of different axonal populations, establishing the appropriate segregation of ipsi- and contralateral inputs. It is difficult to predict what interactions the different combinations of Eph and ephrins will promote, since it has been proposed that ephrin activation may be localized to different subcellular compartments, recruiting different cascades in different neurons (Eberhart et al. 2004). The specific patterns of expression of Eph/ephrins in the NM–NL circuit and the presence of aberrant connections after EphA4 disruption suggest these expression patterns play a crucial role in the definition of innervation territories needed for proper ITD coding.

The transformation of the NM–NL barn owl circuit described in Section 3.2.5 has major implications for sound localization, since the emergence of the apomorphic pattern also results in a reorganization of the delay lines. This reorganization of delay lines and of NL cell morphology may underlie the acuity seen in barn owls for ITD coding at high frequencies. This is supported by the recent findings that a similar arrangement is seen in songbirds with wider frequency hearing ranges (Kubke, unpublished observations). Mammals such as bats with no low-frequency hearing also show an MSO where the cells do not show the characteristic laminar organization found in low-frequency hearing mammals (Grothe 2000). It is therefore not possible to discard the possibility that the modified NL found in songbirds may serve other temporal processing parameters than those used for sound localization by barn owls.

3.4.4 Development of Inhibitory Synaptic Inputs

Inhibitory inputs are also important for binaural computation. Two main inhibitory neurotransmitters are found in the auditory system, γ-aminobutyric acid (GABA) and glycine. At least in mammals, the distribution of both is affected by development and experience.

In young mammals both glycine and its receptors are poorly expressed, and

GABA exerts a more dominant effect (Piechotta et al. 2001; Turecek and Trussell 2002). In older animals, activation of ionotropic glycine receptors has been shown to potentiate glutamate release from the endbulbs of Held in the MNTB. At P11 in rats, before onset of hearing, endbulbs express ionotropic GABA receptors, but few glycine receptors. GABA agonists enhance excitatory postsynaptic currents in young rats, but this effect is no longer seen after P11, when the glycine receptors begin to be expressed (Piechotta et al. 2001; Turecek and Trussell 2002).

Inhibition appears to be important for ITD detection in mammals. In vivo recordings from the MSO of the Mongolian gerbil suggest that precisely timed glycine-controlled inhibition is a critical part of the mechanism by which the physiologically relevant range of ITDs is encoded in the MSO (Brand et al. 2002). In these and other low-frequency hearing mammals there is an experience-dependent refinement of ionotropic inhibitory inputs that does not develop in mammals that do not use ITDs for sound localization. In mammals with high-frequency hearing, glycine receptors are distributed throughout both the soma and the dendrites, while in low-frequency hearing mammals they are restricted to the soma and proximal dendrites (Friauf et al. 1997; Kapfer et al. 2002). The distribution of glycinergic inputs to the dendrites transiently appears in the low-frequency hearing gerbil during development, and this distribution persists in the adult if the cochlea is ablated (Kapfer et al. 2002). The regulation by afferent input of inhibitory connections may be a general feature of the auditory pathway where there are substantial inhibitory inputs that are developmentally regulated (Shneiderman et al. 1993; Sanes and Friauf 2000). Deafening, for example, results in an increase in the mRNA for the $\alpha1$, $\beta1$, and $\gamma2$ subunits of the GABA(A) receptor in the cochlear nuclei (Marianowski et al. 2000). MNTB neurons of congenitally deaf mice have a higher frequency and smaller amplitude of inhibitory postsynaptic currents, a larger mean single channel conductance, and a slower decay time course (Leao et al. 2004).

Inhibitory inputs are found throughout the ascending auditory system in birds. Inhibitory inputs to NM, NA, and NL originate primarily in the SO (Carr et al. 1989; Lachica et al. 1994; Monsivais et al. 2000). GABAergic cells are found in SO, the lemniscal nuclei and IC, and a small population of GABAergic cells are also seen in the vicinity of NA, NM, and NL. GABAergic terminals are observed, in turn, in NA, NM, and NL, as well as in the lemniscal nucleus and IC (Carr et al. 1989; Code et al. 1989; Adolphs 1993). The appearance of GABAergic terminals in chicken NM and NL coincides with formation of endbulbs (Code et al. 1989). The density of GABAergic terminals begins to increase in the caudolateral NL, establishing a density gradient from rostromedial to caudolateral. In contrast, the expression of the GABA receptor does not show a gradient along the tonotopic axis (Code and Churchill 1991). Glycine terminals are also present in the auditory hindbrain. Although there is little glycine expression in NA, NM, or NL in the embryo, glycine terminals are seen on the somas, but not the dendrites, of NL in post-hatch chickens (Code and Rubel 1989).

3.5 Myelination

Myelination represents a crucial event during the maturation of the auditory system since it will change conduction velocities necessary for the proper control of the time of arrival of bilateral inputs. Myelination is one of the later developmental events and progresses over extended periods of postnatal life (Cant 1998). In the cat embryo, myelination begins at about E57, yet myelination of the auditory nerve is not complete until after the second year of life. In the trapezoid body, myelination begins during the first week of life and is not complete at P46.

Myelination is essential for sound localization in barn owls, whose head grows over the first month post-hatch (Haresign and Moiseff 1988). Myelination of the delay lines in barn owls occurs later than myelination in the rest of the hindbrain, and at the time when auditory cues stabilize and the head grows to adult size (Haresign and Moiseff 1988; Carr et al. 1998). There is a delayed myelination of the axonal segments of NM that traverse NL (and provide most of the delay to map contralateral space) with respect to the portions of the axons that lie outside of NL. This delayed myelination may underlie the difference in internodal distance seen in the axonal segments within NL, which is of functional significance because the amount of delay mapped in NL depends on the length and conduction velocity of the myelinated NM afferents. It is not known whether there is a similar regulation of myelination in other birds. In barn owls almost all NM axons within NL have been myelinated by P20 (although most do not have many myelin wrappings), and myelination almost complete at P30. Thus, the delay lines acquire a functional appearance toward the end of the first month of life. Further increases in the degree of myelination continue into the second month post-hatch. The maturation of this circuit coincides with the time at which the head stops growing, and the bird is no longer subject to changing interaural time cues.

4. Biochemical Differentiation of the Auditory Brainstem

The morphological development of the auditory pathways is accompanied by biochemical maturation that includes changes in neurotransmitter receptors and ion channels. The auditory nerve activates glutamate receptors in the cochlear nuclei. In particular, the α-amino-3-hydroxy-5-methyl-4-isoxazolepropionic acid (AMPA)-type glutamate receptor mediates rapid excitatory synaptic transmission in many regions of the nervous system including the auditory nuclei. AMPA receptors in the auditory brainstem differ greatly from receptors in adjacent neurons. The molecular composition of receptors is controlled in a pathway specific basis during development and changes during auditory development. AMPA receptors in auditory neurons are characterized by rapid responses and some permeability to Ca^{2+}. Calcium binding proteins have been proposed to buffer the intracellular levels of Ca^{2+}. Similarly, developmental changes in potassium currents underlie the maturation of the shape of the action potential, and the expression of potassium channel genes changes during synaptic development. The development of glutamate receptors, calcium binding proteins and potassium channels will be reviewed.

4.1 Glutamate Receptors

An amino acid like glutamate mediates excitatory synaptic transmission in the auditory system, and AMPA-type glutamate receptor subtypes with distinct functional properties are expressed in auditory neurons (Parks 2000; Petralia et al. 2000). N-Methyl-D-aspartate (NMDA) receptors are also found in the cochlear nuclei of both birds and mammals (Zhou and Parks 1991; Zhang and Trussell 1994; Petralia et al. 2000; Sato et al. 2000). NMDA receptors have been shown to be important in the early development of the visual system (Kleinschmidt et al. 1987), and may play a similar role early in the development of the auditory system (Walsh et al. 1993; Feldman et al. 1996; Sato and Momose-Sato 2003; Tang and Carr 2004).

The functional properties of AMPA receptors are determined by the relative composition of the different subunits termed GluR1-4 (or GluRA-D). GluRs1-4 can be subjected to posttranslational modifications through mRNA editing and alternative exon splicing that also will result in changes in receptor properties (Borges and Dingledine 1998; Parks 2000). For example, incorporation of the mRNA-edited form of the GluR2 subunit suppresses Ca^{2+} permeability (Geiger et al. 1995; Parks 2000). Editing of the R/G site in GluR2-4 yields a receptor with faster recovery from desensitization. Posttranslational modifications giving rise to the *flip* and *flop* variants of AMPA receptors are of particular interest. In general *flip* variants show slower desensitization than the *flop* counterparts do. These *flop* variants of the GluR3 and GluR4 AMPA receptor subtypes are functionally important because they mediate rapid and precise temporal responses that are characteristic of many auditory brainstem neurons and should therefore play a crucial role in the ability of auditory neurons to phase lock to the auditory stimulus (Trussell 1999; Parks 2000). The cochlear nuclei express AMPA receptors with rapid decay kinetics and Ca^{2+} permeability. Levels of GluR1 are generally low, while other subunits show moderate to high expression levels. Auditory neurons show a relatively low abundance of GluR2, less R/G edited forms of GluR2 and GluR3, and express higher levels of the *flop* isoforms of GluR3 and GluR4. Thus, the principal AMPA auditory receptor appears to be composed by the *flop* variants of the GluR3 and GluR4 subunits (Parks 2000).

The levels of expression of the different subunits are developmentally regulated. GluR2 is expressed in early development, with expression levels diminishing with age after the onset of hearing. In the rat AVCN, Caicedo and Eybalin (1999) saw a transient high expression of GluR1 and GluR2 in the first 2 postnatal weeks with GluR4 detected thereafter. The levels of expression of GluR1 decrease between P16 and P30 until it gradually disappears. In contrast, GluR2, GluR2/3/4c, and GluR4 immunoreactivities increase gradually after birth to reach stable levels between P12 and P16. Similar expression patterns are seen in birds, in which the maturation of the expression of AMPA receptor types occurs during the time of synapse formation, sculpting of dendritic architecture, and onset of rhythmic spontaneous activity (Rubel and Parks 1988; Kubke and Carr 2000). In the adult barn owl, the nuclei of the auditory brainstem contain very high levels of GluR2/3/4c and GluR4 immunoreactivity (Levin et al. 1997). At the time of hatching, levels of GluR1 immunoreactivity in barn owls are very low or absent in all hindbrain auditory nuclei (Kubke and Carr 1998). Immunoreactivity for GluR2/3/4c and GluR4 in NM increases in parallel with the development of the auditory nerve projections to the cochlear nuclei, such that adult patterns of expression are attained between P14 and P21, when the endbulbs have acquired their mature form (Carr and Boudreau 1996; Kubke and Carr 1998). P14 is an important landmark in the young barn owl's development, coinciding with the opening of the eyes, the earliest time at which immature ITD responses can be recorded from NL, and the acquisition of the temporal patterns characteristic of adult auditory brainstem responses (Haresign and Moiseff 1988; Carr et al. 1998).

Less is known about the functional development of NA and SO, but in these two nuclei, the patterns of expression of the AMPA subtypes are also comparable to the adult by the end of the third week. A characteristic observed in NM and NL is a reduction in the overall neuropil staining level over time, which can be attributed in part to a reduction in cell density, increase in cell size, and reduction in dendritic length (Carr et al. 1998). There may be an early expression of other glutamate receptor subtypes associated with the early formation of synaptic structures, since auditory evoked responses can be recorded as early as P4, when the levels of expression of AMPA receptors is relatively low (Carr et al. 1998).

A more detailed study of the development of the expression of different subunits and posttranslational variants of the AMPA-type receptors has been carried out in chickens. In the adult chicken, specialized AMPA receptors are expressed in auditory neurons with characteristics that include rapid desensitization and high Ca^{2+} permeability (Lawrence and Trussell 2000; Ravindranathan et al. 2000). The faster *flop* variants of the GluR3 and GluR4 isoforms are expressed in auditory neurons at higher levels than in motoneurons of the glossopharyngeal and vagus nerves, which express instead the *flip* variants characterized by slower kinetics and low permeability to divalent ions (Raman et al. 1994; Ravindranathan et al. 2000). GluR1 is generally expressed at low levels in the auditory brainstem, and there are no significant changes in the levels of expression during development. The levels of expression of GluR2 decrease between E13-P2, in contrast to GluR3 and GluR4 whose levels remain high in NM over development (Ravindranathan et al. 2000; Sugden et al. 2002). By E10, the *flop* variants predominate in auditory neurons followed by a decrease in the GluR2 subunit in NM and an increase in $GluR3_{flop}$ and $GluR4_{flop}$ variants. By E17 the expression the $GluR4c_{flop}$ variant becomes predominant in auditory structures and the *flop* variants have increased in NM by P2 (Ravindranathan et al. 2000; Sugden et al. 2002).

The developmentally regulated AMPA receptors mediate transmission between VIIIth nerve and NM (Jackson et al. 1985) and between NM and NL (Zhou and Parks 1991). In the adult, rapid desensitization and Ca^{2+} permeability of AMPA receptors reflect the predominance of the $GluR3_{flop}$ and $GluR4_{flop}$, and the scarcity of GluR2 subunit. Cobalt accumulation via AMPA-type glutamate receptors is pronounced in NA, NM, and NL in chicken embryos (Zhou et al. 1995) and after synapse formation in NM, the rate of desensitization increases three-fold in NM neurons which, together with changes in other properties, indicate a decrease in the GluR2 isoform (Geiger et al. 1995; Mosbacher et al. 1995; Lawrence and Trussell 2000). These developmental changes in the properties of AMPA receptors appear to require cellular interactions (Lawrence and Trussell 2000). That afferent input may regulate the expression of different AMPA receptor isoforms is supported by observations by Marianowski et al. (2000) that after rats are deafened there is a decrease in the *flop* variants of the receptor.

4.2 Calcium Binding Proteins (CaBPs)

Calretinin (CR) is a member of the troponin-C family of calcium binding proteins (CaBP), which includes other soluble EF-hand proteins such as parvalbumin and calbindin (Celio et al. 1996). These proteins have a wide and heterogeneous distribution in the brain of vertebrates and are prominent in auditory and electrosensory systems (Braun 1990; Celio et al. 1996; Friedman and Kawasaki 1997). The avian auditory system is characterized by intense and distinct CR immunoreactivity, demarcating the terminal field of the neurons of NL along the ascending time pathway (lemniscal nuclei and IC) whereas it is either not as prominent or absent in the targets of NA (level pathway) (Takahashi et al. 1987; Puelles et al. 1994; Parks et al. 1997).

CaBPs have been proposed to act in the control and regulation of intracellular Ca^{2+} levels (Baimbridge et al. 1992) although their specific role remains unknown. They begin to be expressed early during development during the onset of synaptogenesis and the subsequent maturation of synaptic structures (Friauf 1993; Lohmann and Friauf 1996; Parks et al. 1997; Kubke et al. 1999). It has been hypothesized that these proteins may play a role in buffering the Ca^{2+} that enters the cells during synaptic activation. CR-expressing neurons in chicken cochlear nucleus exhibit Ca^{2+} currents upon glutamate receptor activation (Otis et al. 1995) and CR (but not calbindin) becomes localized beneath the plasma membrane at the onset of spontaneous activity (Hack et al. 2000).

The auditory nuclei of barn owls begin to express CR quite early during development (Kubke et al. 1999). Expression begins in regions that map high-BF in the adult progressing toward lower BF regions over a period of several weeks, encompassing the entire frequency axis by the time the bird opens its eyes. Similar gradients of expression are found in chickens (Parks et al. 1997). CR mRNA and CR immunoreactivity first appear in the cochlear nuclei of chickens at E9, and in NA is restricted to the dorsal and lateral (high-BF) regions, the gradient disappearing by P1 (Parks et al. 1997). The patterns of expression of CR are highly correlated with the future tonotopic axis, suggesting that mechanisms comparable to those that underlie the establishment of the morphological gradients of development may also be responsible for the onset of the expression of CR (see Section 5).

The association between the expression of CaBPs and synaptogenesis described in the rat are not seen with CR in birds (Friauf 1994; Lohmann and Friauf 1996). In chickens, CR is first detected when the first physiological responses can be recorded but its expression does not encompass the whole extent of NA until later in development (Parks et al. 1997). If synaptogenesis were to trigger the expression of CR, the development of synapses and CR would be expected to show similar time courses. In barn owls, the spatiotemporal patterns of expression of CR and synaptogenesis suggests that they do not bear a causal relationship (Kubke et al. 1999) and it has been concluded that synaptic activity does not influence the basal expression of CR in chickens (Parks et al. 1997).

Other studies have investigated the significance of afferent input in the regulation of the expression of CaBPs. In mammals, removal of afferent input by cochlea ablation or treatment with glutamate antagonists is correlated with changes in the levels of expression of CaBPs in the brainstem auditory nuclei (Winsky and Jacobowitz 1995; Caicedo et al. 1997; Alvarado et al. 2004). In contrast, in chickens, reduction of auditory input after the formation of synapses by collumela removal does not alter the patterns of expression of CR (although there may be a small increase in mRNA levels) (Parks et al. 1997). Furthermore, when otocyst removal prevents the formation of contacts between afferents with their central targets in chickens, no changes in the patterns of expression of CR are observed (Parks et al. 1997). Thus, at least in birds, the onset and maintenance of expression of CR may not depend on normal synaptic function. This may represent a difference between avian and mammalian auditory systems.

4.3 Potassium Channels

The intrinsic membrane properties of neurons undergo dynamic changes during development and their maturation is determined in part by the differentiation of various potassium currents that contribute toward the outward current. At least two potassium conductances underlie phase-locked responses in auditory neurons: a low threshold conductance (LTC) and a high threshold conductance (HTC). The LTC activates at potentials near rest and is largely responsible for the outward rectification and nonlinear current voltage relationship seen in many auditory neurons with precise responses to temporal stimuli (Oertel 1999).

4.3.1 Potassium Conductances Active Near Rest

Activation of the LTC shortens the membrane time constant so that the effects of excitation are brief and do not summate in time (Bal and Oertel 2000). Only large excitatory postsynaptic potentials (EPSPs) reaching threshold before significant activation of the LTC would produce spikes. Blocking the LTC, moreover, elicits multiple spiking in response to depolarizing current injection (Trussell 1999). The K^+ channels underlying the LTC appear to be largely composed of Kv1.1 and Kv1.2 subunits. In expression systems, Kv1.1 can produce channel tetramers with properties similar to those of the LTC, including sensitivity to dendrotoxin (Brew and Forsythe 1995). Both subunits are expressed in auditory neurons (Grigg et al. 2000). Kv1.1 is strongly expressed in neurons with low threshold conductances including auditory neurons of the mammalian MNTB, and the neurons of the avian NM and NL (Lu et al. 2004).

Modeling and physiological studies have shown the importance of the LTC in coincidence detection (Gerstner et al. 1996; Agmon-Snir et al. 1998; Kempter et al. 1998). Developmental studies from Kuba et al. (2002) have shown that an increase in the LTC in chicken NL is accompanied by an improvement in coincidence detection between E16–17 and P2–7. During this time, the membrane time constant of NL neurons is reduced while membrane conductance increases five-fold. Improvement of coincidence detection is correlated with the acceleration of the EPSP time course that results from the increase in these conductances.

4.3.2 High-Threshold Potassium Conductance

The HTC is characterized by an activation threshold around -20 mV and by fast kinetics (Rathouz and Trussell 1998; Wang et al. 1998; Rudy and McBain 2001). These features of the HTC result in fast spike repolarization and a large but brief afterhyperpolarization without influencing input resistance, threshold or action potential rise time. Thus, the HTC can keep action potentials brief without affecting action potential generation. In addition, it minimizes Na^+ channel inactivation allowing cells to reach firing threshold sooner, and facilitates high-frequency firing. Currents produced by Kv3 channels share many characteristics of the HTC and, most likely, underlie it. Many auditory neurons express high levels of Kv3 mRNA and protein (Perney and Kaczmarek 1997; Parameshwaran et al. 2001; Lu et al. 2004). Interestingly, in several auditory nuclei including avian NM and NL (Parameshwaran et al. 2001) and rat MNTB (Perney and Kaczmarek 1997), Kv3.1 protein expression varies along the tonotopic map such that mid- to high-BF neurons are most strongly immunopositive while neurons with very low BF are only weakly immunopositive. These results suggest that the electrical properties of higher-order auditory neurons may vary with frequency tuning.

Changes in the functional role of Kv3.1 may be mediated by two splice variants, a and b, which are differentially regulated during development. Kv3.1a is expressed from early in development (Zhou et al. 2001), while Kv3.1b specific staining appears after the onset of synaptogenesis. Early in auditory development, suppression of a high-threshold Kv3.1-like current with tetraethylammonium (TEA) and 4-aminopyridine (4AP) results in a reversible block of migration in cultured neuroblasts (Hendriks et al. 1999). During the period from E14 to E18, when the levels of Kv3.1 immunoreactivity increase in chicken auditory nuclei (Parameshwaran et al. 2001; Zhou et al. 2001), the magnitude of a Kv3.1-like high threshold current in NM neurons in culture also increases (Hendriks et al. 1999). The maturation of tone-elicited and electrically elicited responses from the brainstem auditory nuclei occurs during this time. Response latencies are reduced, the responses become less fatigable, thresholds decrease, and progressively higher frequencies become effective (Saunders et al. 1973; Jackson and Parks 1982).

4.3.3 Kv3.1b Expression in Barn Owls

The onset of Kv3.1b expression in chickens and barn owls NM and NL coincides with synapse maturation and progresses from the more medial high-BF region to the lateral low-BF regions. In barn owl NL, Kv3.1b expression reaches detectable levels between E21 and E26 when synaptic connections between NM and NL are morphologically mature (Carr and Boudreau 1996; Kubke et al. 2002). Expression of Kv3.1b in the time coding neurons follows a time line similar to that of CR and AMPA-type glutamate receptor expression (see Sections 4.1 and 4.2). Kv3.1b expression appears to begin 1 to 2 days later than CR because faint levels can first be detected in rostral NL at E21, reaching approximate adult levels after P6.

The emergence of the Kv3.1b gradient in both chickens and barn owls during development may reflect in part the general rostromedial to caudolateral maturational gradient (see Section 5). The persistence of the gradient in the adult birds, however, indicates that this differential distribution along the tonotopic axis is not merely a developmental phenomenon. The Kv3.1 gradient in the adult is more pronounced in barn owls than in chickens, but is present in both species (Parameshwaran et al. 2001).

5. Development and the Tonotopic Axis

Many of the developmental changes described in the previous section progress in a gradient along the future tonotopic axis, with high-BF regions preceding lower-BF regions in their maturation (Rubel 1978; Kubke and Carr 2000). In the adult, some of these gradients persist as differences in morphological features along the tonotopic axis and in the levels of expression of biochemical markers such as CR, AMPA receptors, or K^+ channels (see Section 4). Thus, the original developmental gradients may reflect a sequence of maturational timing, but may also serve as means of establishing the gradients that persist in the adult. Rubel and Fritzsch (2002) point out that an understanding of the gradients of molecules along topographic axis is an important and general problem in developmental neurobiology, and particularly accessible in the auditory pathways.

In the peripheral auditory system, there is a general trend by which the basal region of the auditory organ matures before the apical region. Some of these maturational changes may be regulated by neurotrophins since the spatiotemporal expression domains of brain derived neurotrophic factor (BDNF) and neurotrophin-3 (NT-3) are essential for the normal development of the ear (Fariñas et al. 2001; Fritzsch 2003). Levels of NT-3 mRNA appear to parallel the basal to apical maturational gradient seen in the peripheral auditory system and disruption of the NT-3 receptor TrkC results in a reduction in the innervation to both inner hair cells (IHCs) and outer hair cells (OHCs), particularly in the basal turn of the cochlea (Fritzsch et al. 1997). The basal to apical maturation gradient seen in the ear is also seen in ganglion cells and neurotrophin receptor expression overlaps both spatially and temporally throughout the early neonatal period (Fritzsch et al. 1997).

Gradients of development that progress along the future tonotopic axis may typify the developing auditory system since numerous morphological and biochemical features of the brainstem auditory nuclei also develop along a spatial gradient that corresponds to the tonotopic axis (Smith and Rubel 1979; Rubel and Parks 1988; Kubke and Carr 2000). For example, innervation of the DCN in mammals begins in the dorsomedial (high-BF) region before the ventrolateral (low-BF) region of the nucleus (Schweitzer and Cant 1984). The morphological and biochemical maturation of cells and synaptic structures within NM and NL in birds also commences in regions that will map high frequencies in the adult and progresses caudolaterally toward lower-BF regions (see Section 3.2.6). Most (if not all) of the maturational processes described so far for barn owls progress along the future tonotopic axis, albeit with different time courses.

The development of the peripheral auditory structures progresses in a similar gradient, suggesting a possible influence of auditory experience on the emergence of the developmental gradients seen in the central auditory system. At least some (if not most) of the aspects of the maturational process of the central auditory nuclei do not, however, depend on auditory input. In chickens, the acquisition of the characteristic morphological gradient in NL cells along the frequency axis persists after otocyst removal (Parks 1981; Parks et al. 1987). The patterns of expression of CR in chickens develop rather normally in the absence of any direct contact with the auditory nerve (Parks et al. 1997). It is still possible that other synaptic input may regulate CR expression. For example, the cochlear nuclei receive synaptic input from sources other than the auditory nerve, and after otocyst removal in chicken an aberrant connection is formed between NM and NM (Jackson and Parks 1988). Nevertheless, Parks et al. (1997) have concluded that synaptic activity does not influence the basal expression of CR, and an early determination program appears to be the more plausible explanation for these gradients.

Rubel and Parks (1988) hypothesized that the spatial gradient along the tonotopic axis in the lower brainstem auditory nuclei arises independently of the emergence of tonotopic organization in the peripheral structures and the data available for chickens and barn owls support this proposal. This suggests that an early determination of the cells may program them to follow a specific developmental timetable along the frequency axis, since several features of the auditory system continue to exhibit such gradients in the absence of normal afferent input.

6. Onset and Development of Auditory Function

It seems likely that synapse formation and some electrochemical transmission begin soon after initial contacts are established. New optical detection methods using voltage-sensitive dyes in chicken embryos have revealed electrical responses evoked by cochlear or vestibular nerve stimulation as early as E7 (Sato and Momose-Sato 2003). Early electrical events have also been observed in mice (Sanes and Walsh 1998). Thus, as is also the case for the visual system, synaptic transmission may precede the time at which the central nervous system processes external acoustic stimuli. This section reviews the onset of function and physiological development of the central auditory system.

6.1 Onset of Hearing

Several methods have been used to establish the onset and development of hearing, including recordings from the cochlear microphonic (CM), the compound action potential (CAP), and auditory brainstem responses (ABR). In birds and mammals, the later stages of development occur over varying time periods of postnatal life (Cant 1998). The CM is a good indicator of active transduction, since it is probably generated by OHCs in the cochlea, and may, in addition, reflect some aspects of middle ear maturation (Woolf and Ryan 1988). The CAP can be first recorded at the time that at which reliable measurements of CMs are first obtained consistent with the notion that the onset of function occurs simultaneously in the IHC and OHCs (Rübsamen and Lippe 1998). Development is characterized by a steady increase in the amplitude and a decrease in the threshold and response latency to the CAP. It has been suggested that the increase in amplitude in the CAP is the result of increased synchronization, as indicated by the improvement in the phase-locked response of afferent nerve fibers that occur over the same period of development.

6.2 ABR Studies

The ABR reflects synchronous activity along the ascending auditory pathway and provides a useful noninvasive method to evaluate the development of auditory function. The emergence of ABR peaks reflects the maturation of the middle and inner ear and development of synchrony within nuclei, while changes in latencies may follow the myelination of ascending tracts. Thus, the development of ABRs is characterized by shortening of latencies and lowering of thresholds (Walsh et al. 1986, 1992; Wang et al. 1993).

The development of ABRs in birds has been studied in several species. In chickens, Saunders (1974) showed that the first ABRs appear at E11 coinciding with the early postsynaptic responses in NM (Jackson et al. 1982). There is a lowering of threshold in higher frequency ranges, and a reduction in the latency to the first peak of the ABR from E15 to P2, but not beyond that age (Saunders et al. 1973; Katayama 1985). This coincides with the reduction in the latency to the EPSP in NM neurons evoked by cochlear nerve stimulation, which occurs between E13 and P4 and which accompanies a reduction in the number of VIIIth nerve terminals innervating each NM neuron (Jackson and Parks 1982).

Studies in altricial birds suggest that the maturation of sensitivity in the region of low and middle frequencies continues to develop throughout a large part of their nesting period (Aleksandrov and Dmitrieva 1992; Brittan-Powell and Dooling 2004). This is in contrast to what is seen in precocial species, in which low to middle frequency responses can already be seen by the time of hatching, while high-frequency responses continue to mature after this time (Saunders et al. 1973).

6.3 Physiology in the Brainstem

Onset of synaptic activity occurs soon after terminals reach their targets. Sound-evoked responses are seen in chickens almost immediately after synaptogenesis but they may be delayed in rodents, probably because of immaturity of the inner ear structures (Sanes and Walsh 1998).

Postsynaptic responses mediated by NMDA receptors emerge in chicken brainstem nucleus at about E6 (Sato and Momose-Sato 2003). This chronological sequence of the emergence of postsynaptic function differs from the previous electrophysiological observations that NM neurons are responsive to VIIIth nerve stimulation from E10 to E11 (Jackson and Parks 1982; Pettigrew et al. 1988). As Jackson et al. (1982) pointed out, this discrepancy may be due to differences in sensitivity of the measurement systems. Prior to the studies with voltage-sensitive dyes, evidence supported the onset of physiological activity in chicken brainstem around E10 to E11, coinciding with the early stages of synaptogenesis, with responses progressing along the future tonotopic axis (Saunders et al. 1973; Jackson et al. 1982; Rubel and Parks 1988). At this time, NM and NL neurons have extended dendritic processes and appear to receive synapses from far more auditory nerve fibers than they retain (Jackson and Parks 1982; Jhaveri and Morest 1982).

Neural activity in the developing brainstem auditory pathway of chicken embryo is dominated by a rhythmic pattern of spontaneous discharge. Tetrodotoxin sensitive rhythmic bursting is present as early as E14 shortly after the onset of functional synaptogenesis, and gives way to more mature, steady level of firing on E19, 2 days prior to hatching (Lippe 1994, 1995; Jones and Jones 2000). Such periods of high activity may facilitate Ca^{2+} influx and regulate the transcription of ion cannels and other proteins (Spitzer 1991; Gu and Spitzer 1995). The increase in the periodicity of bursts and the presence of Ca^{2+} permeable, glutamate-activated AMPA receptor-mediated synaptic currents in chicken NM neurons may necessitate Ca^{2+} buffering ability in auditory neurons (Lippe 1995; Otis et al. 1995; Zhou et al. 1995). CR is expressed early during development and may restrict Ca^{2+} transients (see Section 4.2) (Parks et al. 1997; Kubke et al. 1999; Hack et al. 2000). The rapid depolarization of the action potential mediated by the high threshold conductance may be another mechanism to limit the amount of Ca^{2+} entering the cells (Rudy and McBain 2001). The presence of a systematic relationship between the rate of rhythmic bursting and tonotopic location (Lippe 1995; Jones and Jones 2000) suggests that the spatiotemporal pattern of spontaneous discharges could provide some of the developmental cues for the spatial regulation of gene expression (see Section 5). Around E14, the synapses begin to mature and the first "behavioral" responses to sound are obtained (behavior refers to motility monitored with platinum electrodes inserted beneath the egg shell membrane) (Saunders et al. 1973; Jackson and Rubel 1978; Jackson and Parks 1982).

In barn owls, the anatomical maturation of the delay line axons within the NL coincides with the appearance of sensitivity to ITD. Phase-locked auditory responses in NL produce a large evoked potential termed the neurophonic potential (Sullivan and Konishi 1986). This potential begins to exhibit adultlike sensitivity to ITD after P21 (Carr et al. 1998), while better-developed responses are recorded in a month-old bird. There are also further changes in the neurophonic after P30, with larger peak-trough differences in the neurophonic in a P60 bird. These results suggest that the adultlike sensitivity to ITDs emerges gradually in the first 2 months post-hatch.

In order to examine the onset of responses in the auditory brainstem, Ryan et al. (1982) examined the expression of 2-deoxyglucose in gerbils after stimulation with white noise. 2-deoxyglucose was expressed at P12 only in VCN. Expression was present in all cochlear nuclei, SOC and VNLL by P14 and by P16 it had extended to DNLL, IC, and the auditory thalamus. This suggests that the onset of activity progresses sequentially along the ascending auditory pathway. In rats, the onset of synaptic function occurs at E18 and in mice synaptic transmission can be seen at P4, while neural responses are first observed between P8 and P12, well before the onset of hearing (Cant 1998). Adultlike current-voltage relationships are present in immature auditory neurons, but membrane time constant and input resistance continue to vary with age (Sanes and Walsh 1998). Action potential duration is greater in immature animals and the levels of spontaneous activity is much lower in developing animals than in the adult (Rübsamen and Lippe 1998). In immature mammalian auditory neurons, the amplitude of action potentials near 0 mV is similar to that of the adult, but its duration is two to three times longer (Sanes and Takacs 1993; Kandler and Friauf 1995a; Taschenberger and von Gersdorff 2000). The prolonged duration of the spikes could result in increased levels of Ca^{2+} influx and/or from immature Na^{+} transport mechanisms, which may limit the ability of auditory neurons to respond rapidly to high-frequency stimuli. Immature cells are unable to sustain high firing rates (Wu and Oertel 1987) and synaptic fatigue in the cochlear nuclei is more prominent in immature states (Sanes and Walsh 1998). The duration of the EPSPs and inhibitory postsynaptic potentials (IPSPs) decreases during the first 2 weeks of life in the gerbil LSO.

Immature neurons in the auditory nerve and central nuclei produce low-frequency bursts of action potentials both spontaneously and in response to sound. This pattern of firing becomes continuous if the efferents are severed in neonatal cats, suggesting that cholinergic inhibition of IHCs contributes to the rhythmic activity of the immature auditory pathway. In the cat, the rate of discharge of the auditory nerve shows a rapid increase between P2 and P20, and reaches adultlike characteristics after P40 (Romand 1984; Walsh and McGee 1987, 1988). During the first week postnatal in the kitten phase-locked responses to stimuli below 600 Hz are present, while the high-frequency range is extended and the shapes of response areas of AVCN neurons begin to resemble those seen in the adult. Adult thresholds, maximal discharge rate, and first spike latency achieve adult values between the second and third week postnatal and phase locking is achieved last, between the third and fourth weeks of age (Brugge et al. 1978; Walsh et al. 1993). In the gerbil the cochlear nuclei show spontaneous activity by (but not before) P10, 2 days before the onset of hearing and most AVCN neurons respond to auditory stimulation by P14 and mature evoked responses are seen by P20 (Sanes and Walsh 1998). The third postnatal week is characterized by the improvement of neuronal thresholds, spontaneous firing rates and improvement in phase locking, so that by P18 most neuronal parameters have achieved adult values. Frequency tuning matures from high-BF to low-BF (Woolf and Ryan 1985). Similarly, in the rat, while the onset of hearing is at P12, SOC neurons can generate action potentials as early as E18 (Kandler and Friauf 1995b).

Collaterals begin to invade the LSO anlage at E18 where they can evoke postsynaptic potentials (PSPs) after both ipsi- and contralateral stimulation (Kandler and Friauf 1995b). There is a transition from GABA to glycine transmission in the gerbil LSO during the first 2 weeks postnatal, which is accompanied by a decrease in the postsynaptic localization of the β2,3 subunits of the GABA(A) receptor between P4 and P14 and an increase in the glycine receptor anchoring protein gephryn (Kotak et al. 1998; Korada and Schwartz 1999). In brain slices of developing gerbils the evoked IPSP shows long-term depression when MNTB is activated at low frequencies, and depression can be induced by GABA but not glycine (Kotak and Sanes 2000, 2003). Contralateral glycinergic PSPs are depolarizing between E18 and P4, and can elicit the production of action potentials (Sanes 1993; Kandler and Friauf 1995b). Similar results are seen in the C57Bl/6J mouse (Kullmann and Kandler 2001). These PSPs become hyperpolarizing after P8 in the gerbil and during the first postnatal week in the mouse, and between P9 and P11 in the rat (Sanes 1993; Kandler and Friauf 1995b). This shift in the responses to glycine appears to be due to the underlying maturation of the mechanisms that regulate intracellular Cl^-, which may themselves be under afferent regulation (Ehrlich et al. 1999; Kakazu et al. 1999; Vale and Sanes 2000). During the time at which GABA and glycine elicit depolarizing responses, they also produce increases in intracellular Ca^{2+}, whereas at ages when the responses are hyperpolarizing there is a small decrease in intracellular Ca^{2+} (Kullmann et al. 2002). During this period, there is a transient expression of CaBPs in LSO. This suggests that CaBP expression may be regulated during development in relation to these transient depolarizing responses to inhibitory inputs rather than with glutamate-mediated excitatory inputs as initially proposed (see Section 4.2) (Friauf 1993; Henkel and Brunso-Brechtold 1998). The transition in the response to glycine occurs at the time at which glycinergic connections are refined by activity-dependent mechanisms. Inhibitory inputs are found to be spread out over the frequency axis following functional denervation, suggesting that activity plays a role in the synaptic pruning that gives rise to restricted innervation territories in the normal LSO (Sanes and Siverls 1991; Sanes and Takacs 1993).

In young mammals, EI neurons in the LSO show ILD responses at the earliest time examined. The tonotopic map is already aligned, but ILD responses continue to mature in both dynamic range and resolution (Moore and Irvine 1981; Sanes and Rubel 1988; Blatchley and Brugge 1990). Similar results are obtained in the midbrain. Responses in the IC of C57BL/6J mice show a rapid development between P12 and P17. In younger animals, there is little to no spontaneous activity and many units do not show auditory evoked responses. Spontaneous activity becomes evident between P15 and P17, in parallel with tone-evoked neuronal responses (Shnerson and Willott 1979). In young kittens, ILD sensitivity is seen as early as P8 and ITD sensitivity is present at P12 (Blatchley and Brugge 1990). Thus, although responses to different auditory cues are present in the immature animals, there is a subsequent sharpening of the corresponding receptive fields and a decrease in threshold (Shnerson and Willott 1979; Blatchley and Brugge 1990; Withington-Wray et al. 1990).

6.4 Place Code Shift

One aspect of early development is the expansion of the frequency range over which physiological responses can be observed. Recordings from young animals generally show responses to low-frequency sound (Crowley and Hepp-Reymond 1966; Lippe and Rubel 1985; Hyson and Rudy 1987). This is surprising, since the ear develops from the base to the apex. One early hypothesis put forward to explain the paradoxical relationship between place and best frequency was that of a shifting place code (Lippe and Rubel 1983; Rubel et al. 1984). The idea was that, as the cochlea develops, each cochlear region would respond over time to progressively higher frequencies. Thus, the place of responses to any given frequency would progressively shift toward the apical end. Tests of this hypothesis have instead revealed differences between birds and mammals, and between different parts of the cochlea (see Rübsamen and Lippe 1998 and Manley 1996 for review).

There appears to be only a small frequency shift in the avian papilla (Jones and Jones 1995; Chen et al. 1996), while a large frequency shift occurs in higher frequency regions of the mammalian cochlea (Echteler et al. 1989). Recordings from gerbil spiral ganglion cells at a constant location within the basal cochlea showed significant increases (up to 1.5 octaves) in their best-response frequencies between the second and third weeks of postnatal life, which appear to be largely dependent on maturation of the tectorial membrane (Echteler et al. 1989; Ohlemiller and Echteler 1990). Thus, there is no single comprehensive model for the development of tonotopy in the peripheral auditory system of birds and mammals, and a peripheral place code shift does not appear to be a general mechanism.

It is not yet known how central auditory system responses reflect the tonotopic changes in the peripheral auditory system. There is a tonotopic shift in LSO neurons in Mongolian gerbils, where the tonotopic map changes with age such that the characteristic frequency of neurons in a given anatomical location becomes successively higher during development (Sanes et al. 1989). Behavioral studies also show a shift in the response to frequencies consistent with the place code shift hypothesis (Hyson and Rudy 1987).

7. Experience-Dependent Plasticity

Unlike other sensory systems, auditory space is not mapped in the sensory papilla. Thus, representations of auditory space are the result of neuronal computations that take place along the ascending auditory system. Different direction-dependent features of the auditory stimuli are extracted along the pathway such that auditory receptive fields corresponding to specific locations of the sound source emerge in neurons in the external nucleus of the IC. The existence of these neurons was predicted in barn owls by behavioral data using auditory orienting behavior as an assay for sound localization. Barn owls placed in a dark chamber and presented with relevant stimuli originating from different points in space will either fly toward the sound source or direct their gaze towards it (head saccade) (Konishi 1973b).

The association of specific values of auditory cues with appropriate locations in space occurs during the first 2 months of the barn owl's life (Knudsen 2002). Plasticity in the system is influenced by the age of barn owls, but the definition of a critical period may be less strict than originally thought. The lengths of sensitive and critical periods are influenced by environmental richness as well as by the rate at which cues change. Learning modifies the tuning to sound localization cues in both barn owls and ferrets. In both cases, visual signals shape the development of the auditory space map in the midbrain. Vision is reliable because the visual map is a projection of the retinal map, while the auditory map is computed from binaural cues (Konishi 1986; Knudsen 1999). Furthermore, visual calibration ensures that the neural representations of both sensory modalities share the same topographic organization (King 2002).

Visual calibration of the auditory space map has been carefully examined in barn owls. Barn owls have frontal eyes that do not move much. Therefore, when a sound source is centered in the auditory system, it is also centered in the visual system. Thus, visual cues can be used during development to calibrate the emergence of auditory space specific response so that they are in register with those in the visual field. Studies of how the visual system directs auditory space map formation have become a model for studies of instructive learning (Knudsen 2002).

7.1 Anatomical and Physiological Substrates of Sound Localizing Behavior

Barn owls can localize sounds to about 3° in both azimuth and elevation (Knudsen et al. 1979; Bala et al. 2003). This accuracy has made barn owls a model for studies of sound localization (Makous and Middlebrooks 1990; Knudsen 1999). Barn owl sound localization has also become a model for studies of how experience drives plasticity, because modifying the barn owl's sensory experience reorganizes the map (Knudsen 2002). The neural substrates for sound localization will be reviewed first, followed by studies of experience dependent plasticity.

Behavioral tests make use of barn owl's natural head saccade to center the target in its visual field (Knudsen et al. 1979). The dominant cues used for localizing sounds are binaural differences in the timing and level of sound at the two ears, referred to as ITD and ILD, respectively (Fig. 6.5). ITDs result primarily from the separation between the ears. Sounds originating from the left reach the left ear first, whereas sounds originating from the right reach the right ear first so that ITDs vary primarily with the horizontal (azimuthal) position of the stimulus. In the midbrain, ITDs are associated with appropriate locations in space (Olsen et al. 1989; Brainard and Knudsen 1993). ILDs result primarily from the acoustic collecting effects of the external ears or pinnae and the obstructing effect of the head on sounds propagating to the far ear. For birds with symmetrical ears, ILDs vary mainly with the azimuth of the stimulus (Volman and Konishi 1989). For animals with asymmetrical external ears, such as barn owls, ILDs at high frequencies vary also with the elevation of the stimulus, because they are efficiently collected by the external ears (Fig. 6.5). Encoded values of ILD become associated with appropriate locations in space in the midbrain (Brainard et al. 1992; Mogdans and Knudsen 1994b).

7.1.1 Tuning to Interaural Time Differences

ITD is the principal cue for auditory azimuth representation in birds (Moiseff 1989). There are two stages to ITD computation. First, laminaris neurons act as coincidence detectors to encode interaural phase difference, firing maximally when simultaneously stimulated by inputs from both ears (Carr and Konishi 1990). In this first stage, ambiguities exist with respect to the correspondence between the response in NL and the actual ITD in auditory space. The second

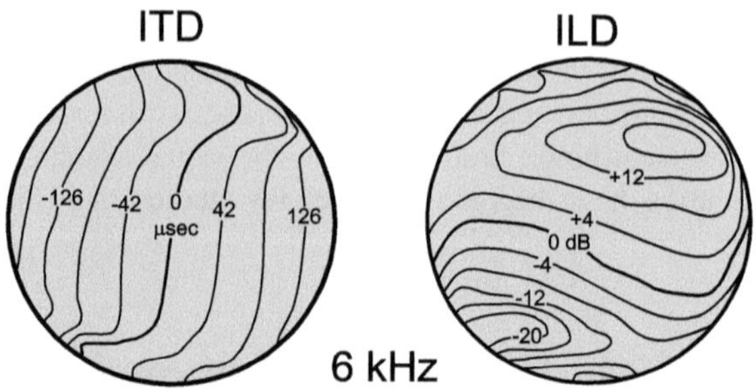

FIGURE 6.5. The relationship between auditory cue values and locations in space in barn owls. Contours in the space around a barn owl representing the locations of equivalent values for ITD and ILD cues at 6 kHz. The changes in ILD are associated with the separation of the ears, with negative values representing left ear leading and positive values representing right ear leading. The different ILD cues result from the asymmetry of the barn owl's ears with negative values representing sounds that are heard louder at the left ear and positive values sounds that are heard louder at the right ear. (From Knudsen 2002 with permission.)

stage of ITD computation occurs in the IC, where across-frequency integration filters phase-ambiguous side peaks, forming neurons that respond mainly to the true ITD (Takahashi and Konishi 1986; Peña and Konishi 2000).

The computation of ITD in the IC begins in the ICCc. In barn owls and in chickens, the core contains sharply frequency tuned ITD sensitive neurons with primary-like response types and similar responses to noise and tones (Coles and Aitkin 1979; Wagner et al. 1987). Barn owls' ICCc is best described as a matrix in which preferred interaural phase difference and frequency covary, so that a single ITD activates all constituent neurons in a set (Wagner et al. 1987). Thus, an ITD is conserved in a population of neurons, not in any single cell (Wagner et al. 1987). Each array projects to ITD and ILD-sensitive neurons in the contralateral lateral shell. These, in turn, project to space-specific neurons in the contralateral ICX, endowing the space-specific neuron with ITD selectivity and, therefore, azimuth coding (Mazer 1997). Ensembles representing ITDs corresponding to the ipsi- and contralateral auditory hemifields are found in the core and lateral shell of the IC, respectively. The lateral shell receives its representation of the contralateral hemifield from the opposite ICCc (Takahashi et al. 1989). The projection from the lateral shell to the ipsilateral ICX forms a map of contralateral space.

7.1.2 Tuning to Level and Interaural Level Difference (ILD) Processing

In barn owls, the vertical asymmetry in ear directionality makes ILD a cue for the vertical coordinate of a target at high frequencies (Fig. 6.5). Level is encoded by cochlear nucleus neurons (Köppl and Carr 2003). ILD sensitivity first emerges in the dorsal nucleus of the lateral lemniscus (LLDp), where neurons are excited by stimulation of the contralateral ear and inhibited by stimulation of the ipsilateral ear (Takahashi and Keller 1992). LLDp neurons thus exhibit EI responses, whose discharge rates are sigmoidal functions of ILD (Manley et al. 1988; Adolphs 1993; Takahashi et al. 1995). The LLDp is therefore similar to the mammalian LSO, except that the excitatory and inhibitory ears are reversed (Tsuchitani 1977; Takahashi et al. 1995). The LLDp appears to project bilaterally to the ICC lateral shell, endowing the neurons there with sensitivity to ILD (Adolphs 1993)

Monaural occlusion during development shifts the range of ILDs experienced by an animal and alters the correspondence of ILDs with source locations. The LLD is one site of plasticity underlying this adaptive adjustment (Mogdans and Knudsen 1994a). Sensitivity of units to the balance of excitatory to inhibitory influences is generally shifted in the adaptive direction. Nevertheless, the adjustment of ILD coding in the LLDp is smaller than expected based on the adjustment of ILD tuning in the OT measured in the same animals. This indicates the involvement of at least one additional site of adaptive plasticity in the auditory pathway (Mogdans and Knudsen 1994a).

LLDp neurons do not unambiguously encode ILD. Although they prefer sound at the contralateral ear, they are also sensitive to changes in average binaural level. ILD tuning gradually emerges in the ICCls. The distributions associated with space specificity are highly correlated with mediolateral position in the lateral shell. Frequency tuning widths broaden with increasing lateral position, while both ITD and ILD tuning widths sharpen. Almost all lateral shell neurons are sensitive to both time and intensity cues (Mazer 1997). Unlike the ITD coding array in the ICCc, a clear topographical representation of ILD is never observed in the ICCls (Mazer 1997). Recordings from the space specific neurons show instead that ILD varies as a function of frequency in a complex manner for any given location (Euston and Takahashi 2002). In ICX, ILD-alone receptive fields are generally horizontal swaths of activity at the elevation of the cell's normal spatial receptive field. An ITD-alone receptive field forms a vertical swath at the azimuth of the cell's normal receptive field, which thus lies at the intersection of the ITD and ILD-alone receptive fields (Euston and Takahashi 2002).

7.1.3 The Map of Auditory Space in the External Nucleus

The ICX space-specific neurons respond to sound only from a particular spatial locus (Knudsen and Konishi 1978; Knudsen and Knudsen 1983; Takahashi and Konishi 1986). The neurons are selective for combinations of ITD and ILD. Driven by noise, they do not show phase ambiguity, and thus differ from the ICCc ITD sensitive cells from which they receive their input (Peña and Konishi 2000). The phase-unambiguous response of space-specific neurons has been explained as follows: They receive inputs via the lateral shell from many ICCc isofrequency laminae (Knudsen 1983), presumably from the ITD-specific arrays (Wagner et al. 1987). These inputs interact at the postsynaptic cell, so that peaks signaling the correct ITD superimpose and add, while secondary, ambiguous peaks cancel by interacting with inhibitory sidebands from other or ambiguous frequencies (Takahashi and Konishi 1986).

The spatially restricted ICX receptive fields are still much larger than the minimum detectable change in sound source location. Comparisons of neuronal activity in the space map with perceptual acuity show that most neurons can reliably signal changes in source location that are smaller than the behavioral threshold. Each source is represented in the space map by a focus of activity in a population of neurons, and source displacement changes the pattern of activity in this population (Bala et al. 2003). This map of contralateral auditory space projects topographically to the OT, whose maps of visual and auditory space are in register (Fig. 6.6) (Knudsen and Knudsen 1983). Tectal activity directs the rapid head movements made by barn owls to auditory and visual stimuli (du Lac and Knudsen 1990).

7.1.4 Experience-Dependent Calibration of Sound Localization

Manipulation of a barn owl's sensory experience reorganizes the map consistent with behavioral learning. Barn owl's accurate auditory orienting behavior measures adaptive adjustment of sound localization after changes in sensory experience, such as disrupting auditory cue values with spatial loci by plugging an ear (Knudsen et al. 1979, 1984a; 1984b; Knudsen 1999). Barn owls with earplugs first err in localizing sounds toward the open ear, then recover accurate orientation. On earplug removal, barn owls make orienting errors in the opposite direction, but these soon disappear with normal experience. Thus, earplugs induce new associations between auditory cues and spatial locations. A second manipulation rearranges the correspondence between visual field and auditory cues by fitting barn owls with prismatic spectacles to displace the visual field (Brainard and Knudsen 1998). Barn owls with prisms learn new associations between auditory and visual cues to recalibrate their auditory and visual worlds.

In the OT, bimodal neurons are driven by visual and auditory stimuli and the functional determination of a neuron's location in the tectum is made accurately with reference to the visual receptive field, which is unaltered. Adaptive changes are centered in the ICX, and depend on changes in axonal projections and adjustments in synaptic strength (Gold and Knudsen 2000; Knudsen 2002). In young barn owls that have experienced either abnormal hearing or prismatic displacement of the visual field, neural tuning of neurons in ICX to sound localization cues is altered adaptively to coordinate the auditory receptive field and the visual receptive field (Brainard and Knudsen 1993). Gradually, these "learned responses" strengthen, while those to the prior ITD range, termed "normal responses," disappear over time.

Work from the Knudsen laboratory has identified three mechanisms to mediate this plasticity: axonal remodeling, and changes in NMDA and GABA receptor expression (Knudsen 2002). Receptive field changes are correlated with axonal remodeling of the topographic projection from the ICC to the ICX. Prism experience appears to induce the formation of learned circuitry in the ICX at least in part through axonal sprouting and synaptogenesis (Fig. 6.6). Normal circuitry also persists, showing that alternative learned and normal circuits can coexist in this network. Both NMDA and GABA receptor changes also appear to mediate plastic changes. NMDA receptors are crucial in the expression of newly learned responses (Feldman and Knudsen 1994). $GABA_A$ receptors also contribute to functional plasticity. In an ICX that is expressing a maximally shifted ITD map, focal application of a GABA antagonist elicits the immediate appearance of normal responses. Thus, in a shifted ITD map, synapses that support normal responses remain patent and coexist with synapses that support learned responses, but responses to the normal synapses are selectively nullified by GABAergic inhibition (Zheng and Knudsen 1999).

7.1.5 The Visual Instructive Signal

Changes in the ICX space map are directed by a visual instructive signal of tectal origin. The dominance of visual input is plausible, in that the pathway's primary function is to initiate gaze towards auditory targets (Knudsen et al. 1993; Wagner 1993). The instructive signal that shapes the auditory space map is a topographic template of visual space. The nature of the instructive signal is shown when a small lesion placed in the tectum eliminates adaptive plasticity in the corresponding portion of the auditory space map in the ICX, while the rest of the auditory map continues to shift adaptively in response to experience (Hyde and Knudsen 2002).

Evidence that vision guides the formation of the space map is threefold. First, barn owls raised blind end up with auditory space maps that are not normal (Knudsen and Brainard 1995). Second, erroneous visual signals shift the space map (Knudsen and Brainard 1991). Third, there is a projection from the tectum to the space map. The brain can generate an auditory space map without vision, but the normal precision and topography of the map depend on visual experience. Not only does vision guide map formation, but vision wins when there is a mismatch between visual and auditory cues. In a classic experiment, Knudsen and Brainard (1991) raised young barn owls with displacing prisms mounted in spectacle frames in front of the eyes. The prisms provided erroneous visual signals that caused the perfectly good auditory map to shift. Thus, vision exerts an overriding influence on auditory map organization.

Because visual experience calibrates the auditory space map, there must be an instructive signal from the visual system. Recent reports have demonstrated a point-to-point feedback projection from the OT to the auditory space map in the external nucleus (Luksch et al. 2000; Hyde and Knudsen 2001). This projection forms even before barn owls hatch, and the projecting neurons have dendrites that extend into the superficial tectal layers, which receive direct input from the retina, and others that extend into the deep tectal layers, which receive feedforward auditory input from the auditory space map, and visual input from

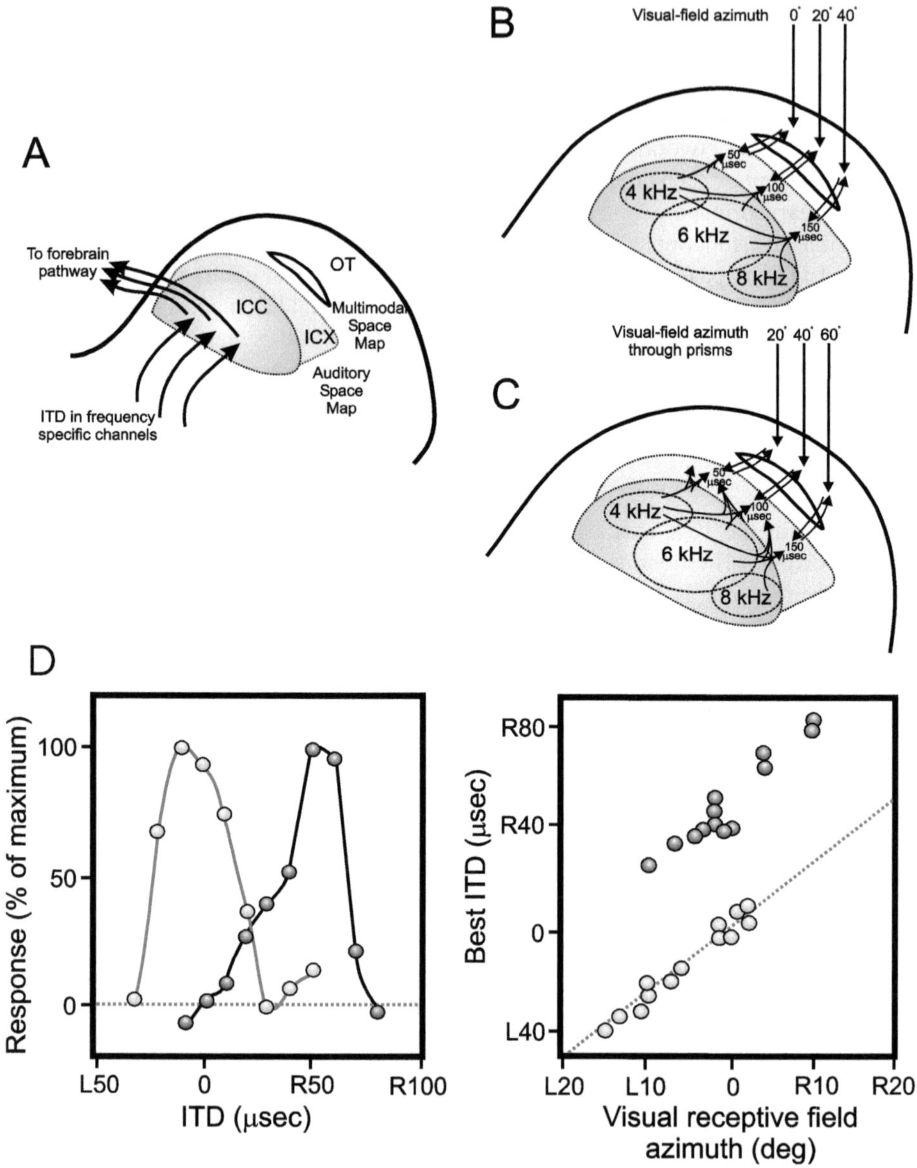

FIGURE 6.6. The map of auditory space in the external nucleus before (**B**) and after (**C**) shifts produced by early prism experience. (**A**) Information proceeds from the ICC to the ICX in the midbrain pathway. In the projection from the ICC to the ICX, information about cue values, including ITD, converges across frequency channels, resulting in spatially restricted auditory receptive fields and a map of space. (**B**) The auditory map of space merges with a visual map of space in the OT, and receives an instructive signal back from the tectum. (**C**) *Arrows* mark the schematic representation of the change in the anatomical projection from the ICC to the ICX that results from early prism

the forebrain. The dominant instructive signal that shapes the auditory space map is a topographic template of visual space. The power and precision of the instructive signal are shown when a small lesion placed in the tectum eliminates adaptive plasticity in the corresponding portion of the auditory space map in the ICX, while the rest of the auditory map continues to shift adaptively in response to experience (Hyde and Knudsen 2002).

Despite the demonstrated existence of a projection from tectum to space map, it proved difficult to unmask physiologically. Strong visual responses, appropriate to guide auditory plasticity, were only revealed in the ICX when inhibition was blocked in the OT (Gutfreund et al. 2002). The visual instructive signal may act as follows: Visual activity would not excite ICX neurons if they have just been activated strongly by an auditory stimulus. Thus, the visual activity arising from a bimodal stimulus does not interfere with auditory processing as long as sound localization cues are represented correctly in the space map. If the visual and auditory maps in the ICX were misaligned, visual activity from bimodal stimuli could excite ICX neurons. This excitation could cause strengthening of auditory inputs that were active in the recent past and thereby adjust the representation of auditory localization cues in the ICX to match the visual representation of space. Since the instructive visual signals are normally gated off in anesthetized barn owls, Gutfreund et al. (2002) suggest that cognitive and/or attentional processes may control the gate.

7.1.6 Plasticity in Other Auditory Pathways

In barn owls' forebrain localization pathway, cue information is combined across frequency channels beyond the level of the primary auditory field to create clustered representations of space (Cohen and Knudsen 1999). Both auditory and visual manipulations cause adaptive changes in the tuning of these forebrain neurons to sound localization cues (Miller and Knudsen 2001). This plasticity emerges in parallel with that observed in the space map and does not derive from it (Cohen et al. 1998).

The auditory thalamus of juvenile barn owls is also plastic (Miller and Knudsen 2003). Juvenile barn owls that experience chronic abnormal ITD cues ex-

FIGURE 6.6. *Continued*
experience. For each direction of prismatic displacement, an abnormal rostralward projection of ICC axons appears on one side of the brain and a caudalward projection appears on the other. In addition, the normal projection persists. (Based on Knudsen 2002.) (**D**) ITD tuning of tectal neurons before and after prismatic visual displacement for neurons with receptive fields centered at 0° azimuth. Before prism experience (*light symbols*), the neuron's maximal response is centered at 0 ms ITD. Barn owls raised with prisms that displace the visual field by L23° for 8 weeks adjust their ITD tuning in the auditory space map (*dark symbols*). (**C**) The relationship between best ITD and visual receptive field azimuth is also shifted (*dark symbols*) systematically from normal (*light symbols*). (Modified from Knudsen et al. 2000. Copyright 2000 National Academy of Sciences, USA.)

hibit adaptive, frequency-dependent shifts in the tuning of thalamic neurons to ITDs (Miller and Knudsen 2003). Abnormal hearing does not alter ITD tuning in the central nucleus of the IC, the primary source of input to the auditory thalamus. Therefore, the thalamus is the earliest stage in the forebrain pathway in which this plasticity is expressed. A visual manipulation similar to those described above, where prisms cause displacement of the visual field, and which leads to adaptive changes in ITD tuning at higher levels in the forebrain, has no effect on thalamic ITD tuning. The results demonstrate that, during the juvenile period, auditory experience shapes neuronal response properties in the thalamus in a frequency-specific manner.

The plasticity in the auditory thalamus is not a result of the plasticity in the space map (Feldman and Knudsen 1998; Zheng and Knudsen 1999; DeBello et al. 2001). Experience with prism spectacles causes large adaptive changes in ITD tuning in the space map. In contrast, experience with prism spectacles does not affect ITD tuning in the nucleus ovoidalis (nOv). Therefore, ITD tuning in the nOv is calibrated independently of ITD tuning in the ICX. This is consistent with the failure of anatomic studies to find connections from the ICX to the nOv (Proctor and Konishi 1997; Cohen et al. 1998) and with previous results, indicating that these two pathways analyze and interpret sound localization cues in parallel (Wagner 1993; Knudsen and Knudsen 1996; Cohen et al. 1998).

7.2 Development of Sound Localization in Mammals

Barn owls are not the only animal to show experience-dependent plasticity of its auditory maps. Ferrets raised and tested with one ear plugged learn to localize as accurately as control animals (King et al. 2001), which is consistent with previous findings that the representation of auditory space in the midbrain can accommodate abnormal sensory cues during development. Adaptive changes in behavior are also observed in adults, particularly if they were provided with regular practice in the localization task. Together, these findings suggest that the neural circuits responsible for sound localization can be recalibrated throughout life (King et al. 2001).

Work in ferrets also shows that instructive visual signals guide the development of an auditory map (King et al. 1998; Doubell et al. 2000). The neural substrates are not identical to the owl, although similar algorithms apply. Projections from the superficial layers of the superior colliculus provide the source of visual signals to guide the development of the auditory space map in the deeper layers of the superior colliculus. Lesions of visual map in young ferrets lead to degraded auditory representations, while equivalent lesions in adult ferrets do not alter the topographic order in the auditory representation, suggesting that visual activity in these layers may be involved in aligning the different sensory maps in the developing superior colliculus. Additional evidence comes from experiments in which surgical deviation of one eye in infancy leads to a compensatory shift in the auditory representation in the opposite superior colliculus, thereby preserving the alignment of the two maps (King et al. 1988).

Developmental sharpening of spatial tuning is also found in auditory cortex, but appears to depend largely on the maturation of the head and ears (Mrsic-Flogel et al. 2003). In young ferrets, spatial response fields of cortical neurons are broader, and transmit less information about stimulus direction, than in older ferrets. However, when infant neurons are stimulated through virtual ears of adults, the spatial receptive field sharpen significantly and the amount of transmitted information increases. Thus, the sharpening of spatial tuning in auditory cortex may depend on peripheral rather than by central factors (Mrsic-Flogel et al. 2003).

7.3 Summary

Developmental studies in barn owls have revealed anatomical and pharmacological changes correlated with experience dependent plasticity. The major results of developmental studies, largely from the Knudsen laboratory, have shown an inherent advantage of innate neuronal connections over connections that are acquired with learning, a decline in learning with age, and an increased capacity for learning in adults that have had appropriate experience as juveniles (Knudsen 2002).

8. Summary and Conclusions

The developmental processes that lay down the basic plan of the auditory circuit are followed by plastic modifications that fine-tune these connections to suitably adapt to the experience of each individual. Birds and mammals share common features in the basic developmental plan by which the morphological and biochemical features of the auditory system emerge generally following the future tonotopic axis. The neuronally computed auditory space map results from the association of binaural cues with specific locations in space of the sound source. Since binaural information will be modified by the characteristics of each individual (head size and shape, for example) the auditory system must rely on the ability of the nervous system to adapt the basic connectivity plan to the characteristics of each individual. As such, accurate associations between binaural cues and space positions can only be assigned after these characteristics have developed, a mechanism that involves experience-dependent plasticity.

Abbreviations

ABR	auditory brain stem response
AMPA	α-amino-3-hydroxy-5-methyl-4-isoxazolepropionic acid

AVCN	anterior ventral cochlear nucleus
BDNF	brain derived neurotrophic factor
BF	best frequency
CaBP	calcium binding proteins
CAP	compound action potential
CM	cochlear microphonic
CR	calretinin
DCN	dorsal cochlear nucleus
DNLL	dorsal nucleus of the lateral lemniscus
EPSP	excitatory postsynaptic potential
GABA	γ-aminobutyric acid
HTC	high threshold conductance
IC	inferior colliculus
ICC	central nucleus of the inferior colliculus
ICCc	core of the central nucleus of the inferior colliculus
ICCls	lateral shell of the central nucleus of the inferior colliculus
ICX	external nucleus of the inferior colliculus
IHC	inner hair cells
ILD	interaural level differences
IPSP	inhibitory postsynaptic potential
ITD	interaural time differences
LLDa	anterior portion of the dorsal nucleus of the lateral lemniscus
LLDp	posterior portion of the dorsal nucleus of the lateral lemniscus
LLI	intermediate nucleus of the lateral lemniscus
LSO	lateral superior olive
LTC	low threshold conductance
MNTB	medial nucleus of the trapezoid body
MSO	medial superior olive
NA	nucleus angularis
NL	nucleus laminaris
NLL	nuclei of the lateral lemniscus
NM	nucleus magnocellularis
NMDA	*N*-methyl-D-aspartate
NT-3	neurotrophic factor 3
nOv	nucleus ovoidalis
OHC	outer hair cells
OT	optic tectum
PVCN	posterior ventral cochlear nucleus
SO	superior olive
SOC	superior olivary complex
VCN	ventral cochlear nucleus
VNLL	ventral nucleus of the lateral lemniscus

References

Adolphs R (1993) Bilateral inhibition generates neuronal responses tuned to interaural level differences in the auditory brainstem of the barn owl. J Neurosci 13:3647–3668.

Agarwala S, Ragsdale CW (2002) A role for midbrain arcs in nucleogenesis. Development 129:5779–5788.

Agmon-Snir H, Carr CE, Rinzel J (1998) The role of dendrites in auditory coincidence detection. Nature 393:268–272.

Aleksandrov LI, Dmitrieva LP (1992) Development of auditory sensitivity of altricial birds: absolute thresholds of the generation of evoked potentials. Neurosci Behav Physiol 22:132–137.

Alladi PA (2002) Prenatal auditory stimulation: effects on programmed cell death and synaptogenesis in chick auditory nuclei [PhD Thesis]. Delhi: All India Institute of Medical Sciences.

Altman J, Bayer SA (1980) Development of the brain stem in the rat. III. Thymidine-radiographic study of the time of origin of neurons of the vestibular and auditory nuclei of the upper medulla. J Comp Neurol 194:877–904.

Alvarado JC, Fuentes-Santamaria V, Henkel CK, Brunso-Bechtold JK (2004) Alterations in calretinin immunostaining in the ferret superior olivary complex after cochlear ablation. J Comp Neurol 470:63–79.

Ariëns Kappers CU, Huber GC, Crosby EC (1936) The comparative anatomy of the nervous system of vertebrates, including man. New York: The Macmillan Company.

Baimbridge KG, Celio MR, Rogers JH (1992) Calcium-binding proteins in the nervous system. Trends Neurosci 15:303–308.

Bajo VM, Merchan MA, Lopez DE, Rouiller EM (1993) Neuronal morphology and efferent projections of the dorsal nucleus of the lateral lemniscus in the rat. J Comp Neurol 334:241–262.

Baker CV, Bronner-Fraser M (2001) Vertebrate cranial placodes. I. Embryonic induction. Dev Biol 232:1–61.

Bal R, Oertel D (2000) Hyperpolarization-activated, mixed-cation current (I(h)) in octopus cells of the mammalian cochlear nucleus. J Neurophysiol 84:806–817.

Bala AD, Spitzer MW, Takahashi TT (2003) Prediction of auditory spatial acuity from neural images on the owl's auditory space map. Nature 424:771–774.

Bally-Cuif L, Wassef M (1995) Determination events in the nervous system of the vertebrate embryo. Curr Opin Genet Dev 5:450–458.

Bell CC (2002) Evolution of cerebellum-like structures. Brain Behav Evol 59:312–326.

Blatchley BJ, Brugge JF (1990) Sensitivity to binaural intensity and phase difference cues in kitten inferior colliculus. J Neurophysiol 64:582–597.

Book KJ, Morest DK (1990) Migration of neuroblasts by perikaryal translocation: role of cellular elongation and axonal outgrowth in the acoustic nuclei of the chick embryo medulla. J Comp Neurol 297:55–76.

Borges K, Dingledine R (1998) AMPA receptors: molecular and functional diversity. Prog Brain Res 116:153–170.

Brainard MS, Knudsen EI (1993) Experience-dependent plasticity in the inferior colliculus: a site for visual calibration of the neural representation of auditory space in the barn owl. J Neurosci 13:4589–4608.

Brainard MS, Knudsen EI (1998) Sensitive periods for visual calibration of the auditory space map in the barn owl optic tectum. J Neurosci 18:3929–3942.

Brainard MS, Knudsen EI, Esterly SD (1992) Neural derivation of sound source location: resolution of spatial ambiguities in binaural cues. J Acoust Soc Am 91:1015–1027.
Brand A, Behrend O, Marquardt T, McAlpine D, Grothe B (2002) Precise inhibition is essential for microsecond interaural time difference coding. Nature 417:543–547.
Braun K (1990) Calcium binding proteins in avian and mammalian central nervous system: localization, development and possible functions. Prog Histochem Cytochem 21: 1–62.
Brew HM, Forsythe ID (1995) Two voltage-dependent K^+ conductances with complementary functions in postsynaptic integration at a central auditory synapse. J Neurosci 15:8011–8022.
Brittan-Powell EF, Dooling RJ (2004) Development of auditory sensitivity in budgerigars (*Melopsittacus undulatus*). J Acoust Soc Am 115:3092–3102.
Brugge JF, Javel E, Kitzes LM (1978) Signs of functional maturation of peripheral auditory system in discharge patterns of neurons in anteroventral cochlear nucleus of kitten. J Neurophysiol 41:1557–1559.
Caicedo A, Eybalin M (1999) Glutamate receptor phenotypes in the auditory brainstem and midbrain of the developing rat. Eur J Neurosci 11:51–74.
Caicedo A, d'Aldin C, Eybalin M, Puel J-L (1997) Temporary sensory deprivation changes calcium-binding proteins levels in the auditory brainstem. J Comp Neurol 378:1–15.
Cambronero F, Puelles L (2000) Rostrocaudal nuclear relationships in the avian medulla oblongata: a fate map with quail chick chimeras. J Comp Neurol 427:522–545.
Cant NB (1998) Structural development of the mammalian central auditory pathways. In: Rubel EW, Popper AN, Fay RR (eds), Development of the Auditory System, New York: Springer-Verlag, pp. 315–413.
Cant NB, Benson CG (2003) Parallel auditory pathways: projection patterns of the different neuronal populations in the dorsal and ventral cochlear nuclei. Brain Res Bull 60:457–474.
Carr CE, Boudreau RE (1993) Organization of the nucleus magnocellularis and the nucleus laminaris in the barn owl: encoding and measuring interaural time differences. J Comp Neurol 334:337–355.
Carr CE, Boudreau RE (1996) Development of the time coding pathways in the auditory brain stem of the barn owl. J Comp Neurol 373:467–483.
Carr CE, Konishi M (1990) A circuit for detection of interaural time differences in the brainstem of the barn owl. J Neurosci 10:3227–3246.
Carr CE, Fujita I, Konishi M (1989) Distribution of GABAergic neurons and terminals in the auditory system of the barn owl. J Comp Neurol 286:190–207.
Carr CE, Kubke MF, Massoglia DP, Cheng SM, Rigby L, Moiseff A (1998) Development of temporal coding circuits in the barn owl. In: Palmer AR, Rees A, Summerfield AQ, Meddis R, (eds), Psychophysical and Physiological Advances in Hearing. London: Whurr, pp. 344–351.
Celio MR, Pauls T, Schwaller B (1996) Guidebook to the Calcium-Binding Proteins. New York: Oxford University Press.
Chen L, Trautwein PG, Shero M, Salvi RJ (1996) Tuning, spontaneous activity and tonotopic map in chicken cochlear ganglion neurons following sound-induced hair cell loss and regeneration. Hear Res 98:152–164.
Code RA, Churchill L (1991) $GABA_A$ receptors in auditory brainstem nuclei of the chick during development and after cochlea removal. Hear Res 54:281–295.

Code RA, Rubel EW (1989) Glycine-immunoreactivity in the auditory brain stem of the chick. Hear Res 40:167–172.
Code RA, Burd GD, Rubel EW (1989) Development of GABA immunoreactivity in brainstem auditory nuclei of the chick: ontogeny of gradients in terminal staining. J Comp Neurol 284:504–518.
Cohen YE, Knudsen EI (1999) Maps versus clusters: different representations of auditory space in the midbrain and forebrain. Trends Neurosci 22:128–135.
Cohen YE, Miller GL, Knudsen EI (1998) Forebrain pathway for auditory space processing in the barn owl. J Neurophysiol 79:891–902.
Coles R, Aitkin L (1979) The response properties of auditory neurones in the midbrain of the domestic fowl (*Gallus gallus*) to monaural and binaural stimuli. J Comp Physiol 134:241–251.
Conlee JW, Parks TN (1983) Late appearance and deprivation-sensitive growth of permanent dendrites in the avian cochlear nucleus (nuc. magnocellualris). J Comp Neurol 217:216–226.
Cramer K, Fraser S, Rubel E (2000) Embryonic origins of auditory brain-stem nuclei in the chick hindbrain. Dev Biol 224:138–151.
Cramer KS, Karam SD, Bothwell M, Cerretti DP, Pasquale EB, Rubel EW (2002) Expression of EphB receptors and EphrinB ligands in the developing chick auditory brainstem. J Comp Neurol 452:51–64.
Cramer KS, Bermingham-McDonogh O, Krull CE, Rubel EW (2004) EphA4 signaling promotes axon segregation in the developing auditory system. Dev Biol 269:26–35.
Crowley D, Hepp-Reymond M (1966) Development of cochlear function in the ear of the infant rat. J Comp Physiol Psychol 62:427–432.
D'Amico-Martel A, Noden D (1983) Contribution of placodal and neural crest cells to to avian cranial peripheral ganglia. Am J Anat 166:445–468.
DeBello WM, Feldman DE, Knudsen EI (2001) Adaptive axonal remodeling in the midbrain auditory space map. J Neurosci 21:3161–3174.
Doubell TP, Baron J, Skaliora I, King AJ (2000) Topographical projection from the superior colliculus to the nucleus of the brachium of the inferior colliculus in the ferret: convergence of visual and auditory information. Eur J Neurosci 12:4290–4308.
du Lac S, Knudsen EI (1990) Neural maps of head movement vector and speed in the optic tectum of the barn owl. J Neurophysiol 63:131–146.
Eberhart J, Barr J, O'Connell S, Flagg A, Swartz ME, Cramer KS, Tosney KW, Pasquale EB, Krull CE (2004) Ephrin-A5 exerts positive or inhibitory effects on distinct subsets of EphA4-positive motor neurons. J Neurosci 24:1070–1078.
Echteler SM, Arjmand E, Dallos P (1989) Developmental alterations in the frequency map of the mammalian cochlea. Nature 341:147–149.
Edmonds JL Jr, Hoover LA, Durham D (1999) Breed differences in deafferentation-induced neuronal cell death and shrinkage in chick cochlear nucleus. Hear Res 127: 62–76.
Ehrlich I, Lohrke S, Friauf E (1999) Shift from depolarizing to hyperpolarizing glycine action in rat auditory neurones is due to age-dependent Cl^- regulation. J Physiol 520 Pt 1:121–137.
Euston DR, Takahashi TT (2002) From spectrum to space: the contribution of level difference cues to spatial receptive fields in the barn owl inferior colliculus. J Neurosci 22:284–293.
Fariñas I, Jones KR, Tessarollo L, Vigers AJ, Huang E, Kirstein M, de Caprona DC,

Coppola V, Backus C, Reichardt LF, Fritzsch B (2001) Spatial shaping of cochlear innervation by temporally regulated neurotrophin expression. J Neurosci 21:6170–6180.

Fekete DM, Wu DK (2002) Revisiting cell fate specification in the inner ear. Curr Opin Neurobiol 12:35–42.

Feldman DE, Knudsen EI (1994) NMDA and non-NMDA glutamate receptors in auditory transmission in the barn owl inferior colliculus. J Neurosci 14:5939–5958.

Feldman DE, Knudsen EI (1998) Experience-dependent plasticity and the maturation of glutamatergic synapses. Neuron 20:1067–1071.

Feldman DE, Brainard MS, Knudsen EI (1996) Newly learned auditory responses mediated by NMDA receptors in the owl inferior colliculus. Science 271:525–528.

Ferreiro B, Skoglund P, Bailey A, Dorsky R, Harris WA (1992) XASH1, a *Xenopus* homolog of achaete-scute: a proneural gene in anterior regions of the vertebrate CNS. Mech Dev 40:25–36.

Friauf E (1993) Transient appearance of calbindin-D28k-positive neurons in the superior olivary complex of developing rats. J Comp Neurol 334:59–74.

Friauf E (1994) Distribution of calcium-binding protein calbindin-D28K in the auditory system of adult and developing rats. J Comp Neurol 349:193–211.

Friauf E, Kandler K (1990) Auditory projections to the inferior colliculus of the rat are present by birth. Neurosci Lett 120:58–61.

Friauf E, Hammerschmidt B, Kirsch J (1997) Development of adult-type inhibitory glycine receptors in the central auditory system of rats. J Comp Neurol 385:117–134.

Friedman MA, Kawasaki M (1997) Calretinin-like immunoreactivity in mormyrid and gymnarchid electrosensory and electromotor systems. J Comp Neurol 387:341–357.

Fritzsch B (2003) Development of inner ear afferent connections: forming primary neurons and connecting them to the developing sensory epithelia. Brain Res Bull 60: 423–433.

Fritzsch B, Silos-Santiago I, Bianchi LM, Fariñas I (1997) The role of neurotrophic factors in regulating the development of inner ear innervation. Trends Neurosci 20: 159–164.

Fritzsch B, Beisel KW, Jones K, Fariñas I, Maklad A, Lee J, Reichardt LF (2002) Development and evolution of inner ear sensory epithelia and their innervation. J Neurobiol 53:143–156.

Gabriele ML, Brunso-Bechtold JK, Henkel CK (2000a) Plasticity in the development of afferent patterns in the inferior colliculus of the rat after unilateral cochlear ablation. J Neurosci 20:6939–6949.

Gabriele ML, Brunso-Bechtold JK, Henkel CK (2000b) Development of afferent patterns in the inferior colliculus of the rat: projection from the dorsal nucleus of the lateral lemniscus. J Comp Neurol 416:368–382.

Geiger JRP, Melcher T, Koh D-S, Sakmann B, Seeburg PH, Jonas P, Monyer H (1995) Relative abundance of subunit mRNAs determines gating and Ca^{2+} permeability of AMPA receptors in principal neurons and interneurons of rat CNS. Neuron 15:193–204.

Gerstner W, Kempter R, van Hemmen JL, Wagner H (1996) A neuronal learning rule for sub-millisecond temporal coding. Nature 383:76–81.

Gold JI, Knudsen EI (2000) Abnormal auditory experience induces frequency-specific adjustments in unit tuning for binaural localization cues in the optic tectum of juvenile owls. J Neurosci 20:862–877.

Grigg JJ, Brew HM, Tempel BL (2000) Differential expression of voltage-gated potassium channel genes in auditory nuclei of the mouse brainstem. Hear Res 140:77–90.

Grothe B (2000) The evolution of temporal processing in the medial superior olive, an auditory brainstem structure. Prog Neurobiol 61:581–610.

Gu X, Spitzer NC (1995) Distinct aspects of neuronal differentiation encoded by frequency of spontaneous Ca^{2+} transients. Nature 375:784–787.

Gutfreund Y, Zheng W, Knudsen EI (2002) Gated visual input to the central auditory system. Science 297:1556–1559.

Hack NJ, Wride MC, Charters KM, Kater SB, Parks TN (2000) Developmental changes in the subcellular localization of calretinin. J Neurosci 20:RC67.

Haresign T, Moiseff A (1988) Early growth and development of the common barn-owl's facial ruff. The Auk 105:699–705.

Harkmark W (1954) Cell migrations from the rhombic lip to the inferior olive, the nucleus raphe and the pons. A morphological and experimental investigation on chick embryos. J Comp Neurol 100:115–209.

Hausler UH, Sullivan WE, Soares D, Carr CE (1999) A morphological study of the cochlear nuclei of the pigeon (*Columba livia*). Brain Behav Evol 54:290–302.

Hendriks R, Morest DK, Kaczmarek LK (1999) Shaw-like potassium currents in the auditory rhombencephalon throughout embryogenesis. J Neurosci Res 58:791–804.

Henkel C, Brunso-Brechtold JK (1998) Calcium-binding proteins and GABA reveal spatial segregation of cell types within the developing lateral superior olivary nucleus of the ferret. Microsc Res Tech 41:234–245.

Hyde PS, Knudsen EI (2001) A topographic instructive signal guides the adjustment of the auditory space map in the optic tectum. J Neurosci 21:8586–8593.

Hyde PS, Knudsen EI (2002) The optic tectum controls visually guided adaptive plasticity in the owl's auditory space map. Nature 415:73–76.

Hyson RL, Rudy JW (1987) Ontogenetic change in the analysis of sound frequency in the infant rat. Dev Psychobiol 20:189–207.

Jackson H, Parks TN (1982) Functional synapse elimination in the developing avian cochlear nucleus with simultaneous reduction in cochlear nerve axon branching. J Neurosci 2:1736–1743.

Jackson H, Parks T (1988) Induction of aberrant functional afferents to the chick cochlear nucleus. J Comp Neurol 271:106–114.

Jackson H, Rubel EW (1978) Ontogeny of behavioral responsiveness to sound in the chick embryo as indicated by electrical recordings of motility. J Comp Physiol Psychol 92:682–696.

Jackson H, Hackett JT, Rubel EW (1982) Organization and development of brain stem auditory nuclei in the chick: ontogeny of postsynaptic responses. J Comp Neurol 210: 80–86.

Jackson H, Nemeth EF, Parks TN (1985) Non-*N*-methyl-*d*-aspartate receptors mediating synaptic transmission in the avian cochlear nucleus: effects of kynurenic acid, dipicolinic acid and streptomycin. Neuroscience 16:171–179.

Jasoni CL, Walker MB, Morris MD, Reh TA (1994) A chicken achaete-scute homolog (CASH-1) is expressed in a temporally and spatially discrete manner in the developing nervous system. Development 120:769–783.

Jeffress L (1948) A place theory of sound localization. J Comp Physiol Psychol 41:35–39.

Jhaveri S, Morest DK (1982) Sequential alterations of neuronal architecture in nucleus magnocellularis of the developing chicken: a Golgi study. Neuroscience 7:837–853.

Jones SM, Jones TA (1995) The tonotopic map in the embryonic chicken cochlea. Hear Res 82:149–157.

Jones TA, Jones SM (2000) Spontaneous activity in the statoacoustic ganglion of the chicken embryo. J Neurophysiol 83:1452–1468.

Kakazu Y, Akaike N, Komiyama S, Nabekura J (1999) Regulation of intracellular chloride by cotransporters in developing lateral superior olive neurons. J Neurosci 19: 2843–2851.

Kandler K, Friauf E (1993) Pre- and postnatal development of efferent connections of the cochlear nucleus in the rat. J Comp Neurol 328:161–184.

Kandler K, Friauf E (1995a) Development of electrical membrane properties and discharge characteristics of superior olivary complex neurons in fetal and postnatal rats. Eur J Neurosci 7:1773–1790.

Kandler K, Friauf E (1995b) Development of glycinergic and glutamatergic synaptic transmission in the auditory brainstem of perinatal rats. J Neurosci 15:6890–6904.

Kapfer C, Seidl AH, Schweizer H, Grothe B (2002) Experience-dependent refinement of inhibitory inputs to auditory coincidence-detector neurons. Nat Neurosci 5:247–253.

Karten H (1967) The organization of the ascending auditory pathway in the pigeon (*Columba livia*). I. Diencephalic projections of the inferior colliculus (nucleus mesencephali lateralis, pars dorsalis). Brain Res 6:409–427.

Katayama A (1985) Postnatal development of auditory function in the chicken revealed by auditory brain-stem responses (ABRs). Electroencephalogr Clin Neurophysiol 62: 388–398.

Kempter R, Gerstner W, van Hemmen JL (1998) How the threshold of a neuron determines its capacity for coincidence detection. Biosystems 48:105–112.

Keynes R, Lumsden A (1990) Segmentation and the origin of regional diversity in the vertebrate central nervous system. Neuron 4:1–9.

Kil J, Kageyama GH, Semple MN, Kitzes LM (1995) Development of ventral cochlear nucleus projections to the superior olivary complex in gerbil. J Comp Neurol 353: 317–340.

King AJ (2002) Neural plasticity: how the eye tells the brain about sound location. Curr Biol 12:R393–395.

King AJ, Hutchings ME, Moore DR, Blakemore C (1988) Developmental plasticity in the visual and auditory representations in the mammalian superior colliculus. Nature 332:73–76.

King AJ, Schnupp JW, Thompson ID (1998) Signals from the superficial layers of the superior colliculus enable the development of the auditory space map in the deeper layers. J Neurosci 18:9394–9408.

King AJ, Kacelnik O, Mrsic-Flogel TD, Schnupp JW, Parsons CH, Moore DR (2001) How plastic is spatial hearing? Audiol Neurootol 6:182–186.

Kleinschmidt A, Bear MF, Singer W (1987) Blockade of "NMDA" receptors disrupts experience-dependent plasticity of kitten striate cortex. Science 238:355–358.

Knowlton VY (1967) Correlation of the development of membranous and bony labyrinths, acoustic ganglia, nerves, and brain centers of the chick embryo. J Morphol 121:179–208.

Knudsen EI (1983) Subdivisions of the inferior colliculus in the barn owl (*Tyto alba*). J Comp Neurol 218:174–186.

Knudsen EI (1999) Mechanisms of experience-dependent plasticity in the auditory localization pathway of the barn owl. J Comp Physiol A 185:305–321.

Knudsen EI (2002) Instructed learning in the auditory localization pathway of the barn owl. Nature 417:322–328.
Knudsen EI, Brainard MS (1991) Visual instruction of the neural map of auditory space in the developing optic tectum. Science 253:85–87.
Knudsen EI, Brainard MS (1995) Creating a unified representation of visual and auditory space in the brain. Annu Rev Neurosci 18:19–43.
Knudsen EI, Knudsen PF (1983) Space-mapped auditory projections from the inferior colliculus to the optic tectum in the barn owl (*Tyto alba*). J Comp Neurol 218:187–196.
Knudsen EI, Knudsen PF (1996) Disruption of auditory spatial working memory by inactivation of the forebrain archistriatum in barn owls. Nature 383:428–431.
Knudsen EI, Konishi M (1978) A neural map of auditory space in the owl. Science 200: 795–797.
Knudsen EI, Blasdel GG, Konishi M (1979) Sound localization by the barn owl (*Tyto alba*) measured with the search coil technique. J Comp Physiol 133:1–11.
Knudsen EI, Esterly SD, Knudsen PF (1984a) Monaural occlusion alters sound localization during a sensitive period in the barn owl. J Neurosci 4:1001–1011.
Knudsen EI, Knudsen PF, Esterly SD (1984b) A critical period for the recovery of sound localization accuracy following monaural occlusion in the barn owl. J Neurosci 4: 1012–1020.
Knudsen EI, Knudsen PF, Masino T (1993) Parallel pathways mediating both sound localization and gaze control in the forebrain and midbrain of the barn owl. J Neurosci 13:2837–2852.
Knudsen EI, Zheng W, DeBello WM (2000) Traces of learning in the auditory localization pathway. Proc Natl Acad Sci USA 97:11815–11820.
Konishi M (1973a) Development of auditory neuronal responses in avian embryos. Proc Natl Acad Sci USA 70:1795–1798.
Konishi M (1973b) How the owl tracks its prey. Am Sci 61:414–424.
Konishi M (1986) Centrally synthesized maps of sensory space. Trends Neurosci 9:163–168.
Köppl, C (1994) Auditory nerve terminals in the cochlear nucleus magnocellularis: differences between low and high frequencies. J Comp Neurol 339:438–446.
Köppl C, Carr CE (2003) Computational diversity in the cochlear nucleus angularis of the barn owl. J Neurophysiol 89:2313–2329.
Korada S, Schwartz IR (1999) Development of GABA, glycine, and their receptors in the auditory brainstem of gerbil: a light and electron microscopic study. J Comp Neurol 409:664–681.
Kotak VC, Sanes DH (2000) Long-lasting inhibitory synaptic depression is age- and calcium-dependent. J Neurosci 20:5820–5826.
Kotak VC, Sanes DH (2003) Gain adjustment of inhibitory synapses in the auditory system. Biol Cybern 89:363–370.
Kotak VC, Korada S, Schwartz IR, Sanes DH (1998) A developmental shift from GABAergic to glycinergic transmission in the central auditory system. J Neurosci 18: 4646–4655.
Krumlauf R, Marshall H, Studer M, Nonchev S, Sham MH, Lumsden A (1993) Hox homeobox genes and regionalisation of the nervous system. J Neurobiol 24:1328–1340.
Kuba H, Koyano K, Ohmori H (2002) Development of membrane conductance improves coincidence detection in the nucleus laminaris of the chicken. J Physiol 540:529–542.

Kubke MF, Carr CE (1998) Development of AMPA-selective glutamate receptors in the auditory brainstem of the barn owl. Microsc Res Tech 41:176–186.
Kubke MF, Carr CE (2000) Development of the auditory brainstem of birds: comparison between barn owls and chickens. Hear Res 147:1–20.
Kubke MF, Gauger B, Basu L, Wagner H, Carr CE (1999) Development of calretinin immunoreactivity in the brainstem auditory nuclei of the barn owl (*Tyto alba*). J Comp Neurol 415:189–203.
Kubke MF, Massoglia DP, Carr CE (2002) Developmental changes underlying the formation of the specialized time coding circuits in barn owls (*Tyto alba*). J Neurosci 22:7671–7679.
Kubke MF, Massoglia DP, Carr CE (2004) Bigger brains or bigger nuclei? Regulating the size of auditory structures in birds. Brain Behav Evol 63:169–180.
Kudo M (1981) Projections of the nuclei of the lateral lemniscus in the cat: an autoradiographic study. Brain Res 221:57–69.
Kudo M, Kitao Y, Okoyama S, Moriya M, Kawano J (1996) Crossed projection neurons are generated prior to uncrossed projection neurons in the lateral superior olive of the rat. Brain Res Dev Brain Res 95:72–78.
Kudo M, Sakurai H, Kurokawa K, Yamada H (2000) Neurogenesis in the superior olivary complex in the rat. Hear Res 139:144–152.
Kullmann PH, Kandler K (2001) Glycinergic/GABAergic synapses in the lateral superior olive are excitatory in neonatal C57Bl/6J mice. Brain Res Dev Brain Res 131:143–147.
Kullmann PH, Ene FA, Kandler K (2002) Glycinergic and GABAergic calcium responses in the developing lateral superior olive. Eur J Neurosci 15:1093–1104.
Lachica EA, Rübsamen R, Rubel EW (1994) GABAergic terminals in nucleus magnocellularis and laminaris orginate from the superior olivary nucleus. J Comp Neurol 348:403–418.
Lawrence JJ, Trussell LO (2000) Long-term specification of AMPA receptor properties after synapse formation. J Neurosci 20:4864–4870.
Leake P, Snyder G, Hradek G (2002) Postnatal refinement of auditory nerve projections to the cochlear nucleus in cats. J Comp Neurol 448:6–27.
Leao RN, Oleskevich S, Sun H, Bautista M, Fyffe RE, Walmsley B (2004) Differences in glycinergic mIPSCs in the auditory brain stem of normal and congenitally deaf neonatal mice. J Neurophysiol 91:1006–1012.
Levi-Montalcini R (1949) The development of the acoustico-vestibular centers in the chick embryo in the absence of the afferent root fibers and of descending fiber tracts. J Comp Neurol 91:209–262.
Levin MD, Kubke MF, Schneider M, Wenthold R, Carr CE (1997) Localization of AMPA-selective glutamate receptors in the auditory brainstem of the barn owl. J Comp Neurol 378:239–253.
Limb C, Ryugo DK (2000) Development of primary axosomatic endings in the anteroventral cochlear nucleus of mice. JARO 1:103–119.
Lippe WR (1994) Rhythmic spontaneous activity in the developing avian auditory system. J Neurosci 14:1486–1495.
Lippe WR (1995) Relationship between frequency of spontaneous bursting and tonotopic position in the developing avian auditory system. Brain Res 703:205–213.
Lippe W, Rubel EW (1983) Development of the place principle: tonotopic organization. Science 219:514–516.
Lippe W, Rubel EW (1985) Ontogeny of tonotopic organization of brain stem auditory

nuclei in the chicken: implications for development of the place principle. J Comp Neurol 237:273–289.
Liu M, Pereira FA, Price SD, Chu M-j, Shope C, Himes D, Eatock RA, Brownell WE, Lysakowski A, Tsai M-J (2000) Essential role of *BETA2/NeuroD1* in development of the vestibular systems. Genes Dev 14:2839–2854.
Lohmann C, Friauf E (1996) Distribution of the calcium-binding proteins parvalbumin and calretinin in the auditory brainstem of adult and developing rats. J Comp Neurol 367:90–109.
Lu Y, Monsivais P, Tempel BL, Rubel EW (2004) Activity-dependent regulation of the potassium channel subunits Kv1.1 and Kv3.1. J Comp Neurol 470:93–106.
Luksch H, Gauger B, Wagner H (2000) A candidate pathway for a visual instructional signal to the barn owl's auditory system. J Neurosci 20:RC70.
Makous JC, Middlebrooks JC (1990) Two-dimensional sound localization by human listeners. J Acoust Soc Am 87:2188–2200.
Manley G (1996) Ontogeny of frequency mapping in the peripheral auditory system of birds and mammals: a critical review. Audit Neurosci 3:199–214.
Manley GA, Köppl C, Konishi M (1988) A neural map of interaural intensity differences in the brain stem of the barn owl. J Neurosci 8:2665–2676.
Marianowski R, Liao WH, Van Den Abbeele T, Fillit P, Herman P, Frachet B, Huy PT (2000) Expression of NMDA, AMPA and GABA(A) receptor subunit mRNAs in the rat auditory brainstem. I. Influence of early auditory deprivation. Hear Res 150:1–11.
Marín F, Puelles L (1995) Morphological fate of rhombomeres in chick/quail chimeras: a segmental analysis of hindbrain nuclei. Eur J Neurosci 7:1714–1738.
Martin MR, Rickets C (1981) Histogenesis of the cochlear nucleus of the mouse. J Comp Neurol 197:169–184.
Masterton B, Thompson GC, Bechtold JK, RoBards MJ (1975) Neuroanatomical basis of binaural phase-difference analysis for sound localization: a comparative study. J Comp Physiol Psychol 89:379–386.
Mazer J (1997) The integration of parallel processing streams in the sound localization system of the barn owl. Comput Neurosci Trends Res 5:735–739.
McAlpine D, Grothe B (2003) Sound localization and delay lines—do mammals fit the model? Trends Neurosci 26:347–350.
McConnell SK, Kaznowski CE (1991) Cell cycle dependence of laminar determination in developing neocortex. Science 254:282–285.
Meininger V, Baudrimont M (1981) Postnatal modifications of the dendritic tree of cells in the inferior colliculus of the cat. A quantitative Golgi analysis. J Comp Neurol 200:339–355.
Miller GL, Knudsen EI (2001) Early auditory experience induces frequency-specific, adaptive plasticity in the forebrain gaze fields of the barn owl. J Neurophysiol 85: 2184–2194.
Miller GL, Knudsen EI (2003) Adaptive plasticity in the auditory thalamus of juvenile barn owls. J Neurosci 23:1059–1065.
Mogdans J, Knudsen EI (1994a) Site of auditory plasticity in the brain stem (VLVp) of the owl revealed by early monaural occlusion. J Neurophysiol 72:2875–2891.
Mogdans J, Knudsen EI (1994b) Representation of interaural level difference in the VLVp, the first site of binaural comparison in the barn owl's auditory system. Hear Res 74:148–164.
Moiseff A (1989) Bi-coordinate sound localization by the barn owl. J Comp Physiol A 164:637–644.

Molea D, Rubel EW (2003) Timing and topography of nucleus magnocellularis innervation by the cochlear ganglion. J Comp Neurol 466:577–591.
Monsivais P, Yang L, Rubel EW (2000) GABAergic inhibition in nucleus magnocellularis: implications for phase locking in the avian auditory brainstem. J Neurosci 20: 2954–2963.
Moore DR, Irvine DR (1981) Development of responses to acoustic interaural intensity differences in the car inferior colliculus. Exp Brain Res 41:301–309.
Mosbacher J, Schoepfer R, Monyer H, Burnashev N, Seeburg PH, Ruppersberg JP (1995) A molecular determinant for submillisecond desensitization in glutamate receptors. Science 266:1059–1062.
Mrsic-Flogel TD, Schnupp JW, King AJ (2003) Acoustic factors govern developmental sharpening of spatial tuning in the auditory cortex. Nat Neurosci 6:981–988.
Myat A, Henrique D, Ish-Horowicz D, Lewis J (1996) A chick homologue of *Serrate* and its relationship with *Notch* and *Delta* homologues during central neurogenesis. Dev Biol 174:233–247.
Nieder B, Wagner H, Luksch H (2003) Development of output connections from the inferior colliculus to the optic tectum in barn owls. J Comp Neurol 464:511–524.
Oertel D (1999) The role of timing in the brain stem auditory nuclei of vertebrates. Annu Rev Physiol 61:497–519.
Ohlemiller KK, Echteler SM (1990) Functional correlates of characteristic frequency in single cochlear nerve fibers of the Mongolian gerbil. J Comp Physiol A 167:329–338.
Oliver DL, Beckius GE, Bishop DC, Kuwada S (1997) Simultaneous anterograde labeling of axonal layers from lateral superior olive and dorsal cochlear nucleus in the inferior colliculus of cat. J Comp Neurol 382:215–229.
Olsen JF, Knudsen EI, Esterly SD (1989) Neural maps of interaural time and intensity differences in the optic tectum of the barn owl. J Neurosci 9:2591–2605.
Otis TS, Raman IM, Trussell LO (1995) AMPA receptors with high Ca^{2+} permeability mediate synaptic transmission in the avian auditory pathway. J Physiol 482 (Pt 2): 309–315.
Overholt EM, Rubel EW, Hyson RL (1992) A circuit for coding interaural time differences in the chick brainstem. J Neurosci 12:1698–1708.
Palmgren A (1921) Embryological and morphological studies on the mid-brain and cerebellum of vertebrates. Acta Zool 2:1–94.
Panin V, Irvine K (1998) Modulators of Notch signaling. Cell Dev Biol 9:609–617.
Parameshwaran S, Carr CE, Perney TM (2001) Expression of the Kv3.1 potassium channel in the avian auditory brainstem. J Neurosci 21:485–494.
Parks TN (1979) Afferent influences on the development of the brain stem auditory nuclei of the chicken: otocyst ablation. J Comp Neurol 183:665–678.
Parks TN (1981) Changes in the length and organization of nucleus laminaris dendrites after unilateral otocyst ablation in chick embryos. J Comp Neurol 202:47–57.
Parks TN (2000) The AMPA receptors of auditory neurons. Hear Res 147:77–91.
Parks TN, Jackson H (1984) A developmental gradient of dendritic loss in the avian cochlear nucleus occuring independently of primary afferents. J Comp Neurol 227: 459–466.
Parks TN, Rubel EW (1975) Organization and development of brain stem auditory nucleus of the chicken: organization of projections from N. magnocellularis to N. laminaris. J Comp Neurol 164:435–448.
Parks TN, Gill SS, Jackson H (1987) Experience-independent development of dendritic organization in the avian nucleus laminaris. J Comp Neurol 260:312–319.

Parks TN, Code RA, Taylor DA, Solum DA, Strauss KI, Jacobowitz DM, Winsky L (1997) Calretinin expression in the chick brainstem auditory nuclei develops and is maintained independently of cochlear nerve input. J Comp Neurol 383:112–121.

Peña JL, Konishi M (2000) Cellular mechanisms for resolving phase ambiguity in the owl's inferior colliculus. Proc Natl Acad Sci USA 97:11787–11792.

Perney TM, Kaczmarek LK (1997) Localization of a high threshold potassium channel in the rat cochlear nucleus. J Comp Neurol 386:178–202.

Person AL, Pat Cerretti D, Pasquale EB, Rubel EW, Cramer KS (2004) Tonotopic gradients of Eph family proteins in the chick nucleus laminaris during synaptogenesis. J Neurobiol 60:28–39.

Petralia RS, Rubio ME, Wang YX, Wenthold RJ (2000) Differential distribution of glutamate receptors in the cochlear nuclei. Hear Res 147:59–69.

Pettigrew AG, Ansselin AD, Bramley JR (1988) Development of functional innervation in the second and third order auditory nuclei of the chick. Development 104:575–588.

Piechotta K, Weth F, Harvey RJ, Friauf E (2001) Localization of rat glycine receptor alpha1 and alpha2 subunit transcripts in the developing auditory brainstem. J Comp Neurol 438:336–352.

Proctor L, Konishi M (1997) Representation of sound localization cues in the auditory thalamus of the barn owl. Proc Natl Acad Sci USA 94:10421–10425.

Puelles L, Martinez-de-la-Torre M (1987) Autoradiographic and Golgi study on the early development of n. isthmi principalis and adjacent grisea in the chick embryo: a tridimensional viewpoint. Anat Embryol (Berl) 176:19–34.

Puelles L, Robles C, Martinez de la Torre M, Martinez S (1994) New subdivision schema for the avian torus semicircularis: neurochemical maps in the chick. J Comp Neurol 340:98–125.

Pujol R, Lavigne-Rebillard M, Lenoir M (1998) Development of sensory and neural structures in the mammalian cochlea. In: Rubel EW, Popper AN, Fay RR (eds), Development of the Auditory System. New York: Springer, pp. 146–192.

Rakic P (1974) Neurons in rhesus monkey visual cortex: systematic relation between the time of origin and eventual disposition. Science 183:425–427.

Raman I, Zhang S, Trussell LO (1994) Pathway-specific variants of AMPA receptors and their contribution to neuronal signaling. J Neurosci 14:4998–5010.

Rathouz M, Trussell L (1998) Characterization of outward currents in neurons of the avian nucleus magnocellularis. J Neurophysiol 80:2824–2835.

Ravindranathan A, Donevan SD, Sugden SG, Greig A, Rao MS, Parks TN (2000) Contrasting molecular composition and channel properties of AMPA receptors on chick auditory and brainstem motor neurons. J Physiol 523 Pt 3:667–684.

Rogowski BA, Feng AS (1981) Normal postnatal development of medial superior olivary neurons in the albino rat: a Golgi and Nissl study. J Comp Neurol 196:85–97.

Romand R (1984) Functional properties of auditory-nerve fibers during postnatal development in the kitten. Exp Brain Res 56:395–402.

Romand R, Avan P (1997) Anatomical and functional aspects of the cochlear nucleus. In: Ehret G, Romand R (eds), The Central Auditory System. New York: Oxford University Press.

Rubel EW (1978) Ontogeny of structure and function in the vertebrate auditory system. In: Handbook of Sensory Physiology, Vol. IX: Development of Sensory Systems. Ed. M. Jacobsen. New York: Springer-Verlag, pp. 135–237.

Rubel EW, Fritzsch B (2002) Auditory system development: primary auditory neurons and their targets. Annu Rev Neurosci 25:51–101.
Rubel EW, Parks TN (1988) Organization and development of the avian brain-stem auditory system. In: Edelman GM, Einar Gall W, Maxwell Cowan W (eds), Brain Function. New York: John Wiley & Sons, pp. 3–92.
Rubel EW, Smith DJ, Miller LC (1976) Organization and development of brain stem auditory nuclei of the chicken: ontogeny of n. magnocellularis and n. laminaris. J Comp Neurol 166:469–490.
Rubel EW, Lippe WR, Ryals BM (1984) Development of the place principle. Ann Otol Rhinol Laryngol 93:609–615.
Rübsamen R, Lippe W (1998) The development of cochlear function. In: Rubel EW, Popper AN, Fay RR (eds), Development of the Auditory System. New York: Springer-Verlag.
Rudy B, McBain CJ (2001) K^+ channels: voltage-gated K+ channels designed for high-frequency repetitive firing. Trends Neurosci 24:517–526.
Ryan AF, Woolf NK, Sharp FR (1982) Functional ontogeny in the central auditory pathway of the Mongolian gerbil. A 2-deoxyglucose study. Exp Brain Res 47:428–436.
Ryugo DK, Fekete DM (1982) Morphology of primary axosomatic endings in the anteroventral cochlear nucleus of the cat: a study of the endbulbs of Held. J Comp Neurol 210:239–257.
Ryugo DK, Parks TN (2003) Primary innervation of the avian and mammalian cochlear nucleus. Brain Res Bull 60:435–456.
Ryugo DK, Sento S (1991) Synaptic connections of the auditory nerve in cats: relationship between endbulbs of held and spherical bushy cells. J Comp Neurol 305:35–48.
Ryugo DK, Wu MM, Pongstaporn T (1996) Activity-related features of synapse morphology: a study of endbulbs of Held. J Comp Neurol 365:141–158.
Ryugo DK, Pongstaporn T, Huchton D, Niparko J (1997) Ultrastructural analysis of primary endings in deaf white cats: morphologic alterations in endbulbs of Held. J Comp Neurol 385:230–244.
Ryugo DK, Rosenbaum B, Kim P, Niparko J, Saada A (1998) Single unit recordings in the auditory nerve of congenitally deaf white cats: morphological correlates in the cochlea and cochlear nucleus. J Comp Neurol 397:532–548.
Sanes DH (1993) The development of synaptic function and integration in the central auditory system. J Neurosci 13:2627–2637.
Sanes DH, Friauf E (2000) Development and influence of inhibition in the lateral superior olivary nucleus. Hear Res 147:46–58.
Sanes DH, Rubel EW (1988) The ontogeny of inhibition and excitation in the gerbil lateral superior olive. J Neurosci 8:682–700.
Sanes DH, Siverls V (1991) Development and specificity of inhibitory terminal arborizations in the central nervous system. J Neurobiol 22:837–854.
Sanes DH, Takacs C (1993) Activity-dependent refinement of inhibitory connections. Eur J Neurosci 5:570–574.
Sanes DH, Walsh EJ (1998) Development of auditory processing. In: Rubel EW, Popper AN, Fay RR (eds), Development of the Auditory System. New York: Springer-Verlag. pp. 271–314.
Sanes DH, Merickel M, Rubel EW (1989) Evidence for an alteration of the tonotopic map in the gerbil cochlea during development. J Comp Neurol 279:436–444.
Sanes DH, Goldstein NA, Ostad M, Hillman DE (1990) Dendritic morphology of central

auditory neurons correlates with their tonotopic position. J Comp Neurol 294:443–454.
Sanes DH, Song J, Tyson J (1992) Refinement of dendritic arbors along the tonotopic axis of the gerbil lateral superior olive. Brain Res Dev Brain Res 67:47–55.
Sato K, Momose-Sato Y (2003) Optical detection of developmental origin of synaptic function in the embryonic chick vestibulocochlear nuclei. J Neurophysiol 89:3215–3224.
Sato K, Shiraishi S, Nakagawa H, Kuriyama H, Altschuler RA (2000) Diversity and plasticity in amino acid receptor subunits in the rat auditory brain stem. Hear Res 147:137–144.
Saunders JC (1974) The development of auditory evoked responses in the chick embryo. Min Oto 24:221–229.
Saunders JC, Coles RB, Gates GR (1973) The development of auditory evoked responses in the cochlea and cochlear nuclei of the chick. Brain Res 63:59–74.
Schweitzer L, Cant NB (1984) Development of the cochlear innervation of the dorsal cochlear nucleus of the hamster. J Comp Neurol 225:228–243.
Schweitzer L, Cant NB (1985a) Development of oriented dendritic fields in the dorsal cochlear nucleus of the hamster. Neuroscience 16:969–978.
Schweitzer L, Cant NB (1985b) Differentiation of the giant and fusiform cells in the dorsal cochlear nucleus of the hamster. Brain Res 352:69–82.
Shneiderman A, Henkel CK (1987) Banding of lateral superior olivary nucleus afferents in the inferior colliculus: a possible substrate for sensory integration. J Comp Neurol 266:519–534.
Shneiderman A, Oliver DL, Henkel CK (1988) Connections of the dorsal nucleus of the lateral lemniscus: an inhibitory parallel pathway in the ascending auditory system? J Comp Neurol 276:188–208.
Shneiderman A, Chase MB, Rockwood JM, Benson CG, Potashner SJ (1993) Evidence for a GABAergic projection from the dorsal nucleus of the lateral lemniscus to the inferior colliculus. J Neurochem 60:72–82.
Shnerson A, Willott JF (1979) Development of inferior colliculus response properties in C57BL/6J mouse pups. Exp Brain Res 37:373–385.
Smith Z (1981) Organization and development of brain stem auditory nuclei of the chicken: dendritic development in n. laminaris. J Comp Neurol 203:309–333.
Smith ZDJ, Rubel EW (1979) Organization and development of brain stem auditory nuclei of the chicken: dendritic gradients in nucleus laminaris. J Comp Neurol 186:213–239.
Smith ZD, Gray L, Rubel EW (1983) Afferent influences on brainstem auditory nuclei of the chicken: n. laminaris dendritic length following monaural conductive hearing loss. J Comp Neurol 220:199–205.
Soares D, Carr CE (2001) The cytoarchitecture of the nucleus angularis of the barn owl (*Tyto alba*). J Comp Neurol 429:192–205.
Solum D, Hughes D, Major MS, Parks TN (1997) Prevention of normally occurring and deafferentation-induced neuronal death in chick brainstem auditory neurons by periodic blockade of AMPA/kainate receptors. J Neurosci 17:4744–4751.
Spitzer NC (1991) A developmental handshake: neuronal control of ionic currents and their control of neuronal differentiation. J Neurobiol 22:659–673.
Sugden SG, Zirpel L, Dietrich CJ, Parks TN (2002) Development of the specialized AMPA receptors of auditory neurons. J Neurobiol 52:189–202.

Sullivan WE, Konishi M (1986) Neural map of interaural phase difference in the owl's brainstem. Proc Natl Acad Sci USA 83:8400–8404.
Takahashi TT, Keller CH (1992) Commissural connections mediate inhibition for the computation of interaural level difference in the barn owl. J Comp Physiol A 170: 161–169.
Takahashi T, Konishi M (1986) Selectivity for interaural time difference in the owl's midbrain. J Neurosci 6:3413–3422.
Takahashi TT, Carr CE, Brecha N, Konishi M (1987) Calcium binding protein-like immunoreactivity labels the terminal field of nucleus laminaris of the barn owl. J Neurosci 7:1843–1856.
Takahashi TT, Wagner H, Konishi M (1989) Role of commissural projections in the representation of bilateral auditory space in the barn owl's inferior colliculus. J Comp Neurol 281:545–554.
Takahashi TT, Barberini CL, Keller CH (1995) An anatomical substrate for the inhibitory gradient in the VLVp of the owl. J Comp Neurol 358:294–304.
Tang YZ, Carr CE (2004) Development of NMDA R1 expression in chicken auditory brainstem. Hear Res 191:79–89.
Taschenberger H, von Gersdorff H (2000) Fine-tuning an auditory synapse for speed and fidelity: developmental changes in presynaptic waveform, EPSC kinetics, and synaptic plasticity. J Neurosci 20:9162–9173.
Trussell LO (1999) Synaptic mechanisms for coding timing in auditory neurons. Annu Rev Physiol 61:477–496.
Tsuchitani C (1977) Functional organization of lateral cell groups of cat superior olivary complex. J Neurophysiol 40:296–318.
Turecek R, Trussell LO (2002) Reciprocal developmental regulation of presynaptic ionotropic receptors. Proc Natl Acad Sci USA 99:13884–13889.
Vale C, Sanes DH (2000) Afferent regulation of inhibitory synaptic transmission in the developing auditory midbrain. J Neurosci 20:1912–1921.
Volman SF, Konishi M (1989) Spatial selectivity and binaural responses in the inferior colliculus of the great horned owl. J Neurosci 9:3083–3096.
Wadhwa S, Roy A, Anand P (1997) Quantitative study of cellular and volumetric growth of magnocellular and laminar auditory nuclei in the developing chick brainstem. J Biosci 22:407–417.
Wagner H (1993) Sound-localization deficits induced by lesions in the barn owl's auditory space map. J Neurosci 13:371–386.
Wagner H, Takahashi T, Konishi M (1987) Representation of interaural time difference in the central nucleus of the barn owl's inferior colliculus. J Neurosci 7:3105–3116.
Walsh EJ, McGee J (1987) Postnatal development of auditory nerve and cochlear nucleus neuronal responses in kittens. Hear Res 28:97–116.
Walsh EJ, McGee J (1988) Rhythmic discharge properties of caudal cochlear nucleus neurons during postnatal development in cats. Hear Res 36:233–247.
Walsh EJ, McGee J, Javel E (1986) Development of auditory-evoked potentials in the cat. II. Wave latencies. J Acoust Soc Am 79:725–744.
Walsh EJ, Gorga M, McGee J (1992) Comparisons of the development of auditory brainstem response latencies between cats and humans. Hear Res 60:53–63.
Walsh EJ, McGee J, Fitzakerley JL (1993) Development of glutamate and NMDA sensitivity of neurons within the cochlear nuclear complex of kittens. J Neurophysiol 69: 201–218.

Wang LY, Gan L, Forsythe ID, Kaczmarek LK (1998) Contribution of the Kv3.1 potassium channel to high-frequency firing in mouse auditory neurones. J Physiol 509 (Pt 1):183–194.

Wang Z, Liou L, Li DJ, Liou WZ (1993) Early-stage development of auditory center: an experimental study of auditory evoked electrophysiologic recordings from fetal and newborn guinea pigs. Ann Otol Rhinol Laryngol 102:802–804.

Webster DB (1992) An overview of mammalian auditory pathways with an emphasis on humans. In: Webster DB, Popper AN, Fay RR (eds), The Mammalian Auditory Pathway: Neuroanatomy, New York: Springer-Verlag, pp. 1–22.

Webster DB, Fay RR, Popper AN (1992a) The Evolutionary Biology of Hearing. New York: Springer-Verlag.

Webster DB, Popper AN, Fay RR (1992b) The Mammalian Auditory Pathway: Neuroanatomy. New York: Springer-Verlag.

Wild JM (1987) Nuclei of the lateral lemniscus project directly to the thalamic auditory nuclei in the pigeon. Brain Res 408:303–307.

Wild JM, Kubke MF, Carr CE (2001) Tonotopic and somatotopic representation in the nucleus basalis of the barn owl, *Tyto alba*. Brain Behav Evol 57:39–62.

Winsky L, Jacobowitz DM (1995) Effects of unilateral cochlea ablation on the distribution of calretinin mRNA and immunoreactivity in the guinea pig ventral cochlar nuclei. J Comp Neurol 354:564–582.

Withington-Wray DJ, Binns KE, Keating MJ (1990) The developmental emergence of a map of auditory space in the superior colliculus of the guinea pig. Brain Res Dev Brain Res 51:225–236.

Woolf NK, Ryan AF (1985) Ontogeny of neural discharge patterns in the ventral cochlear nucleus of the mongolian gerbil. Brain Res 349:131–147.

Woolf NK, Ryan AF (1988) Contributions of the middle ear to the development of function in the cochlea. Hear Res 35:131–142.

Wu SH, Oertel D (1987) Maturation of synapses and electrical properties of cells in the cochlear nuclei. Hear Res 30:99–110.

Yang L, Monsivais P, Rubel EW (1999) The superior olivary nucleus and its influence on nucleus laminaris: a source of inhibitory feedback for coincidence detection in the avian auditory brainstem. J Neurosci 19:2313–2325.

Young SR, Rubel EW (1983) Frequency-specific projections of individual neurons in chick brainstem auditory nuclei. J Neurosci 3:1373–1378.

Young SR, Rubel EW (1986) Embryogenesis of arborization pattern and topography of individual axons in N. Laminaris of the chicken brain stem. J Comp Neurol 254:425–459.

Zhang S, Trussell LO (1994) A characterization of excitatory postsynaptic potentials in the avian nucleus magnocellularis. J Neurophysiol 72:705–718.

Zheng W, Knudsen EI (1999) Functional selection of adaptive auditory space map by GABAA-mediated inhibition. Science 284:962–965.

Zhou N, Parks TN (1991) Pharmacology of excitatory amino acid neurotransmission in nucleus laminaris of the chick. Hear Res 52:195–200.

Zhou N, Taylor DA, Parks TN (1995) Cobalt-permeable non-NMDA receptors in developing chick brainstem auditory nuclei. NeuroReport 6:2273–2276.

Zhou X, Baier C, Hossain WA, Goldenson M, Morest DK (2001) Expression of a voltage-dependent potassium channel protein (Kv3.1) in the embryonic development of the auditory system. J Neurosci Res 65:24–37.

7

Interaural Correlation as the Basis of a Working Model of Binaural Processing: An Introduction

CONSTANTINE TRAHIOTIS, LESLIE R. BERNSTEIN, RICHARD M. STERN, AND THOMAS N. BUELL

1. Introduction

Our goal in writing this chapter is to present and to integrate at an introductory level several recent empirical and theoretical findings concerning binaural processing that can be understood within the general rubric of interaural correlation. Specifically, this chapter was designed to discuss interaural correlation in a manner that facilitates an intuitive understanding of the relevant phenomena, their explanation, and their import. Toward that end, and in order to make the material more accessible to the novice, the presentation minimizes the use of jargon and formal mathematical analyses. It is the authors' intent that this chapter will serve as an entrée for readers who wish to gain a deeper understanding via the primary literature and/or several professional-level book-chapters targeted to individuals who work in the field (e.g., Stern and Trahiotis 1995, 1997; Colburn 1996; Wightman and Kistler 1997; Yost 1997).

The discussion will begin by defining indices of interaural correlation and their application to binaural perception. After that, the concept of the three-dimensional cross-correlation function is introduced and it is shown how one can understand binaural perception both intuitively and quantitatively in terms of sophisticated types of "pattern processing" operations based on this function. Finally, neurophysiological evidence is presented that verifies the existence of neural mechanisms that appear to function as elements of the cross-correlation mechanism postulated a half-century ago by Jeffress (1948).

This effort is dedicated to Lloyd A. Jeffress who, along with his colleagues, was largely responsible for the success and popularity of interaural correlation as a framework for the explanation of data obtained in binaural experiments. His many contributions encompassing a variety of topics were fundamental, and typically included elegant and informative mixtures of data and theory. It is our hope that this chapter will serve to motivate readers currently unfamiliar with binaural hearing to explore and to appreciate the work of Lloyd Jeffress and his colleagues.

2. What Is Meant by "Interaural Correlation?"

In general, interaural correlation refers to the similarity/dissimilarity between a pair of signals presented simultaneously to the left and right ears. Interaural correlation has been expressed using two different types of functions. The interpretation that has been the most commonly encountered in the literature is

$$\rho_{cov} = \frac{\int (L(t) - m_L)(R(t) - m_R)}{\sqrt{\int (L(t) - m_L))^2}\sqrt{\int (R(t) - m_R)^2}} \tag{7.1}$$

where $L(t)$ and $R(t)$ represent the time waveforms or signals to the left and right ears, respectively, and m_L and m_R represent their mean (or average) values. This formula quantifies what is commonly known as the Pearson product-moment correlation. It is the index of correlation with which most people are familiar and the one typically encountered in many undergraduate courses including statistics and psychology. The Pearson-product-moment is sometimes called the *correlation coefficient* and is known more formally as the *normalized covariance*.

A second type of correlation index is obtained by comparing the instantaneous values of the left- and right-ear time-waveforms $L(t)$ and $R(t)$ directly:

$$\rho_{corr} = \frac{\int L(t)R(t)}{\sqrt{\int L(t)^2}\sqrt{\int R(t)^2}} \tag{7.2}$$

This formula specifies what is formally referred to as the *normalized correlation*. Later, we will show why this distinction is important. Typically, we think of the left- and right-ear signals as being AC electrical waveforms, such as sine waves, that have a mean value of zero. That is, over time, the values of the waveforms in the positive and negative directions "balance out." When this is true, the normalized correlation is equivalent to the normalized covariance. Historically, interaural correlation has been used successfully to account for a wide variety of data obtained in investigations that employed low-frequency waveforms (e.g., Robinson and Jeffress 1963; Osman 1971). In some investigations, the index of correlation is explicitly (e.g., Pollack and Trittipoe 1959a,b; McFadden 1968; Gabriel and Colburn 1981; Jain et al. 1991) taken to be the normalized covariance (correlation coefficient) and in others that identification is implicit (e.g., Durlach et al. 1986).

In contrast, it is well known that, for high-frequency complex stimuli, binaural timing information is conveyed by the envelopes of the waveforms and the "fine-structure-based" timing information is not encoded by the auditory periphery. Figure 7.1 shows a high-frequency narrow-band noise that is delayed in the right ear relative to its counterpart in the left ear. The line drawn through the individual positive peaks of the more rapidly changing fine-structure represents the envelope (or time-varying amplitude) of the stimulus. Note that the values of the envelope are never negative. Thus, the envelope function is not mean-

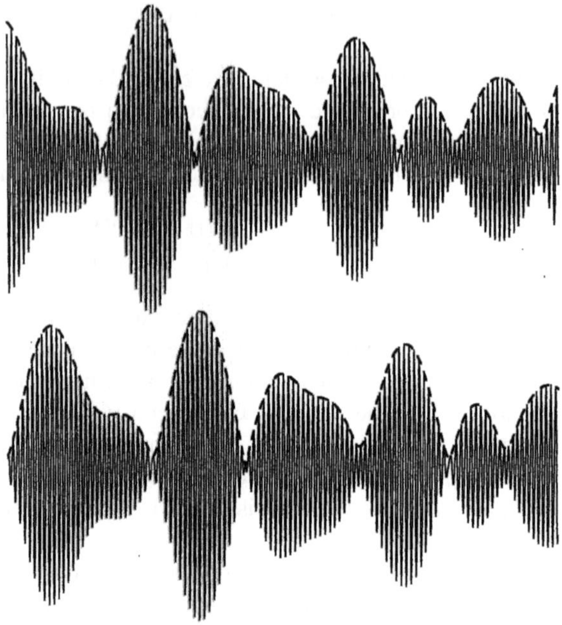

FIGURE 7.1. A high-frequency narrow-band noise that is delayed to the right ear with respect to the left ear. The *dashed lines* drawn through the positive peaks of the waveforms represent the envelopes.

zero. These characteristics are captured by stating that the time-varying envelope has a "positive DC offset." One would expect that, if the DC offset of the envelope were encoded neurally and affected perception, then the appropriate index of interaural correlation would be the normalized correlation and *not* the normalized covariance. Only the former incorporates the DC offset.

In order to investigate these notions, Bernstein and Trahiotis (1996a) measured listeners' performance in a special binaural discrimination task. The task required listeners to discriminate between two different types of tonal signals added to a background of narrow-band noise centered at 4 kHz. The noise itself was identical at the two ears (i.e., diotic). Therefore, the noise (and its envelope) had an interaural correlation of 1.0. One type of signal was a 4-kHz tone that was also presented diotically. Adding such a signal to the noise resulted in a signal-plus-noise stimulus also having an interaural correlation of 1.0. The second type of signal was a 4-kHz tone that was interaurally phase-reversed. Adding this signal to the diotic background noise resulted in a signal-plus-noise stimulus whose envelope, which carried the relevant timing information, had an interaural correlation that was less than 1.0. The actual decrease in interaural envelope correlation depended on the actual level of the phase-reversed signal.

Figure 7.2a plots the interaural correlation of the envelope resulting from the addition of the phase-reversed 4-kHz tonal signal to the diotic noise, as a func-

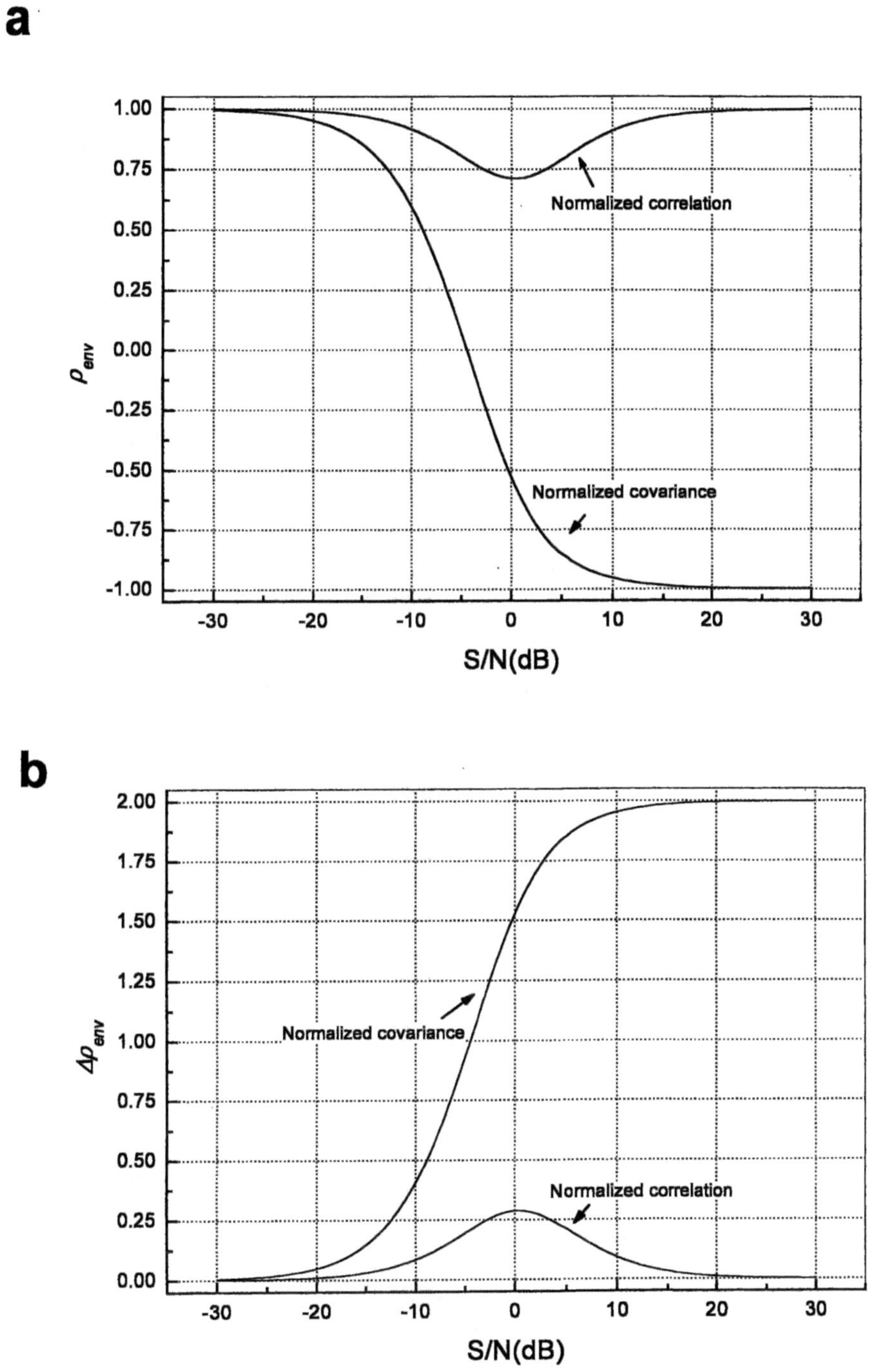

FIGURE 7.2. (**a**) Interaural correlation, ρ, as a function of S/N in dB computed with two different indices: the normalized correlation and the normalized covariance. (**b**) The measures in panel a expressed in terms of the change in interaural correlation, Δρ (from 1.0). (From Bernstein and Trahiotis 1996a. Reprinted with permission.)

tion of the signal-to-noise ratio (S/N) in dB. The line labeled "normalized covariance" indicates the interaural correlation computed when the DC offset of the envelope is removed. Note that, for very low S/Ns, the normalized covariance is essentially 1.0, meaning that the phase-reversed signal is so weak that it does not appreciably reduce the 1.0 normalized covariance of the envelope of the background noise. In stark contrast, at very high S/Ns, the addition of the phase-reversed signal to the background noise results in values of normalized covariance of the envelope that approach -1.0.

The line in Figure 7.2a labeled "normalized correlation indicates the interaural correlation of the envelope computed when the DC offset of the envelope is retained. Note that, at very low values of S/N, the normalized correlation, like the normalized covariance, is essentially 1.0. That, however, is where the similarity ends. The minimum of the normalized correlation occurs for S/Ns near 0 dB and at very high S/Ns the normalized correlation returns to 1.0!

Figure 2b shows the information depicted in Figure 7.2a replotted in terms of *decreases* in the interaural correlation ($\Delta\rho$) of the noise brought about by the addition of the interaurally phase-reversed tonal signal. Increases in $\Delta\rho$ would be expected to lead to increases in listeners' abilities to discriminate which interval contained the phase-reversed signal. Thus, at the higher S/Ns, listeners' performance would be expected to be poor if the *normalized correlation* captures auditory processing because, for that metric, $\Delta\rho$ becomes very small and approaches zero. On the other hand, performance at high S/Ns would be expected to be excellent if the normalized covariance captures auditory processing because, for that metric, $\Delta\rho$ becomes very large.

Figure 7.3 shows the results of Bernstein and Trahiotis's (1996a) study. The graph plots the listeners' sensitivity (in terms of the metric, d') as a function of S/N in dB. Note that the ability to discriminate between the addition of diotic and phase-reversed signals follows the form of the function expected from the assumption that the normalized correlation is the appropriate index. By extension, this outcome provides strong evidence that the DC offset of the stimulus envelopes is retained through peripheral auditory processing and affects behavioral performance. This conclusion is bolstered by the demonstration that the normalized interaural correlation of the envelope can be used to explain classical data obtained by Nuetzel and Hafter (1981). Nuetzel and Hafter measured listeners' sensitivities to interaural time delay (ITD) conveyed by the envelopes of sinusoidally amplitude modulated (SAM) tones centered at 4kHz and modulated at 300 Hz. The parameter of interest was the depth of modulation of the SAM tones. The top panel of Figure 7.4 illustrates two such SAM tones, one modulated at 100% ($m = 1$) and one modulated at 20% ($m = 0.2$). Note that the SAM tone with the lower depth of modulation has a "shallower" envelope. From our discussion and this picture, it should be clear that changes in depth of modulation are changes in the relative DC offset of the envelope of the SAM tone. Because the normalized correlation incorporates the DC offset of the envelope, changes in sensitivity to ITD (threshold-ITD) that occur with changes in depth of modulation should be predictable by casting Nuetzel and Hafter's

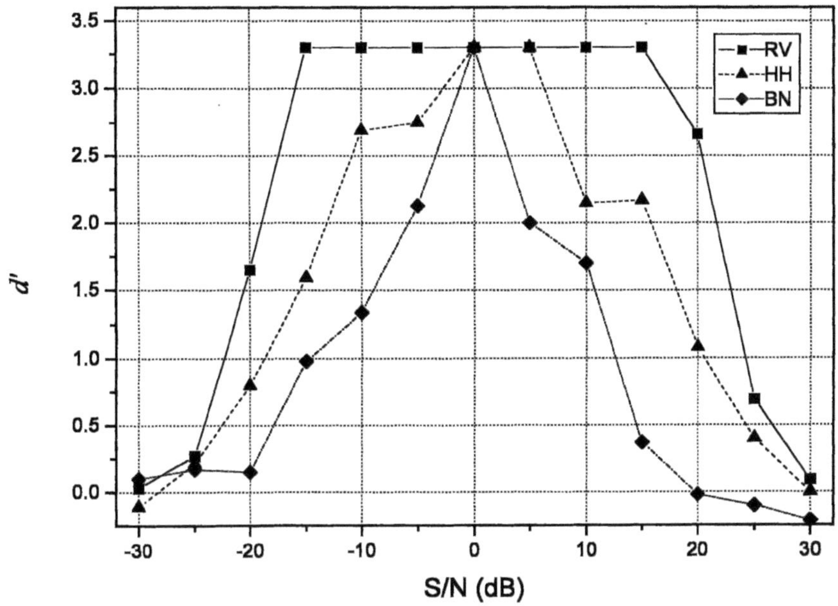

FIGURE 7.3. d' as a function of S/N in dB for the diotic versus phase-reversed discrimination task (see text). The data are plotted separately for each listener. (From Bernstein and Trahiotis 1996a. Reprinted with permission.)

threshold-ITDs as changes in the normalized correlation (of the envelope). The assumption is that, at threshold sensitivity, a listener requires a constant value of $\Delta\rho$, that is, a constant decrease from a correlation of 1.0. The bottom panel of Figure 7.4 verifies the expectations. The normalized correlation does, indeed, provide an excellent fit to the data, especially when it is considered that the threshold-ITD changed from about 50 ms to about 950 ms as depth of modulation was reduced from 1.0 to 0.1. Successful predictions were also obtained for data gathered by McFadden and Pasanen (1976) in a similar experiment employing high-frequency two-tone complexes (Bernstein and Trahiotis 1996a). In addition to noting the success of the normalized correlation, the reader should also appreciate that the normalized covariance, because it removes the DC offsets of the envelopes, *cannot* explain the data. Specifically, the normalized covariance would be *unaffected* by changes in the depth of modulation and would predict *constant threshold-ITDs* as a function of that parameter. Further implications of these findings are discussed in detail in Bernstein and Trahiotis (1996a).

At this point, it should be clear that the normalized correlation is the index of choice for characterizing the interaural correlation of the envelopes of high-frequency stimuli. As it turns out, it is a useful index for characterizing interaural correlation across the audible spectrum. To understand why this is so, it

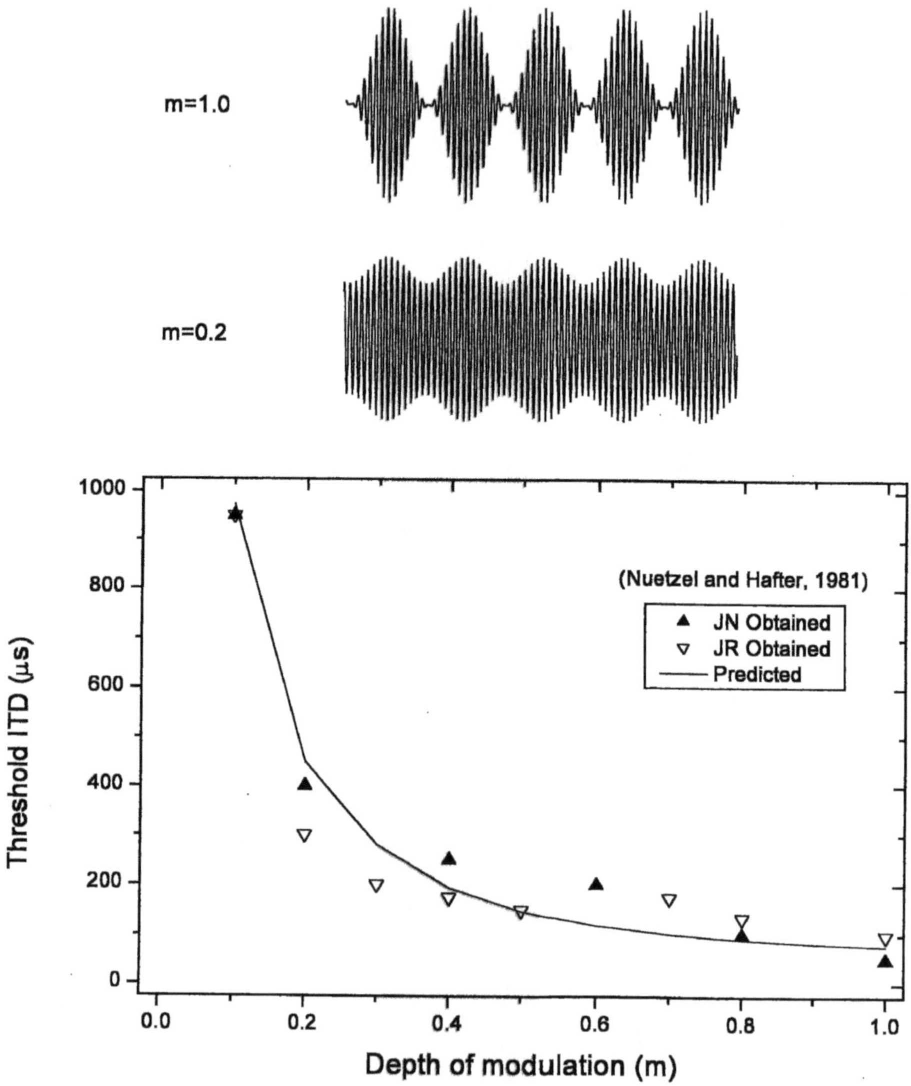

FIGURE 7.4. (*Top*) A pair of SAM tones, one modulated at 100% ($m = 1.0$), the other modulated at 20% ($m = 0.2$). (*Bottom*) Threshold ITD (in μs) as a function of depth of modulation for a high-frequency SAM tone (carrier, 4 kHz; modulator, 300 Hz). The data were transcribed from Nuetzel and Hafter's (1981) Figure 4 for their listeners JN (upward triangles) and JR (downward triangles). The *solid line* represents the predicted thresholds based on the assumption that the two listeners require the same constant change in normalized envelope correlation for detection to occur. (From Bernstein and Trahiotis 1996a. Reprinted with permission.)

is necessary to consider the basic stages of peripheral auditory processing, as depicted in Figure 7.5, for a single auditory receptor (hair cell) that monitors a single frequency "channel" along the cochlear partition.

The three stages are bandpass filtering, half-wave rectification, and low-pass filtering. Bandpass filtering comes about largely as a result of the mechanics of the inner ear and is manifest in the finding that any given auditory hair cell/nerve-fiber complex only responds to, or is "tuned" to, a limited range of frequencies. Half-wave rectification occurs because hair cells tend to respond to only one direction of displacement of the stimulating waveform. Finally, low-pass filtering is a process which acts to "smooth over" rapid changes in the waveform. It describes the finding that the synchronized discharge of auditory nerve fibers to the fine-structure of the waveform diminishes as the frequency of stimulation is increased (for a review, see Ruggero and Santos-Sacchi 1997). The result of all of this processing is shown in the bottom row of Figure 7.5. At low frequencies (well below the cutoff of the low-pass filter), the result of peripheral processing is the rectified *waveform*, and neural discharges are synchronized to it. At high frequencies (well above the cutoff of the low-pass filter),

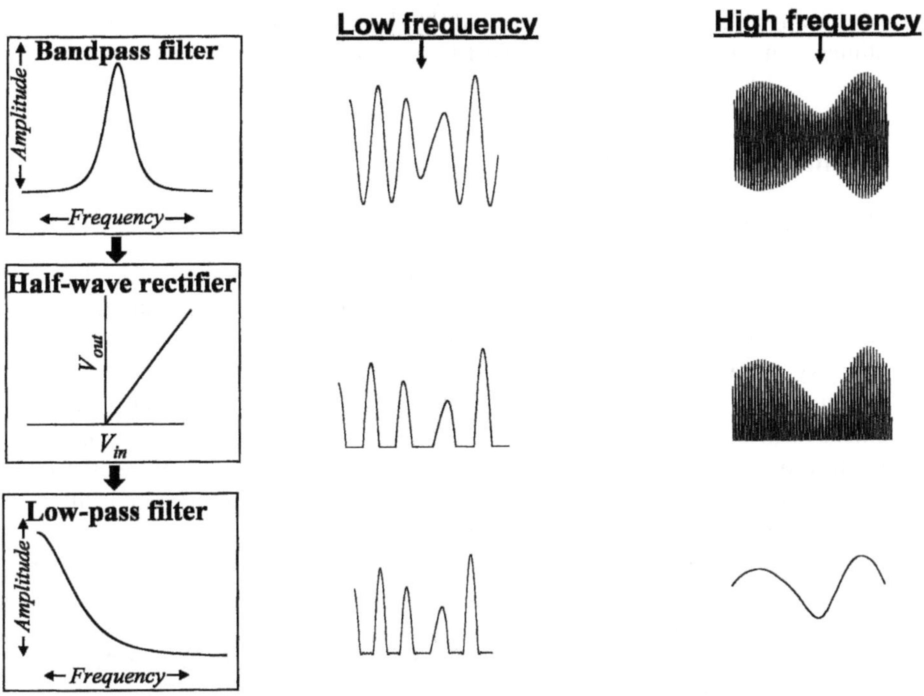

FIGURE 7.5. The effects of the three basic stages of peripheral auditory processing (bandpass filtering, rectification, and low-pass filtering) on a low-frequency and a high-frequency narrow band of noise.

however, the result of peripheral processing is that only the *envelope* emerges and neural discharges are synchronized to it.

For "intermediate" frequencies there occurs a transition such that neural discharges are synchronized to some degree to both aspects of the stimuli as processed, depending on the relative spectral "distance" between center frequency and the cutoff of the low-pass filter. The very crucial observation to be made is that *all* stimuli processed via the peripheral auditory system have a DC offset resulting from the rectification operation performed by the auditory hair cells. Therefore, one would expect that the normalized correlation, which includes DC offsets, would be the metric to account for discriminability of interaural correlation at *all* frequencies.

Figure 7.6, taken from Bernstein and Trahiotis (1996b), shows data obtained in the same discrimination task described above for stimuli centered at several center frequencies spanning the range from 500 Hz, where neural impulses are synchronized to the waveforms of the stimuli, to 2000 Hz, where neural impulses are synchronized to the envelopes of the stimuli. The figure clearly demonstrates that predictions (solid lines) using a model that computes changes in normalized correlation subsequent to peripheral processing, can, independent of frequency, account quite well for binaural detection data. This conclusion is bolstered by the results of a recent study (Bernstein and Trahiotis 2002) in which such a model was able to account for threshold-ITDs when the center frequency of the stimuli that carried the envelope-based ITDs was as high as 10 kHz.

3. Interaural Cross-Correlation in the Auditory System

3.1 The Basics—Responses to Pure Tones and Noise

So far, the discussion has been restricted to measures of the interaural correlation computed on either a pair of waveforms considered as physical stimuli or as those stimuli after processing via the auditory periphery. This approach is satisfactory for explaining and understanding a variety of binaural detection and discrimination experiments. It is the case, however, that the model must be extended in order to capture other basic binaural phenomena including the lateralization of sounds. Lateralization refers to the left/right displacement of sounds "within the head" (intracranial images) produced by interaural differences of time and/or intensity presented via earphones.

The principal extension of the model stems from Jeffress's (1948) postulation of a neural place mechanism that would enable the extraction of interaural timing information. His general idea was that external interaural delays (ITDs) could be coded by central units that record the simultaneous occurrence, or "coincidences," of neural impulses stemming from activity in peripheral nerve fibers originating at the left and right ears, respectively. Each central unit would be responsible for responding to neural coincidences that occur subsequent to the imposition of an *internal* time-delay. The operation of such neural delay-line

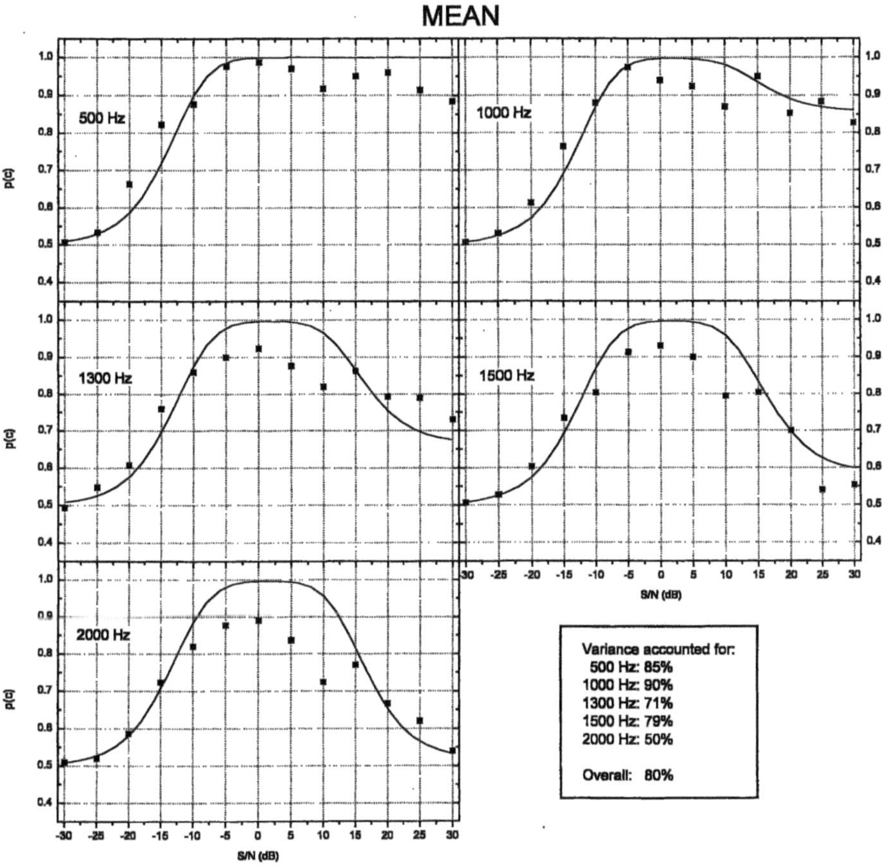

FIGURE 7.6. Proportion correct *p(c)* as a function of signal-to-noise ratio in dB for the data averaged across four listeners in the diotic versus phase-reversed discrimination task (see text). *Squares* represent the empirical data. *Solid lines* represent predicted values of *p(c)* using a model that computes changes in normalized correlation subsequent to peripheral processing. Each panel displays data and predictions for a different center frequency. The lower right-hand corner of the figure displays the amount of variance accounted for by the predictions. (From Bernstein and Trahiotis, 1996b. Reprinted with permission.)

mechanisms, which have since been anatomically and physiologically verified (e.g., Yin et al., 1997), is mathematically equivalent to computing the interaural correlation after each and every internal delay has been applied. Seen in this light, Jeffress's place mechanism enables the computation of the interaural correlation *function*. The *index* of interaural correlation discussed above is a special case in which the interaural correlation is computed after the imposition of an internal delay of 0 μs, that is, after no internal delay. Said differently, the index of interaural correlation described above quantifies activity at only one point along the correlation function, specifically activity at "lag-zero." In this light, it can be understood that an ITD of 0 μs is just as much an ITD as any other value. As will become clear later, an ITD of 0 μs certainly does not signal there being no interaural temporal information to be encoded, to be processed, or to be perceptually relevant.

Figure 7.7 depicts the putative internal cross-correlation function for a 500-Hz tone having an ongoing ITD (which, for a pure tone, is equivalent to a phase-shift) such that the right ear leads by 500 μs. Note that the maximum correlation is 1.0 and occurs at an internal delay of 500 μs. This is so because the *internal delay* of 500 μs effectively compensates, or cancels out, the opposing external delay and, therefore, internally realigns the activity produced by the waveforms. This results in an interaural correlation of 1.0 at the 500-μs "place" and, concomitantly, an interaural correlation of 0.0 at the 0-μs place (no delay). This latter value is the index of interaural correlation described in the previous section.

Adding the dimension of "frequency" to the characterization shown in Figure 7.7, allows one to plot the interaural correlation as a *surface* which is commonly referred to as a "correlogram." The height of the correlogram represents the number of coincidences, or amount of neural activity, at a given time-delay (rather than the value of the normalized correlation plotted in Figure 7.7). The number of coincidences at a given time-delay is represented mathematically by the numerator of Eq. (7.1); which is a cross-product *not normalized* by the denominator of Eq. (7.1). As such, values of the cross-product, or number of

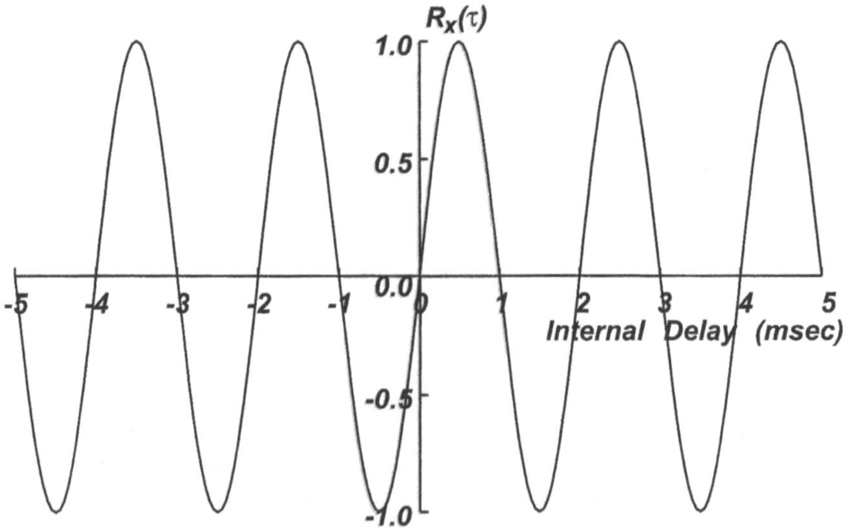

FIGURE 7.7. The putative internal cross-correlation function for a 500-Hz tone having an ongoing ITD such that the right ear leads by 500 μs. Note that the maximum correlation is 1.0 and occurs at an internal delay of 500 μs.

coincidences, computed for a given delay are not restricted to range from 1.0 to −1.0.

Such correlograms are shown in the top row of Figure 7.8. In each case, a band of noise centered at 500 Hz contains a 1.5-ms interaural delay with the right ear leading. The left-hand panel depicts the correlogram computed when the bandwidth is 50-Hz-wide; the right-hand panel depicts the correlogram computed when the bandwidth is 400-Hz-wide. There are important features of these correlograms that must be highlighted. They result from the fact that the computation is performed on the stimuli *as processed peripherally* (i.e., subsequent to bandpass filtering and rectification). First, notice that, as a result of rectification, the height of the correlogram never falls below zero. Second, notice that robust activity occurs across the entire range of frequencies plotted, even for the 50-Hz-wide noise. In order to understand why this is so, note that the "frequency-axis," CF in Hz, represents the center frequencies of the auditory filters through which the noise must pass. As shown, for example, by Kim and Molnar (1979), the responses of the population of auditory-nerve fibers to narrow-band stimuli presented at moderate levels (e.g., 70 dB SPL) are essentially equivalent across single units tuned several octaves apart.

Note that both correlograms contain a peak of activity at the internal delay of 1.5 ms which matches, or compensates for, the imposed external ITD of that magnitude which favors the right ear. Also note that the correlogram of the 50-Hz-wide noise contains two other peaks of activity that are aligned with internal delays of −0.5 ms and −2.5 ms. All of the peaks are separated by multiples of the period of the 500-Hz center frequency of the noise. The correlogram for the 400-Hz-wide noise looks quite different. The trajectories of its peaks, other than the one at 1.5 ms, are not straight but, instead, are curved across frequency. This is so because the 400-Hz-wide stimulus contains energy at frequencies well below and well above the center frequency of 500 Hz. The periods of the lower frequencies are larger than those of the higher frequencies, making the spacing between the trajectories of neighboring peaks of activity larger for them.

3.2 Straightness and Centrality

In order to use information arising from these correlograms to account for or explain how the stimuli are lateralized, two more modifications are necessary. Both of them reflect fundamental aspects of binaural processing. First, what is known as "centrality" must be included. Centrality refers to the general finding that small values of internal delay weigh more heavily in binaural perception than do larger values of internal delay. It is commonly thought that centrality reflects a relatively greater number of neural coincidence detectors "tuned" to smaller delays. In fact, that has been found to be the case in several physiological experiments (e.g., Crow et al. 1978, Kuwada and Yin 1983, Kuwada et al.1987). A number of mathematical functions, derived principally from psychoacoustic data (e.g., Sayers and Cherry 1957, Colburn 1977, Blauert and Cobben 1978, Stern and Colburn 1978, Shackelton et al. 1992, Stern and Shear

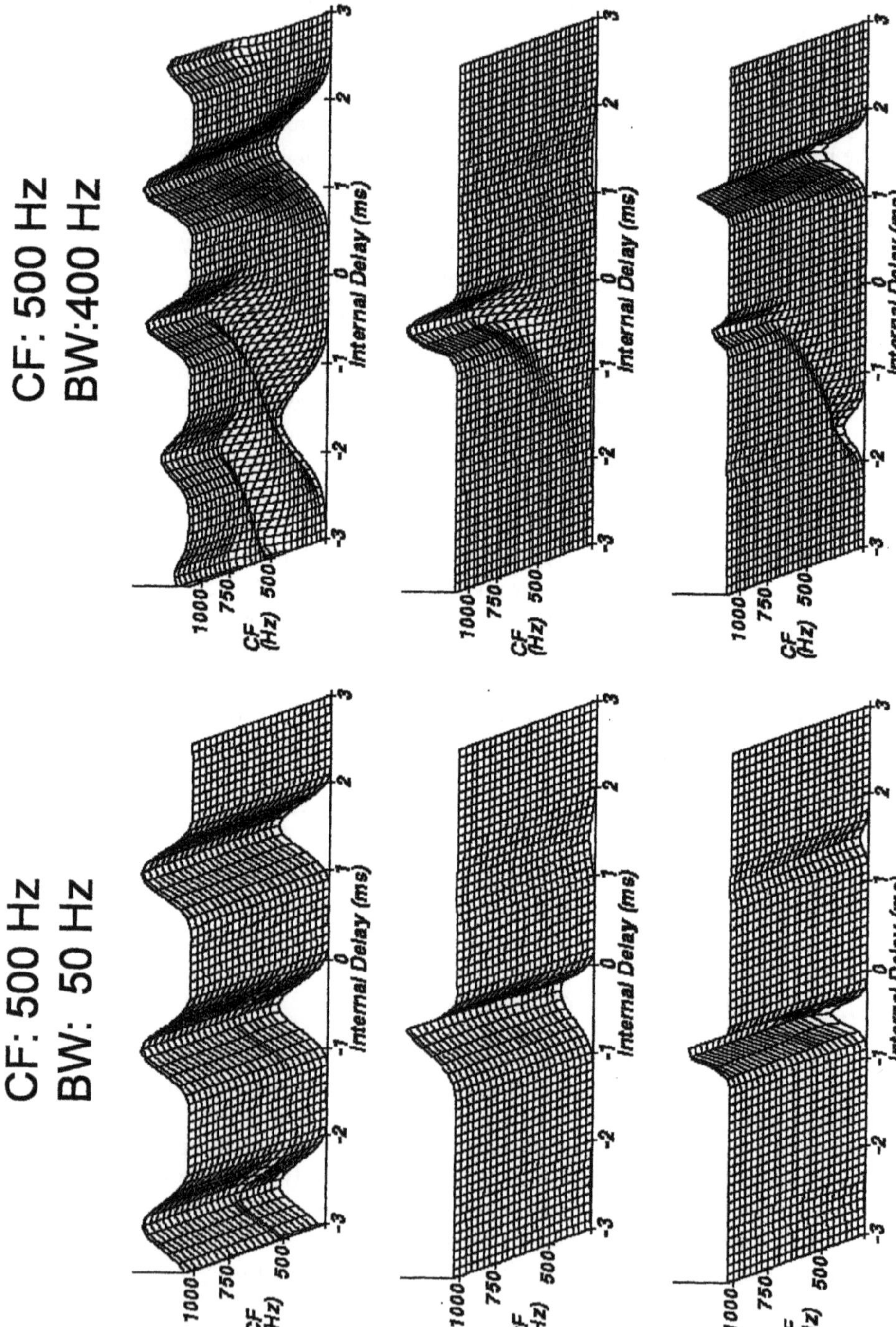
CF: 500 Hz
BW: 50 Hz
CF: 500 Hz
BW:400 Hz
CF (Hz)
1000
750
500
Internal Delay (ms)
-3
-2
-1
0
1
2
3

1996), have been used to enhance the central portion of the correlogram. Correlograms that incorporate a weighting for centrality are shown in the second row of Figure 7.8. As can be seen, centrality weighting is very potent and acts to reduce the heights of trajectories centered on interaural delays of larger magnitude.

The second modification that must be made to the correlograms is one that accounts for the general finding that the extent to which maxima in the interaural cross-correlation of the stimuli are *consistent*, in the sense that they appear at the same internal delay over a range of frequencies, strongly influences binaural perception. This is commonly referred as "straightness," and, as described above for "centrality," mathematical functions derived from psychoacoustic data have also been used to account for it (e.g., Stern et al. 1988).

The bottom row of Figure 7.8 displays correlograms that incorporate both centrality and straightness weightings. A dramatic effect of adding the straightness weighting is to enhance greatly the amount of activity at an internal delay of 1.5 ms for the 400-Hz-wide band of noise (lower right panel). Notice that the centrality weighting highlighted the most central trajectory at the expense of that straight trajectory (compare top-right and middle-right panels), while the straightness weighting caused the straight trajectory to emerge as the dominant one! This makes clear why Stern et al. (1988) conceive of centrality and straightness weightings as two, sometimes conflicting, weighing functions that determine the relative salience of individual peaks of the cross-correlation function.

3.3 Lateralization and Interaural Discrimination of Simple Binaural Stimuli

We are finally in a position to understand how the modified correlogram, which incorporates stages of (monaural) peripheral processing and knowledge of binaural processing, can be linked to binaural perception. Returning to the "final" correlograms in the bottom row of Figure 7.8, note that the strongest activity for the 50-Hz-wide band of noise occurs at an interaural delay of −0.5 ms. This indicates that perceived location would be on the *left* side of the head, despite the fact that the physical stimulus contains an ITD such that the right ear leads by 1.5 ms. In fact, the model correctly predicts that listeners perceive that stimulus configuration on the "wrong" (i.e., lagging) side of the head. In contrast, for the 400-Hz-wide band of noise, the strongest activity is centered at internal delay of 1.5 ms. This indicates that perceived location would be very

Figure 7.8. Correlograms for 50-Hz-wide (*left*) and 400-Hz-wide (*right*) Gaussian noises centered at 500 Hz that contain an ITD of 1.5 ms. (*Top*) Correlograms computed subsequent to peripheral bandpass filtering. (*Middle*) Correlograms computed subsequent to peripheral bandpass filtering and "centrality weighting." (*Bottom*) Correlograms computed subsequent to peripheral bandpass filtering, centrality weighting, and "straightness" weighting.

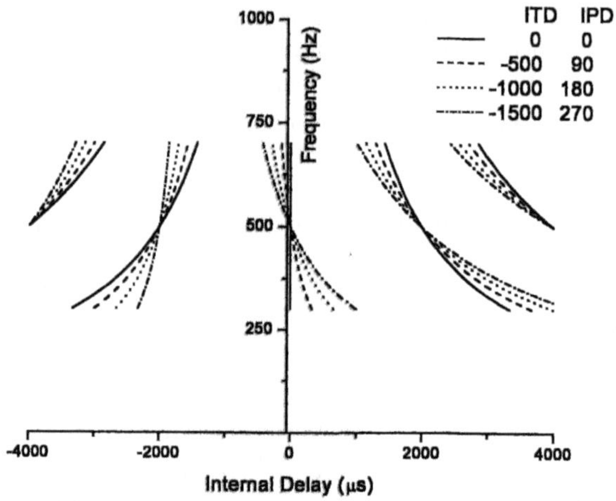
ITD IPD
0 0
-500 90
-1000 180
-1500 270
Frequency (Hz)
1000
750
500
250
-4000
-2000
0
2000
4000
Internal Delay (µs)

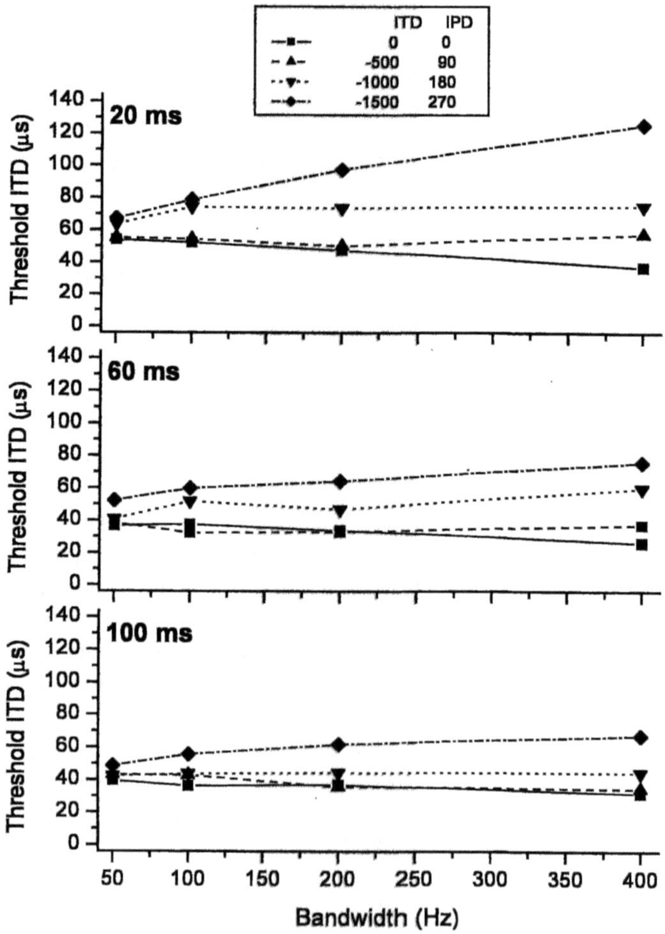
ITD IPD
0 0
-500 90
-1000 180
-1500 270
20 ms
60 ms
100 ms
Threshold ITD (µs)
0
20
40
60
80
100
120
140
50
100
150
200
250
300
350
400
Bandwidth (Hz)

far toward the *right* (leading) side of the head in accord with the actual ITD of the stimulus. Once more, the prediction of the model is correct. Although not shown, it is the case that listeners perceive the binaural image to migrate from near the lagging ear to positions near the midline and, finally, to a position near the leading ear as bandwidth is increased from 50 to 400 Hz. The perceived position of the stimuli depends on bandwidth because, as one increases the bandwidth from 50 to 400 Hz, one increases the relative strength of straightness relative to centrality (for details see Stern et al. 1988; Trahiotis and Stern 1989). Although he did not discuss and quantify such phenomena in terms of the straightness and centrality of patterns within the correlogram, Jeffress (1972) did provide a cogent qualitative account of listeners' perceptions of similar stimuli. Jeffress's discussion focused on the differential amounts of neural activity that would be expected to occur along the internal delay line for narrow-band versus broadband stimuli, each having an ITD of 2.0 ms. As often occurred, his foresight stimulated others to conduct formal experiments and analyses that verified his intuitions.

The relative straightness and curvature of the trajectories within the correlogram have recently been shown to affect listeners' abilities to detect changes in ITD (Trahiotis et al. 2001). Detecting a change in ITD can be thought of as equivalent to detecting a horizontal shift, that is, a translation along the internal delay axis, of the correlogram. To the degree that is the case, then the change in ITD would be expected to be most discernable for the straightest trajectories because they are the ones with the least "variance," or inconsistency, along the internal delay axis. One would also expect, other things being equal, that increases in the amount of curvature would lead to decreases in the ability to discern changes in ITD. This line of argument is consistent with and follows from the now commonly used signal-detection theory approach. Within that approach sensitivity is characterized as a mean-to-sigma ratio. That is, sensitivity is not only related to the amount of "signal" present but, also, to the amount of variability included in the measurements. Here the amount of "signal" is the displacement of a trajectory within the correlogram and the "variance" is related to the degree of dispersion (curvature) of that trajectory along the internal delay axis.

The top panel of Figure 7.9 displays a bird's-eye view of the peaks of the correlograms for four different types of stimuli. The stimuli depicted are 400-Hz-wide bands of noise centered at 500 Hz. The four types of stimuli differ with respect to their relative amounts of ITD and interaural phase-shift (IPD). One of them (indicated by the solid line) had both an ITD of zero and an IPD of zero. As shown, the correlogram for that type of stimulus has a peak that is

FIGURE 7.9. (*Top*) "Bird's-eye view" of the trajectories of the cross-correlation for each of the four reference stimuli (see text). (*Bottom*) Average threshold ITDs obtained with the 20-, 60- and 100-ms-long stimuli plotted as a function of bandwidth. The parameter within each panel is the combination of ITD and IPD that was imposed on the bands of noise that served as the reference. (From Trahiotis et al. 2001. Reprinted with permission.)

straight (consistent) across frequency at an internal delay of zero (midline). The other three types of stimuli had mixtures of ITD and IPD chosen to anchor the 500-Hz component of their most central trajectory at an internal delay of zero, while producing varying degrees of curvature. As discussed above, if one were to reduce the bandwidth of the stimuli, then there would less curvature to be "seen" in the correlogram because the lengths of the trajectories in the vertical dimension (i.e., along the frequency axis) would be diminished. Therefore, one would expect that, for narrow-band stimuli, threshold-ITDs would not vary appreciably across the different combinations of ITD and IPD, but that they would do so as bandwidth is increased.

It is important to note that all four types of stimuli depicted in Figure 7.9 are perceived as being at or near the midline, regardless of their bandwidth. This indicates that the central trajectory dominates the perceived laterality of the four types of reference stimuli. Said differently, the stimuli were constructed so that it was never the case that the weighing of straightness versus centrality produced any substantial influence of a trajectory other than the one at midline.

The bottom portion of Figure 7.9 displays the threshold-ITDs measured with the four types of reference stimuli when the stimuli were 20, 60, or 100 ms long. Each plot displays the data averaged across the same four listeners. As expected, the greater the curvature of the central trajectory of the correlogram, the larger the threshold-ITD. This can readily be seen by noting that, at a bandwidth of 400 Hz, threshold-ITDs increase monotonically with the amount of curvature as produced by the combinations of ITD and IPD. Then, as the bandwidth is decreased, and there is less curvature to "see" along the surface of the correlogram, threshold-ITDs decrease for all four types of stimuli. In general, however, the ordering of the data remains consistent with the idea that increased curvature leads to decreased sensitivity to changes in ITD (i.e., larger threshold-ITDs). With regard to the overall improvement in sensitivity with duration evident in Figure 7.9, suffice it to say that sample-to-sample variability is smaller at the longer durations than at the shorter durations. Successful accounts of the data incorporating the effects of stimulus duration and spectral integration within the context of a quantitative model based on the correlogram computed subsequent to peripheral auditory processing are presented in Trahiotis et al. (2001).

3.4 The Precedence Effect

It is also the case that this same general interaural correlation-based approach can be used to account for another class of findings that fall under the rubric of "the precedence effect." In the classic laboratory study of the "precedence effect," Wallach et al. (1949) used earphones to present listeners with pairs of successive binaural transients (or "clicks"). The clicks were presented sufficiently close in time to one another so that the listeners perceived one auditory event. That is, the pairs of clicks presented in rapid succession yielded one fused acoustic image. The object of the experiment was to determine the relative influence on the position of the auditory image of ITDs conveyed by the first versus the second binaural click. They did so by imposing an ITD on the first pair of clicks and then determining the magnitude of an opposing ITD that had to be conveyed by the second binaural click so that the fused auditory was heard in the center of the head. That is, they determined the ITD of the second binaural click required to "balance out" the lateralization produced by the ITD conveyed by the first click. Wallach et al. found that ITDs that had to be imposed on the second binaural click in order to achieve this balance were much larger than those carried by the first click. That is, the ITD that was imposed on the first, or *preceding*, click was prepotent, or took *precedence*. Since 1949, many investigators have conducted a wide range of similar studies using binaural click pairs in order to study "binaural precedence" (e.g., Zurek 1980; Yost and Soderquist 1984; Aoki and Houtgast 1992; Houtgast and Aoki 1994; Shinn-Cunningham et al. 1995; Tollin and Henning 1998, 1999). The reader is also referred to three comprehensive reviews (Blauert 1983; Zurek 1987; Litovsky et al. 1999).

As mentioned above, the precedence effect, as measured with successive click-pairs, can be accounted for via a cross-correlation of the stimuli as processed by the auditory periphery. It is useful to begin by describing how pairs of binaural clicks, each of which have a broad spectral content, would be transformed by the auditory periphery. The top portion of Figure 7.10 displays the frequency response of an auditory filter centered at 500 Hz. The plot displays the response of a 75-Hz-wide gammatone filter (Glasberg and Moore 1990). The bottom panel illustrates the temporal response of the filter to a single click (i.e., the filter's impulse response). Notice that, in response to the click, the filter's output builds up to a maximum output about 6 ms after the occurrence of the click and, importantly, the filter continues to "ring" for another 15 ms or so. Clearly, if one were to present two clicks separated by 1 ms or so, their responses, at the output of such a filter, would be "smeared" together. In order to show the consequences of this, Figure 7.11 displays the outputs of a *pair* of 500-Hz-centered filters, one for the left ear, the other for the right ear. As shown at the top of the figure, the inputs to the filters were a pair of binaural clicks. The left, center, and right panels depict conditions in which the time by which the second binaural click followed the first (the interclick interval, or ICI) was 1 ms, 2 ms, or 3 ms, respectively. In all cases, the first binaural click carried an ITD of 0 μs and the second binaural click carried an ITD of 200 μs, leading in the left ear. The first row shows the output of the filter at the left ear; the second row shows the output of the filter at the right ear. First, note that although the input to each filter is a discrete pair of clicks, the response of each filter reflects the "smearing together" of the inputs. Also note that the sizes and shapes of the outputs of each filter are heavily dependent upon the ICI. This is so because the ICI determines the degree to which the temporal ringing of the filter that is induced by the second click adds in or out of step, that is, in or out of phase with the ringing produced by the first click.

The third and fourth rows of Figure 7.11 show the *instantaneous values* of

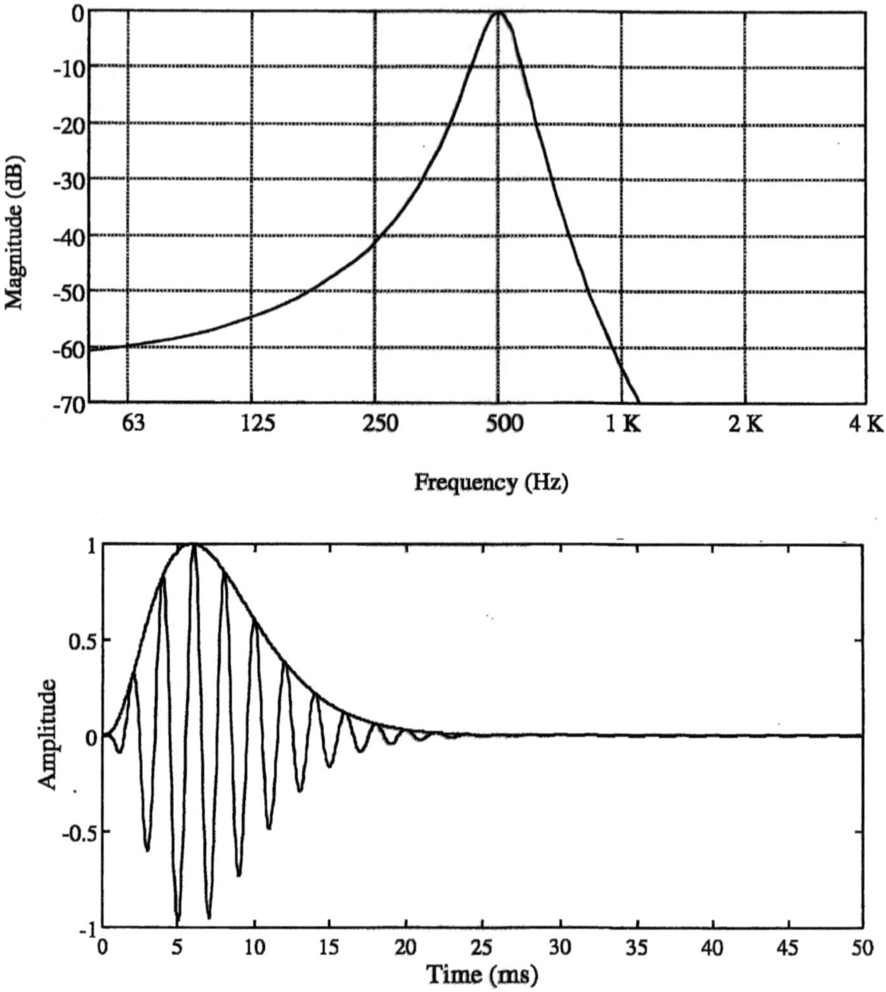

FIGURE 7.10. (*Top*) The logarithmic amplitude of the frequency response of a gammatone-filter centered at 500 Hz and having a 3-dB down bandwidth of 75 Hz. (*Bottom*) The impulse response of this filter. The *solid line* drawn through the positive peaks of the response represents the envelope of the output. (From Hartung and Trahiotis 2001. Reprinted with permission.)

FIGURE 7.11. The outputs of a gammatone filter, like the one shown in Figure 7.10, for inputs that are pairs of binaural transients having monaural interclick intervals (ICIs) of either 1, 2, or 3 ms, as defined the time between the onsets of pairs of monaural inputs. The ITD conveyed by the second pair of inputs was 200 μs. The *upper two rows* of the figure display the outputs of a tandem of "left" and "right" filters when the input to each is the pair of successive transients described at the top of each column. The third and fourth rows of each column show the instantaneous values of ITDs and IIDs as measured after filtering. (From Hartung and Trahiotis 2001. Reprinted with permission.)

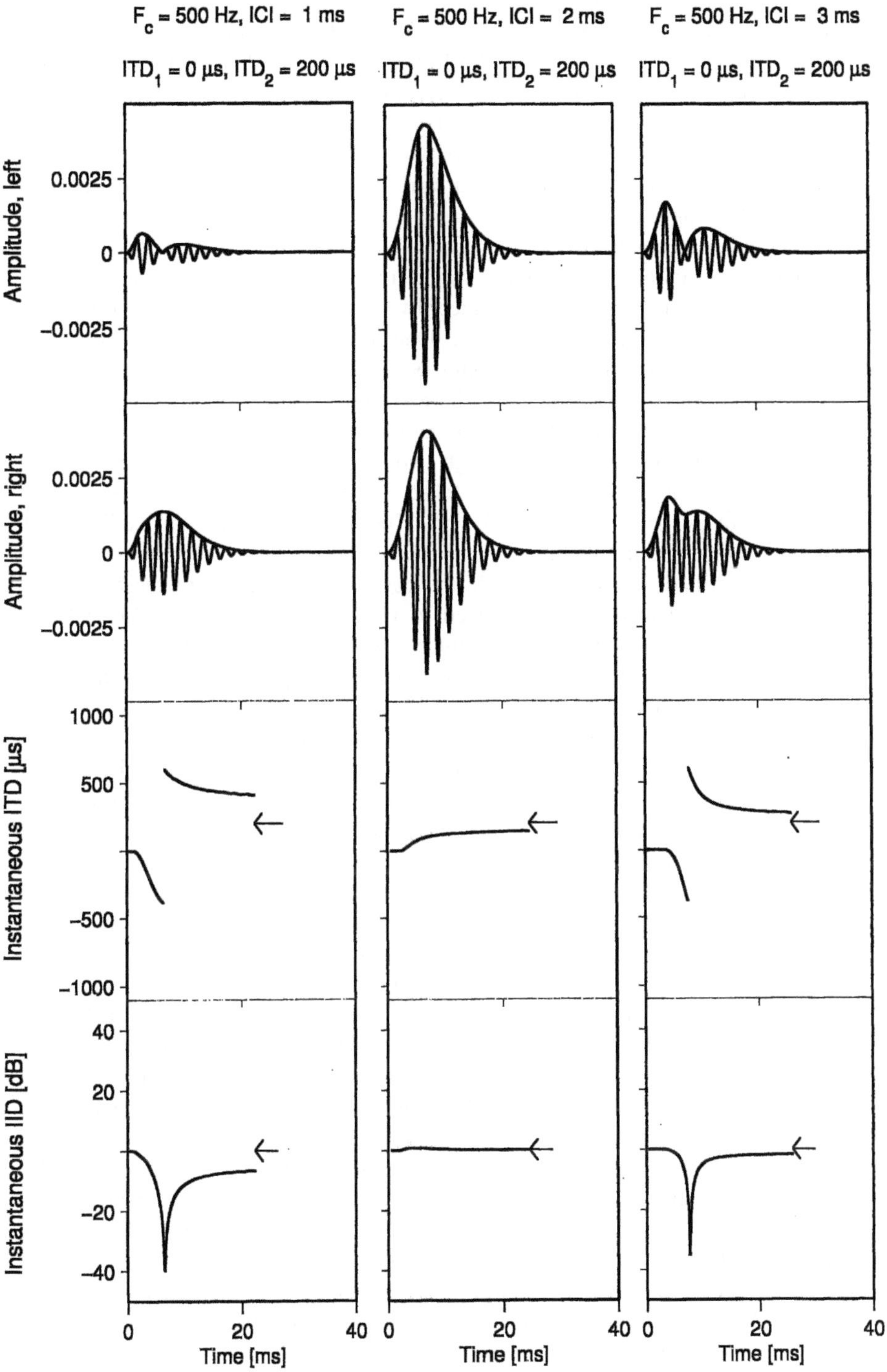

F_c = 500 Hz, ICI = 1 ms
ITD_1 = 0 μs, ITD_2 = 200 μs
F_c = 500 Hz, ICI = 2 ms
ITD_1 = 0 μs, ITD_2 = 200 μs
F_c = 500 Hz, ICI = 3 ms
ITD_1 = 0 μs, ITD_2 = 200 μs
Amplitude, left
0.0025
0
−0.0025
Amplitude, right
0.0025
0
−0.0025
Instantaneous ITD [μs]
1000
500
−500
−1000
Instantaneous IID [dB]
40
20
−20
−40
0
20
40
Time [ms]

ITD and interaural intensitive disparities (IID), respectively, computed by comparing the responses of the left and right filters. The arrows within these panels indicate the *external, physical* ITD (200 μs) and IID (0 dB) imposed on the second click. These plots reveal that the ITDs and IIDs at the outputs of the filters vary over time and, for the most part, are drastically different than their counterparts at the input. For example, when the ICI was either 1 or 3 ms, the instantaneous ITDs and IIDs that occur early in the response of the filter actually favor the lagging ear!

It should be recognized that the outputs of left/right pairs of filters centered at frequencies other than 500 Hz would reflect their own, idiosyncratic, responses, which, in turn, would produce diverging patterns of instantaneous values of ITD and IID across frequency. One might expect that the collection of binaural cues across frequency would be so chaotic as to preclude a useful account of listeners' perceptions of these stimuli. That turns out not to be the case. Hartung and Trahiotis (2001) computed the correlogram for binaural clicks after passing the stimuli through two stages of peripheral processing: (1) a bank of auditory filters similar to those described above which spanned the frequency-range from about 250 Hz to 1700 Hz and (2) a "hair-cell" model (Meddis 1986, 1988; Meddis et al. 1990, Slaney 1993). The hair-cell model provided rectified, low-pass-filtered, and compressed versions of the bandpass-filtered clicks. The reader is referred to Hartung and Trahiotis (2001) for a detailed discussion of the effects produced by the hair-cell model.

In order to make predictions for Wallach et al.'s experimental conditions, the correlogram was averaged across auditory filters (i.e., across frequency). Such averaging effectively emphasizes the consistent patterns of activity within the correlogram. In fact, such across-frequency averaging was postulated by Shackelton et al. (1992) as a way to account for the straightness effects found by Stern et al. (1988) and Trahiotis et al. (2001) discussed earlier.

This resulted in a two-dimensional representation of the correlogram, specifically activity as a function of internal delay. The location, in terms of internal delay, of the most central peak of activity was then taken to be a prediction of the perceived position of the fused intracranial image produced by the pairs of binaural clicks defining each stimulus condition. All that remained was to determine pairings of ITDs (one imposed on the first binaural click [ITD_1], the other imposed on the trailing binaural click [ITD_2] that resulted in the predicted position of the total stimulus being at midline. These pairings of ITD form the ordinate (ITD_1) and the abscissa (ITD_2) of Figure 7.12. The top panel of the figure shows the data obtained by Wallach et al. (1949) from their two listeners (open symbols) along with the predictions of the model (solid diamonds). The bottom panel shows data and predictions obtained in Yost and Soderquist's (1984) replication of Wallach et al.'s study. Note that, overall, the model predicts the data quite well including the somewhat paradoxical nonmonotonicity or "reversals" that occur when very large ITDs are conveyed by the second pair of clicks. Although not shown here, the model was also able to account for precedence data obtained by Shinn-Cunningham et al. (1995) when the stimuli were

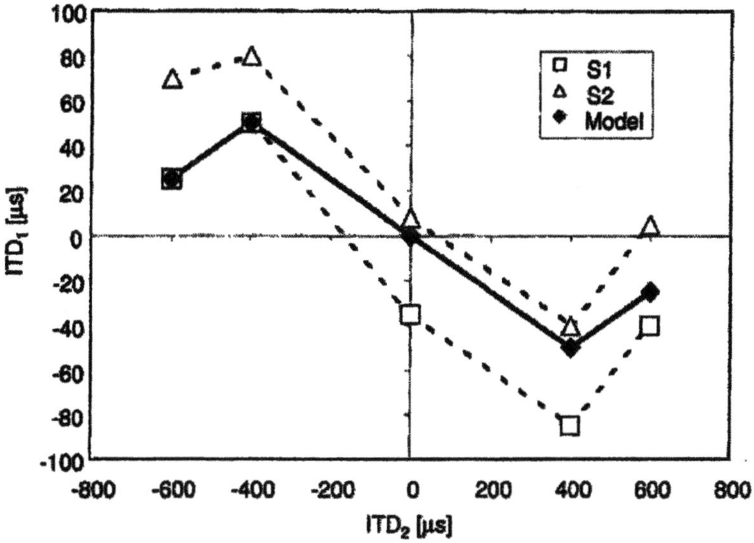

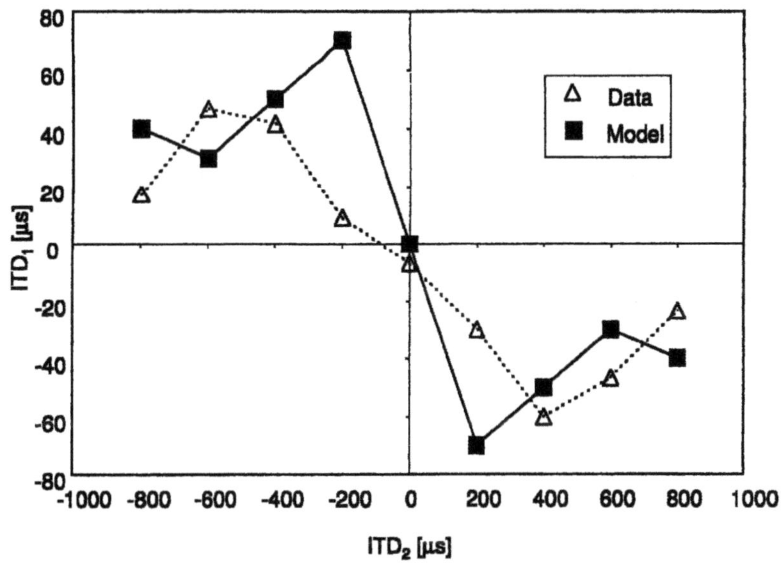

FIGURE 7.12. (*Top*) Combinations of ITD conveyed by the first pair of binaural transients and ITDs conveyed by the second pair of transients required to produce a midline intracranial image for the total stimulus, as reported by Wallach et al. (1949). The *open symbols* represent the data obtained for their two subjects and the *closed diamond* represents the predictions of the model that takes into account peripheral auditory processing, as described in the text. (*Bottom*) Similar data obtained by Yost and Soderquist (1984). The *open symbols* represent the average data obtained for their three subjects and the closed squares represents the predictions of the model. (From Hartung and Trahiotis 2001. Reprinted with permission.)

bursts of band-limited noise rather than clicks. They required listener's to indicate intracranial position for pairs of bursts of noise for a variety of combinations of ITDs conveyed by the first and trailing burst, respectively. Thus, the model can account for precedence data obtained with either clicks or bursts of bandlimited noise for stimulus conditions producing either centered or substantially lateralized intracranial images. The model successfully accounts for the data without invoking neural inhibitory mechanisms like those typically invoked to explain both behavioral (e.g., Lindemann 1986a; Zurek 1987; Litovsky et al. 1999) and physiological (e.g., Yin 1994; Fitzpatrick et al. 1995; Litovsky et al. 1999) measures of the precedence effect, as studied with transient stimuli presented over earphones. This is not to say that the results of other types of studies of the precedence effect, which employ a wide variety of longer stimuli such as speech, are explainable solely on the basis of peripheral processing followed by cross-correlation. Indeed, central, attentive, and perhaps other cognitive factors may also play a role. In our view, the evidence presented above and congruent findings regarding recent physiological measures of precedence (Trahiotis and Hartung 2002) make clear that any role(s) played by those factors can only be evaluated *after* the effects of peripheral auditory processing have been taken into account.

4. Additional Issues

In this section, several further issues concerning the application, generality, and explanatory power of cross-correlation–based models are considered. The objective is to provide a general understanding of their relevance and import and, therefore, details are largely omitted.

4.1 Relating the Two-Dimensional Interaural Cross-Correlation Function to the Interaural Correlation

In the previous section we illustrated how changes in the *locations* of the trajectories along the internal delay axis of the correlogram can account for data obtained in tasks in which changes in subjective lateral position, per se, appear to mediate performance. In Section 2 we described how data obtained with other tasks, for which lateral position does not appear to be the dominant cue, can be accounted for on the basis of changes in the *index* of interaural correlation, Eq. (7.2). Such changes are manifest as changes in the relative *heights* of the trajectories of the correlogram in regions corresponding to the frequency of the signal. For example, Figure 7.13 illustrates how an ensemble of "Jeffress-type" coincidence-counting units responds to typical stimuli that are used for binaural detection experiments. The bottom panel of the figure depicts, after weighting for centrality, the response to a broadband noise masker that is interaurally in phase. The center panel shows the corresponding response when a

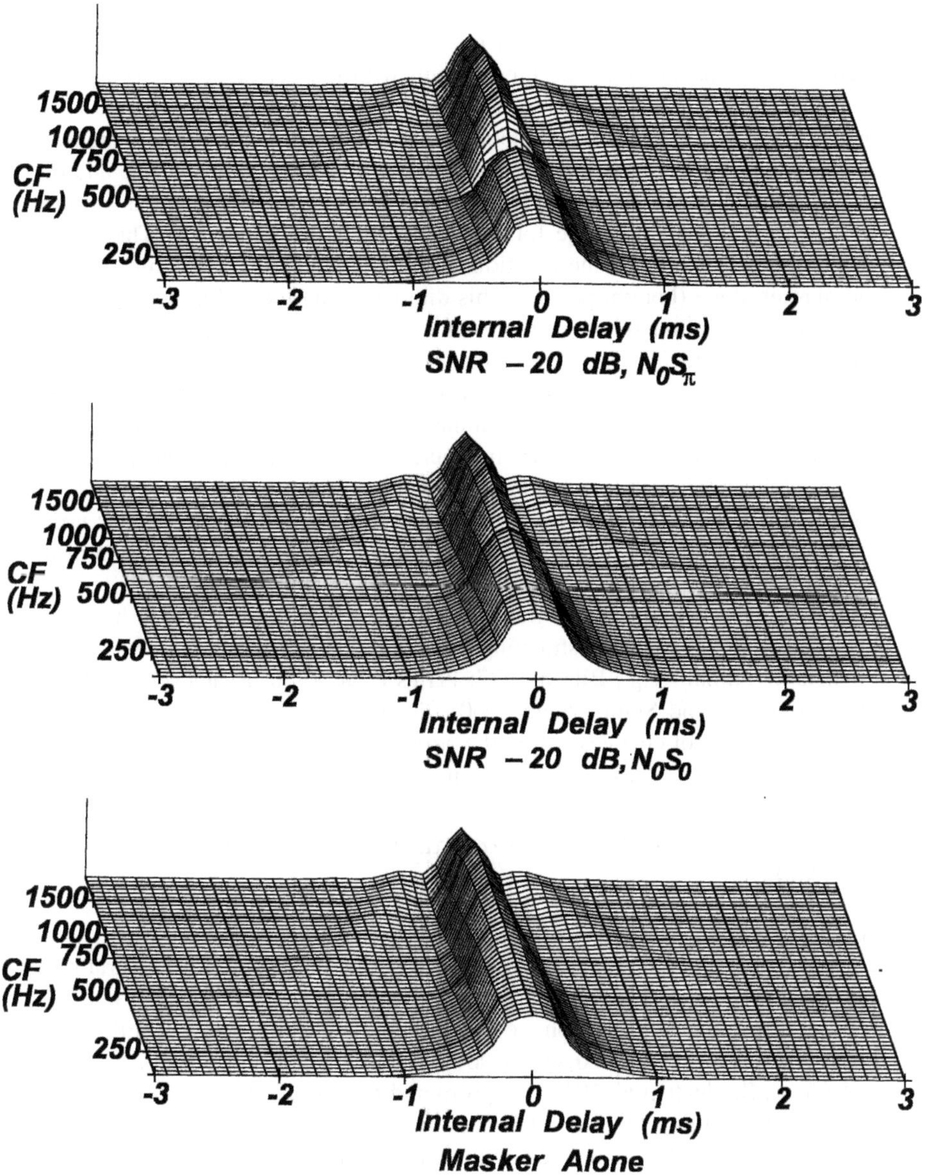

FIGURE 7.13. (*Bottom*) The response to a broadband noise maker that is interaurally in phase, after weighting for centrality. (*Center*) The corresponding response when a 500-Hz target is added to the broadband masker, with both target and masker being interaurally in phase (i.e., in the NoSo configuration). (*Top*) The corresponding response when the target is presented interaurally out of phase (i.e., in the N0Sπ configuration). (From Stern and Trahiotis 1997. Reprinted with permission.)

500-Hz target is added to the broadband masker, with both target and masker being interaurally in phase (i.e., in the NoSo configuration). As can be seen, there is little change from the presentation with masker alone (bottom panel), and this is used to explain why the target is not detected. Finally, the upper panel shows the corresponding response when the target is presented interaurally out of phase (i.e., in the NoSp configuration). It can be seen that the addition of the 500-Hz out-of-phase (Sp) target to the in-phase masker causes a "dimple" to appear in the central ridge for CFs near the target frequency. This dimple accounts for the fact that the addition of the target is easily discriminated from the masker alone (bottom panel). This dimple, being located at lag-zero of the interaural correlation function, represents a decrease in the index of interaural correlation [Eq. (7.2)]. We believe that such changes in the index of interaural correlation successfully describe the results of many binaural detection experiments utilizing diotic maskers. When maskers are not diotic, for example, when the masker contains an ITD or an interaural phase-shift, similar changes in the shape of the interaural correlation function can also reveal the presence of the signal. In such cases, however, the changes that foster detection occur not at lag-zero, but at internal delays, depending on the particular configuration of the stimuli.

The relative heights of the trajectories along the frequency axis have also been shown to be useful in estimating the "central" pitches produced by a variety of dichotic stimuli (e.g., Bilsen 1977; Frijns et al. 1986; Raatgever and Bilsen 1986; Akeroyd and Summerfield 2000). Depending on the nature of the task, differing aspects of changes within the correlogram (i.e., horizontal translations, relative heights, or both) may provide information to the listener.

4.2 The The Range of Interaural Delays Used in Binaural Analysis

A second issue concerns the range of internal time-delays that should be included in correlograms used in models to describe or account for binaural processing. In addition to the experiments discussed above (see Figs. 7.8 and 7.9), it has been known from earphone-based studies that listeners can utilize very large ITDs (up to 10 to 20 ms or so) in lateralization and binaural detection tasks (e.g., Blodgett et al. 1956; Langford and Jeffress 1964; Rabiner et al. 1966; Mossop and Culling 1998). Such data, however, do not force the conclusion that listeners somehow code the stimuli using internal delays of such large magnitudes. It is logically possible that very large values of external time-delays can be processed, albeit somewhat poorly or inefficiently, by neurons tuned to much smaller ITDs. As seen in the correlograms of low-frequency stimuli, the patterns are often quite broad along the internal delay axis and, therefore, there may be sufficient information in the "edges" of activity to support the observed performance. That is, a nominally very large value of external delay might "excite" neural elements tuned to smaller internal delays.

Recent analyses of binaural detection data by van der Heijden and Trahiotis

(1999) bear directly on this issue. Their data and quantitative analyses indicate that human listeners are able to encode, or compensate for, external delays of up to 700 μs or so virtually without error. External delays of 700 μs or so are the largest head-related interaural delays available to humans. In addition van der Heijden and Trahiotis showed that listeners are able to encode and compensate for delays of up to 3 to 4 ms or more with a precision that declines with the magnitude of the delay.

The findings are consistent with the concept of "centrality" (the relatively greater salience of small interaural delays) discussed earlier. Recall that centrality is often attributed to there being an essentially inverse relation between the magnitude of internal delay and the density of neural units "tuned" to those delays. The crucial point is that, in order to account for the behavioral results, one *must* explicitly include the processing of large interaural delays. Said differently, if one were to truncate the correlogram at ±700 μs (the range of ecologically available delays produced by the travel-time of a wavefront around the head), one simply could not account for the data. The argument appears to generalize to other species. Saberi (1999) has recently shown that the binaural perception of barn owls is also affected by the straightness and centrality of the correlogram over a range of internal delays that far exceeds the range of external delays that are naturally available. The patterning of the data Saberi obtained was much like that observed with human listeners.

Physiological investigations have provided evidence that is consistent with the concept of centrality and the conclusion that external delays larger than those encountered naturally are encoded neurally. Centrality is supported by the findings that the majority of neural units encountered that exhibit tuning to interaural delay are tuned to relatively small values (e.g., Crow et al. 1978; Kuwada and Yin 1983). Those investigations reveal that there are, indeed, a relatively small number of neural units tuned to quite large delays. These findings are typified by the measurements shown in Figure 7.14, which was provided by Dr. Shigeyuki Kuwada. The plot displays the distribution of characteristic delays measured from more than 400 single neural units in the inferior colliculus of awake rabbits. There exist units with characteristic delays of 750 μs or more even though the largest ecologically valid head-size–related interaural delay for the rabbit is about 300 μs. This is not to say that distribution of neural units in terms of the internal delays to which they are tuned is the sole factor underlying effects attributed to centrality. Other factors may operate.

All of these recent findings validate the remarkably insightful discussion offered by Jeffress et al. in 1956. Those authors wrote, "We picture, therefore, a system of delay nets with a finely graded series of delays in the region corresponding to the median plane, and coarser and coarser delay steps as the total delay becomes larger. The longer delays are probably provided by chains of synapses. We picture also a considerable mass of nerve tissue in the region representing the median plane and greater sparsity of tissue as the delays become longer" (p. 420). The apparently necessary incorporation of decreasing precision of encoding associated with larger and larger delays found by van der

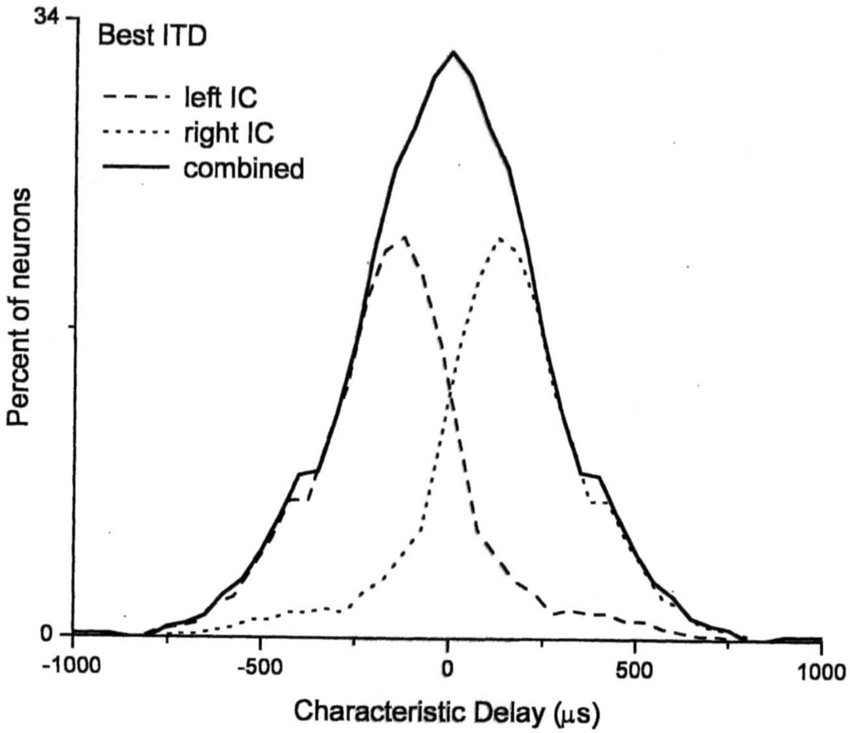

FIGURE 7.14. The distribution of characteristic delays measured from single neural units in the inferior colliculus of awake rabbits, for which the largest ecologically valid head-size related interaural delay is about 300 μs. The plot is a summary of data obtained from more than 400 such units.

Heijden and Trahiotis (1999) some four decades after Jeffress et al.'s conjectures could result from either a more coarse representation of larger internal delays, fewer units tuned to larger delays, or both. The reader is referred to McFadden (1973) for an insightful discussion of how large interaural delays may be processed, be implemented neurally, and affect perception.

4.3 Binaural "Sluggishness" and Temporal Integration

A third general issue concerns the "tracking" or "following" of dynamic, rather than static, changes in interaural delay. In that regard, the binaural system has been found time and time again (pun intended) to be rather "sluggish" in that it appears to "smooth over" or average, rapid fluctuations in interaural disparities (e.g., Licklider et al. 1950; Perrott and Nelson 1969; Perrott and Musicant 1977; Grantham and Wightman 1978, 1979; Grantham 1982, 1984b; for a review, see

Bernstein 1997). Specifically, when the rates of change of interaural temporal cues exceeds about 5 to 10 Hz, listeners are very poor at tracking the changes. The time-constant of this "binaural sluggishness" is an order of magnitude greater than what would be expected on the basis of the temporal responses of the (monaural) auditory filters. On this basis, sluggishness appears to be a central phenomenon.

Several quantitative analyses of data obtained in a wide variety of experiments employing dynamically changing ITDs are now available. For the most part they have focused on the averaging of changes in the binaural cues in order to characterize the shape and time-constants of "temporal windows" that are assumed to mediate the integration of binaural information over time (e.g., Kollmeier and Gilkey 1990; Culling and Summerfield 1998; Holube et al. 1998; Bernstein et al. 2001). In fact, and somewhat remarkably, the same form of temporal window was found to describe the processing of dynamic changes in ITD for barn owls (Wagner 1991) and human listeners (Bernstein et al. 2001) when both species were tested with similar stimuli and comparable paradigms.

One need not be restricted to quantifying sensitivity to dynamic changes in ITD in terms of an averaging of the physical cues themselves. For example, one successful explanation of binaural sluggishness considers the major limiting factor to be "persistence of activity" within the surface of the correlogram itself, or perhaps, within whatever mechanism "reads" that surface (Stern and Bachorski 1983). Within this conception, one envisions the surface of the correlogram as being quite "viscous" and, for that reason, unable to change its shape rapidly in response to changes in ITD (see Bernstein 1997 for a detailed discussion of these issues).

Recall that the correlograms used to account for the precedence effect were averaged across frequency. They were also averaged across a period of time (30 ms). That averaging effectively "smeared" or averaged the dynamically changing binaural information produced by the interaction of the click-stimuli with the auditory filters. This caused the model to predict a stable, rather than time-varying percept that occupied only one intracranial position. Given the success of that model in predicting the precedence results it appears necessary to incorporate binaural sluggishness in accounts of binaural precedence.

One might wonder how interaural intensity differences (IIDs), another major binaural cue, are incorporated within the rubric of interaural correlation. Although not considered here in any detail, it is the case that several lines of evidence point toward the conclusion that IIDs function as simple weights applied to the correlogram (Sayers and Cherry 1957; Blauert and Cobben 1978; Stern and Colburn 1978; Buell et al. 1994). Specifically, IIDs favoring one ear are envisioned as effectively amplifying the activity on the side of the correlogram corresponding to the favored ear. It is interesting to consider that IIDs, regardless of their magnitude, appear to affect binaural perception essentially equivalently across the audible spectrum (e.g., Yost 1981; Grantham 1984a). This is true even though large IIDs are rarely encountered naturally at low frequencies. Readers interested in understanding how both natural and unnatural combinations of ITDs and IIDs affect binaural perception are referred to an excellent recent investigation and analysis by Gaik (1993).

4.4 Closing Remarks

From the foregoing, it should be clear that much can be gained by using the concept of interaural correlation as the basis of a working model of binaural processing. Does that success necessitate the existence of any particular, or even orderly, arrangement of specialized neurophysiological elements? We think not. Rather, it seems prudent to adopt the viewpoint of Huggins and Licklider (1951). They suggested two principles to be used regarding the organization and functioning of the nervous system. The first is the principle of "sloppy workmanship;" the second is the principle of diversity. According to Huggins and Licklider:

> The principle of sloppy workmanship states that it is dangerous to postulate a neural structure that is precisely arranged in detail. The nervous system is the product of a superb architect and a sloppy workman, and in his plans the architect took into account the fact that the workman would not get all the terminal boutons where they belonged. One of the basic facts of neurophysiology is that the nervous system works despite a considerable amount of misarrangement of detail. That fact should be taken into account in constructing theory. . . . it is important to keep in mind that a statistical interpretation of details is required. Thus, in our opinion, the hypothesis that the nervous system computes an exact derivative [in the context of this chapter, a correlation], as by a digital process, is hardly to be taken seriously. But, for example, the hypothesis that the nervous system performs, in its statistical and analogical way, an operation that may be roughly described as differentiation [correlation], and one that we may represent by differentiation [correlation] in a mathematical model, seems to account economically for a considerable range of facts.
>
> The principle of diversity states that the nervous system often hedges: Instead of presenting a single transform of the peripheral stimulation to the higher centers, the auditory tract may present a number of transforms. Given a number of views of the stimulus, the cortex may look them over and take the most useful one. Or it may accept them all and operate upon them all, trying to piece together a consistent picture of the outside world. . . . The principle of diversity is in a sense opposed to the principle of parsimony. It suggests that we look not for the simplest explanation consistent with the data at hand but for an explanation consistent both with the data and with the demands of efficiency in the processing of information. The principle of diversity suggests that a simple description of the auditory process may not be possible because the process may not be

> simple. Theories that appear at first thought to be alternatives may in fact supplement one another. (p. 299)

A recent, comprehensive review by Yin et al. (1997) presents a variety of evidence suggesting that some neural mechanisms postulated by Jeffress within his interaural "delay line" do exist and, in general, do function as the elements of a complex cross-correlator. It is the case, however, that the same review makes clear that some challenges exist. For example, while there is some evidence for an orderly spatial mapping of internal delay within the medial superior olive (MSO), such a spatial mapping has not yet been identified in more central nuclei. In light of the principles of sloppy workmanship and diversity and the great complexity of the auditory system, the fact that the preponderance of evidence, be it behavioral, anatomical, or physiological, supports one or another of the mechanisms envisioned by Jeffress a half-century ago attests to his remarkable achievements.

5. Summary

This chapter presents an introduction to binaural information processing explained or accounted for within the general rubric of interaural correlation. Interaural correlation is described and considered in two different ways. First, interaural correlation is considered in terms of mathematical formulas and indices used to quantify the amount of similarity between a pair of signals presented to the ears. Second, interaural correlation is considered in terms of a putative, internal, cross-correlation function (correlogram) that represents neural activity at specific paired values of frequency and internal delay. The two representations of interaural correlation are shown to account quantitatively for several binaural perceptual phenomena including binaural detection, resolution of ITDs, lateralization, and the precedence effect. An overarching principle that is proposed is that binaural perception is strongly tied to, or explained by, the locations and shapes of features within the correlogram. The generality of the approach is considered both in terms of other perceptual phenomena including binaural "sluggishness" and dichotic pitch and in terms of neurophysiological findings that are consistent with the operation of the mechanisms proposed.

Acknowledgments. The authors thank Dr. Richard L. Freyman for his helpful comments and Dr. Shigeyuki Kuwada for providing the data in Figure 7.14. The preparation of this chapter was supported by research grants NIH DC-04147, DC-04073, and DC-00234 from the National Institute on Deafness and Other Communication Disorders, National Institutes of Health.

References

Akeroyd MA, Summerfield AQ (2000) The lateralization of simple dichotic pitches. J Acoust Soc Am 108:316–334.

Aoki S, Houtgast T (1992) A precedence effect in the perception of inter-aural cross correlation. Hear Res 59:25–30.

Bernstein LR (1997) Detection and discrimination of interaural disparities: modern earphone-based studies. In: Gilkey RH, Anderson T (eds), Binaural and Spatial Hearing. Mahwah, NJ: Lawrence Erlbaum, pp. 117–138.

Bernstein LR and Trahiotis C (1996a) On the use of the normalized correlation as an index of interaural envelope correlation. J Acoust Soc Am 100:1754–1763.

Bernstein LR and Trahiotis C (1996b) The normalized correlation: accounting for binaural detection across center frequency. J Acoust Soc Am 100:3774–3784.

Bernstein LR, Trahiotis C (2002) Enhancing sensitivity to interaural delays at high frequencies by using "transposed stimuli". J Acoust Soc Am 112:1026–1036.

Bernstein LR, Trahiotis C, Akeroyd MA, Hartung K (2001) Sensitivity to brief changes of interaural time and interaural intensity. J Acoust Soc Am 109:1604–1615.

Bilsen FA (1977) Pitch of noise signals: Evidence for a "central spectrum." J Acoust Soc Am 61:150–161.

Blauert, J (1983) Spatial Hearing—The Psychophysics of Human Sound Source Localization. Cambridge, MA: MIT Press.

Blauert J, Cobben W (1978) Some consideration of binaural cross correlation analysis. Acustica 39:96–103.

Blodgett HC, Wilbanks WA, Jeffress LA (1956) Effects of large interaural time differences upon the judgment of sidedness. J Acoust Soc Am 28:639–643.

Buell TN, Trahiotis C, Bernstein LR (1994) Lateralization of bands of noise as a function of combinations of interaural intensitive differences, interaural temporal differences and bandwidth. J Acoust Soc Am 95:1482–1489.

Colburn HS (1977) Theory of binaural interaction based on auditory-nerve data. II: Detection of tones in noise. J Acoust Soc Am 61:525–533.

Colburn HS (1996) Computational models of binaural processing. In: Hawkins HL, McMullen TA, Popper AN, Fay RR (eds), Auditory Computation. New York: Springer Verlag, pp. 332–400.

Crow G, Rupert AL, Moushegian G (1978) Phase locking in monaural and binaural medullary neurons: implications for binaural phenomena. J Acoust Soc Am 64:493–501.

Culling JF, Summerfield Q (1998) Measurements of the binaural temporal window using a detection task. J Acoust Soc Am 103:3540–3553.

Durlach NI, Gabriel KJ, Colburn HS, Trahiotis C (1986) Interaural correlation discrimination: II. Relation to binaural unmasking. J Acoust Soc Am 79:1548–1557.

Fitzpatrick DC, Kuwada S, Batra R, Trahiotis C (1995) Neural responses to simple simulated echoes in the auditory brain stem of the unanesthetized rabbit. J Neurophysiol 74:2469–2486.

Frijns HM, Raatgever J, Bilsen FA (1986) A central spectrum theory of binaural processing: the binaural edge pitch revisited. J Acoust Soc Am 80:442–451.

Gabriel KJ, Colburn HS (1981) Interaural correlation discrimination: I. Bandwidth and level dependence. J Acoust Soc Am 69:1394–1401.

Gaik W (1993) Combined evaluation of interaural time and intensity differences: psychoacoustic results and computer modeling. J Acoust Soc Am 94:98–110.

Glasberg BR, Moore BCJ (1990) Derivation of auditory filter shapes from notched-noise data. Hear Res 47:103–138.

Grantham DW (1982) Detectability of time-varying interaural correlation in narrow-band noise stimuli. J Acoust Soc Am 72:1178–1184.

Grantham DW (1984a) Interaural intensity discrimination: Insensitivity at 1000 Hz. J Acoust Soc Am 75:1191–1194.

Grantham DW (1984b) Discrimination of dynamic interaural intensity differences. J Acoust Soc Am 76:71–76.

Grantham DW, Wightman FL (1978) Detectability of varying interaural temporal differences. J Acoust Soc Am 63:511–523.

Grantham DW, Wightman FL (1979) Detectability of a pulsed tone in the presence of a masker with time-varying interaural correlation. J Acoust Soc Am 65:1509–1517.

Hartung K, Trahiotis C (2001) Peripheral auditory processing and investigations of the 'precedence effect' which utilize successive transient stimuli. J Acoust Soc Am 110: 1505–1513.

Heijden M van der, Trahiotis C (1999) Masking with interaurally delayed stimuli: the use of "internal" delays in binaural detection. J Acoust Soc Am 105:388–399.

Holube I, Kinkle M, Kollmeier B (1998) Binaural and monaural auditory filter bandwidths and time constants in probe tone detection experiments. J Acoust Soc Am 104: 2412–2425.

Houtgast T, Aoki S (1994) Stimulus-onset dominance in the perception of binaural information. Hear Res 72:29–36.

Huggins WH, Licklider JCR (1951) Place mechanisms of auditory frequency analysis. J Acoust Soc Am 23:290–299.

Jain M, Gallagher DT, Koehnke J, Colburn HS (1991) Fringed correlation discrimination and binaural detection. J Acoust Soc Am 90:1918–1926.

Jeffress LA (1948) A place mechanism of sound localization. J Comp Physiol 41:35–39.

Jeffress LA (1972) Binaural signal detection: vector theory. In: Tobias JV (ed), Foundations of Modern Auditory Theory, Vol. 2. New York: Academic Press, pp. 349–368.

Jeffress LA, Blodgett HC, Sandel TT, Wood CL III (1956) Masking of tonal signals. J Acoust Soc Am 28:416–426.

Kim DO, Molnar CE (1979) A population study of cochlear nerve fibers: comparison of spatial distributions of average-rate and phase-locking measures of responses to single tones. J Neurophysiol 42:16–30.

Kollmeier B, Gilkey RH (1990) Binaural forward and backward masking: evidence for sluggishness in binaural detection. J Acoust Soc Am 87:1709–1719.

Kuwada S, Yin YCT (1983) Binaural interaction in low-frequency neurons in the inferior colliculus of the cat. I. Effects of long interaural delays, intensity, and repetition rate on interaural delay function. J Neurophysiol 50:981–999.

Kuwada S, Stanford TR, Batra R (1987) Interaural phase-sensitive units in the inferior colliculus of the unanaesthetized rabbit: effects of changing frequency. J Neurophysiol 57:1338–1360.

Langford TL, Jeffress LA (1964) Effect of noise crosscorrelation on binaural signal detection. J Acoust Soc Am 36:1455–1458.

Licklider JCR, Webster JC, Hedlun JM (1950) On the frequency limits of binaural beats. J Acoust Soc Am 22:468–473.

Lindemann W (1986a) Extension of a binaural cross-correlation model by contralateral

inhibition. I. Simulation of lateralization for stationary signals. J Acoust Soc Am 80: 1608–1622.
Lindemann W (1986b) Extension of a binaural cross-correlation model by contralateral inhibition. II. The law of the first wavefront. J Acoust Soc Am 80:1623–1630.
Litovsky RY, Colburn HS, Yost WA, Guzman SJ (1999) The precedence effect. J Acoust Soc Am 106:1633–1654.
McFadden DM (1968) Masking-level differences determined with and without interaural disparities in masker intensity. J Acoust Soc Am 44:212–223.
McFadden DM (1973) Precedence effects and auditory cells with long characteristic delays. J Acoust Soc Am 54:528–530.
McFadden DM, Pasanen EG (1976) Lateralization at high frequencies based on interaural time differences. J Acoust Soc Am 59:634–639.
Meddis R (1986) Simulation of mechanical to neural transduction in the auditory receptor. J Acoust Soc Am 79:702–711.
Meddis R (1988) Simulation of auditory-neural transduction: further studies. J Acoust Soc Am 83:1056–1063.
Meddis R, Hewitt MJ, Shackleton TM (1990) Implementation details of a computational model of the inner hair-cell/auditory-nerve synapse. J Acoust Soc Am 87:1013–1016.
Mossop JE, Culling JF (1998) Lateralization of large interaural delays. J Acoust Soc Am 104:1574–1579.
Nuetzel JM, Hafter ER (1981) Discrimination of interaural delays in complex waveforms: spectral effects. J Acoust Soc Am 69:1112–1118.
Osman E (1971) A correlation model of binaural masking level differences. J Acoust Soc Am 50:1494–1511.
Perrott DR, Musicant AD (1977) Rotating tones and binaural beats. J Acoust Soc Am 61:1288–1292.
Perrott DR, Nelson MA (1969) Limits for the detection of binaural beats. J Acoust Soc Am 46:1477–1481.
Pollack I, Trittipoe WJ (1959a) Binaural listening and interaural noise cross correlation. J Acoust Soc Am 31:1250–1252.
Pollack I, Trittipoe WJ (1959b) Interaural noise correlations: examination of variables. J Acoust Soc Am 31:1616–1618.
Raatgever J, Bilsen FA (1986) A central spectrum theory of binaural processing. Evidence from dichotic pitch. J Acoust Soc Am 80:429–441.
Rabiner LR, Laurence CL, Durlach NI (1966) Further results on binaural unmasking and the EC model. J Acoust Soc Am 40:62–70.
Robinson DE, Jeffress LA (1963) Effect of varying the interaural noise correlation on the detectability of tonal signals. J Acoust Soc Am 35:1947–1952.
Ruggero MA, Santos-Sacchi J (1997) Cochlear mechanics and biophysics. In: Crocker MJ (ed), Encyclopedia of Acoustics, Vol. 3. New York: John Wiley & Sons, pp. 1357–1369.
Saberi K, Takahashi Y, Farahbod H, Konishi M (1999) Neural bases of an auditory illusion and its elimination in owls. Nat Neurosci 2:656–659.
Sayers B McA, Cherry EC (1957) Mechanism of binaural fusion in the hearing of speech. J Acoust Soc Am 29:973–987.
Shackleton TM, Meddis R, Hewitt, MJ (1992) Across frequency integration in a model of lateralization. J Acoust Soc Am 91:2276–2279.
Shinn-Cunningham BG, Zurek P, Durlach NI, Clifton RK (1995) Cross-frequency interactions in the precedence effect. J Acoust Soc Am 98:164–171.

Slaney M (1993) An efficient implementation of the Patterson–Holdsworth auditory filter bank. Apple Computer Technical Report No. 35.

Stern RM, Bachorski SJ (1983) Dynamic cues in binaural perception. In: Klinke R, Hartmann R (eds), Hearing—Physiological Bases and Psychophysics. New York: Springer-Verlag, pp. 209–215.

Stern RM, Colburn HS (1978) Theory of binaural interaction based on auditory-nerve data. IV. A model for subjective lateral position. J Acoust Soc Am 64:127–140.

Stern RM, Shear GD (1996) Lateralization and detection of low-frequency binaural stimuli: effects of distribution of interaural delay. J Acoust Soc Am 100:2278–2288.

Stern RM, Trahiotis C (1995) Models of binaural interaction, In: Moore BCJ (ed), Handbook of Perception and Cognition, Vol. 6. Hearing. New York: Academic Press, pp. 347–385.

Stern RM, Trahiotis C (1997) Models of binaural perception. In: Gilkey RH, Anderson T (eds), Binaural and Spatial Hearing. Mahwah, NJ: Lawrence Erlbaum, pp. 499–531.

Stern RM, Zeiberg AS, Trahiotis C (1988) Lateralization of complex binaural stimuli: a weighted-image model. J Acoust Soc Am 84:156–165.

Tollin DJ, Henning GB (1998) Some aspects of the lateralization of echoed sound in man. I. The classical interaural delay-based effect. J Acoust Soc Am 104:3030–3038.

Tollin DJ, Henning GB (1999) Some aspects of the lateralization of echoed sound in man. I. The role of stimulus spectrum. J Acoust Soc Am 105:838–849.

Trahiotis C, Hartung K (2002) Peripheral auditory processing, the precedence effect and responses of single units in the inferior colliculus. Hear Res 168:55–59.

Trahiotis C, Stern RM (1989) Lateralization of bands of noise: effects of bandwidth and differences of interaural time and phase. J Acoust Soc Am 86:1285–1293.

Trahiotis C, Bernstein LR, Akeroyd MA (2001) Manipulating the "straightness" and "curvature" of patterns of interaural cross-correlation affects listeners' sensitivity to changes in interaural delay. J Acoust Soc Am 109:321–330.

Wagner H (1991) A temporal window for lateralization of interaural time difference by barn owls. J Comp Physiol A 169:281–289.

Wallach H, Newman EB, Rosenzweig MR (1949) The precedence effect in sound localization. Am J Psychol 52:315–336.

Wightman FL, Kistler DJ (1997) Factors affecting the relative salience of sound localization cues. In: Gilkey RH, Anderson T (eds), Binaural and Spatial Hearing. Mahwah, NJ: Lawrence Erlbaum, pp. 1–23.

Yin TCT (1994) Physiological correlates of the precedence effect and summing localization in the inferior colliculus of the cat. JNeurosci 14:5170–5186.

Yin TCT, Joris PX, Smith PH, Chan JCK (1997) Neuronal processing for coding interaural time disparities. In: Gilkey RH, Anderson T (eds), Binaural and Spatial Hearing. Mahwah, NJ: Lawrence Erlbaum, pp. 427–445.

Yost WA (1981) Lateral position of sinusoids presented with interaural intensive and temporal differences. J Acoust Soc Am 70:397–409.

Yost WA (1997) The cocktail party problem: forty years later. In: Gilkey RH, Anderson T (eds), Binaural and Spatial Hearing. Mahwah, NJ: Lawrence Erlbaum, pp. 329–347.

Yost WA, Soderquist DR (1984) The precedence effect: Revisited. J Acoust Soc Am 76: 1377–1383.

Zurek PM (1980) The precedence effect and its possible role in the avoidance of interaural ambiguities. J Acoust Soc Am 67:952–964.

Zurek PM (1987) The precedence effect. In: Yost WA, Gourevitch(eds), Directional Hearing. New York: Springer-Verlag, pp. 85–105.

8

Models of Sound Localization

H. STEVEN COLBURN AND ABHIJIT KULKARNI

1. Introduction

The process of extracting information about the locations of acoustic sources from the characteristics of two sound pressure waveforms has been addressed from several points of view. A signal-processing approach considers the problem to be estimation of unknown parameters with partial information available about the signal and the environment's transformation of the signal. A physiological approach considers the mechanisms of signal transduction and recoding through the cochlea, brainstem, midbrain, and cortex. A psychophysical approach provides measures of human performance and relates abilities to the characteristics of the signals and the environment. These three approaches are all informed by the physical acoustics of the environment and of the effects of the head and body on the received signals.

The overall problem of sound localization is outlined in Figure 8.1. A sound source emits a partially characterized signal $s(t)$, which propagates through an environment to generate received signals at the localizer. This localizer is typically a human listener with two ears, but may be generalized to a microphone or a microphone array with many inputs. The acoustic environment, the physical structure of the receiver (e.g., the shape of the human head, torso, and external ear), and the relative locations of the source and the receiver in the environment determine the relationship of the received signals to the original signal waveform. In general, the sound localization problem is "How does the localizer interpret the received signals to determine the location of the sound source?" In the process of localizing the sound source, the listener also extracts information about the acoustic environment, makes use of a priori and multimodal information about the original acoustic signal and the environment, and estimates the signal or properties of the signal. Historically, specific aspects or attributes of the received signals that are used in the localization process, called cues for sound localization, have been identified. These cues always share two properties: first, these attributes must vary with the location of the stimulus; and second, the auditory system must be sensitive to these attributes. As we describe

FIGURE 8.1. Block diagram of signal flow in localization.

more fully below, the classic cues for sound localization are the interaural time difference (ITD, roughly the time delay between the signals at the two ears), the interaural intensity difference (IID, the difference between the level of the signals at the two ears), and the spectral shape (which is a general word for the variation in the level of the received signal as a function of frequency). As we shall see, these general categories require further refinement.

Of course, there may be several signal sources and the received signals are in general the sum of the signals that would be received from each source separately. When multiple sources are present, any sources other than the one being localized would be considered as interferers or distracters. As described in later sections, current models of localization are not adequate for describing human localization performance in the presence of reflections or distracters.

The effect of the environment (including the head, ear, etc.) is assumed to be a linear, locally time-invariant transformation of the signal so that the received signals at the ears are linearly related to the source signal. This transformation from source to ear for a given direction and environment is conveniently described by a transfer function in the frequency domain. In anechoic (nonreverberant) environments, the transfer function is called the head-related transfer function (HRTF), and is usually defined empirically as the Fourier transform of the head-related impulse response (HRIR). In a general reverberant (nonanechoic) environment, although the transfer function is sometimes still called an HRTF, the term HRTF is reserved in this chapter for anechoic conditions and the term spatial transfer function (STF) is used for the general case (and is applied to any receiver including microphones in arrays as well as to ears). Finally, note that there is no uniform definition for the reference pressure for STFs (or HRTFs). Some people use the pressure at the location of the center of the head with no head present as a reference for the transfer functions; others find it more direct to relate the received signal to the source signal. This is the convention used here.

Our notation for the spatial transfer function, $H(w,\theta)$, describes the ratio, for any spatial arrangement and acoustic environment, of the Fourier transform $R(w)$ of the received waveform $r(t)$ to the Fourier transform $S(w)$ of the source waveform $s(t)$, which is a function of the location θ of the source within the room. The location of the head in the room is assumed to be fixed and is not explicitly included in our notation. The parameter θ here is a general representation for the location of the source. In some cases, we focus on a single-dimension angle to specify location, such as the azimuthal angle for source location in an anechoic far field, in which case the transfer function is independent of distance and the elevation angle is held fixed. In other cases, θ is a vector parameter and specifies the coordinates of the source. (The orientation of the source, although relevant and interesting to consider, is not addressed here, primarily because there has been little work on this in the context of human localization.)

For two ears, there are two received signals, and and the general relationships of the transforms is given by

$$\begin{aligned} R_L(w) &= S(w)\ H_L(w) \\ R_R(w) &= S(w)\ H_R(w) \end{aligned} \tag{8.1}$$

where the spatial transfer functions $H_L(w)$ and $H_R(w)$ are determined for each source location θ and include all effects of the environment on the sound transformation. As noted above, the received signals and their transforms must depend on the location θ of the source, on the environment, and on the source. In these terms, the problem of a localizer is to determine the location θ (which is generally a three-component vector) of the source using the received signals, given partial knowledge about the source signal, environment, and receiver properties (including past experience in the environment, inputs from other modalities, and a priori information).

This chapter first provides background about different approaches to the modeling of sound localization (Section 2) and then describes computational models of the sound localization process, separated into azimuth alone (Section 3), elevation alone (Section 4), azimuth and elevation together in the far field (Section 5), and full three-dimensional localization models (Section 6). Finally, several complicated issues are discussed briefly (Sections 7 to 9), and concluding comments are given in the last section (Section 10). In the background material in Section 2, sound localization is summarized as an engineering system problem. Specifically, the relationships of sound localization models to ideas from signal processing, physical acoustics, and physiological mechanisms are described. Even though the focus of this chapter is the modeling of human sound-localization behavior, concepts from other approaches provide useful insights for mechanisms and models. In fact, work from all these areas have always been intertwined, and discussion of models of localization is more meaningful when an integrated point of view is maintained.

The reader will note that almost all of the work on explicit models of localization has been limited to single sources in anechoic space. (Most empirical data are also limited this way.) There has been minimal consideration of the effects of reverberation, multiple sources, or dynamic changes in locations or signal characteristics. In terms of models, this means that almost all of the effort has been directed toward the simplest problems. In the simplest case, a single broadband source with a known spectrum in an anechoic environment, there is more than enough information in the received waveforms to specify the location of the source. That is, there are multiple ways to use partial information from the waveforms to localize the source, and experiments indicate that humans can use multiple aspects of the signals to reach a decision. In the more complex cases with multiple sources, reverberant environments, or uncertain source waveforms, localization behavior with only two received waveforms is underdetermined, and localization judgments depend on a combination of top-down and bottom-up processing. Estimated source locations will depend on expectations, previous histories of signals, and evolved solutions to similar cases. These cases, which may be important for determining the actual algorithms that define human performance, are not highly studied, either theoretically or empirically.

Most of the chapter, like most of the work in the literature, is concerned with the simple cases, but the full complexity should be kept in mind by the reader. The focus of this chapter is modeling, and this treatment should be complemented by reviews that summarize available data more explicitly. The review of sound localization by Middlebrooks and Green (1991) and the book on spatial hearing by Blauert (1997) are particularly recommended as excellent introductions to both empirical and theoretical aspects of sound localization.

2. Background

In the first subsection of the background (Section 2.1), sound localization is discussed as a signal-processing problem, and general aspects of processing with one, two, or more received signals are described. In Section 2.2, a brief review of physical acoustics is presented in the context of a discussion of the nature of the signals received at the ears and their dependence on the source location and the environment. In Section 2.3, the relevant aspects of the peripheral physiological transformations are summarized, in part because they constrain the further processing and in part because many aspects of sound localization performance, and essentially all computational models, are consistent with the peripheral physiology as one would expect. Finally, in Section 2.4, a general model of localization is outlined and used to discuss several aspects of the sound localization process, including top-down influences, cross-modality effects, and cognitive influences in general.

2.1 Engineering Aspects of the Sound-Localization Problem

For the discussion in this section, we assume that "the environment is known" in the sense that $H(w,\theta)$ is available to the localizer for each receiver. This implies that the set of STFs are known explicitly for each source location θ. (It is implicitly assumed here that the receiver [ear] is at a known location and that the known STFs are those that correspond to this fixed receiver location.) It is also imagined that the STF is unique for each θ, so that knowledge of the frequency response is enough to determine a specific value of location. In other words, we assume that the w-contours for different directions are sufficiently distinctive for different directions to allow the direction θ to be determined from looking at the contour of w-dependence of $H(w,\theta)$ derived from a given measurement.

2.1.1 Single Receiver

Assume first that a known, broadband source $s(t)$ [with Fourier transform $S(w)$] results in the received signal $r(t)$ [with transform $R(w)$]. Now since $R(w)$ and $S(w)$ determine the shape of the STF $H(w,\theta)$ as a function of w, even when the direction θ is still unknown [i.e., $H(w,\theta) = R(w)/S(w)$], knowledge of the shape of the STF versus w determines the direction when it is assumed that shapes are distinctive. It is important to note that this computational method of localization depends on a priori information about $S(w)$ and a broad bandwidth. In many circumstances, however, $S(w)$ is not fully known or it is of limited bandwidth. Then, even if the dependence of the spatial transfer function on θ is known, some a priori knowledge about the source or source location must be used to determine source location.

If $S(w)$ is not known, then for any $H(w,\theta)$, there exists an $S(w,\theta)$ that would be consistent with the received data $R(w)$, and the uncertainty in $S(w)$ leads to an inability to determine θ from $R(w)$. This is an instance of the "deconvolution problem" stating that the result of convolving two time functions (corresponding to multiplying transforms) cannot be uniquely resolved into the original two time functions without additional knowledge about the time functions. In other words, given only the product of frequency functions, it is not possible to factor them unambiguously unless additional properties or constraints are provided. For example, for each candidate angle θ, since $H(w,\theta)$ is known, one could set $S(w,\theta) = R(w)/H(w,\theta)$ and the received signal would then be consistent with a source $S(w,\theta)$ at the direction θ.

This analysis shows that, with a single receiver (monaural listening), the ability to localize sound (i.e., determine the direction of the source) depends on knowledge of the source waveform and knowledge of the spatial filter for each direction. Since the human localization system has two receivers, the study of two-receiver systems is critical for understanding human performance, including the relevance of knowledge about the source waveform.

2.1.2 Two Receivers (e.g., Two Ears)

For a single source and two receivers (e.g., right and left ears), the source and its direction are represented at both receivers, consistent with Eq. (8.1) above, and cues for localization become available beyond those that are possible with a single receiver. In the binaural case, the ratio of the received spectra provides a measure that is independent of the source at each frequency (as long as there is energy in the source at that frequency), so that there is less dependence on a priori knowledge.

The interaural transfer function $I(w,\theta)$ is defined as follows:

$$I(w,\theta) = \frac{R_L(w)}{R_R(w)} = \frac{H_L(w,\theta)}{H_R(w,\theta)} \tag{8.2}$$

Now, if source location θ is determined by the shape of the frequency dependence of I (as discussed above for H), then θ can be determined independently of $S(w)$ as long as $S(w)$ is sufficiently broadband to allow the computation of $I(w,\theta)$ over a reasonable range of w. Note that $R_L(w)$ and $R_R(w)$ are both zero and their ratio is undefined for any w for which $S(w)$ is zero.

The interaural ratio $I(w,\theta)$, being a ratio of complex functions, has a magnitude and a phase angle for each frequency at which $S(w)$ is non-zero. The magnitude corresponds to the interaural amplitude ratio for a tone at this frequency and the phase angle of $I(w,\theta)$ corresponds to the interaural phase difference (IPD) at frequency w. These two quantities, often called α and ϕ, respectively, are the basis of a large part of binaural hearing analysis and of localization studies. With arg(z) representing the phase angle of z, α and ϕ are defined as

$$\alpha(w,\theta) = |I(w,\theta)| = \left|\frac{H_L(w,\theta)}{H_R(w,\theta)}\right| \tag{8.3}$$

$$\phi = \arg(I(w,\theta)) = \arg\left(\frac{H_L(w,\theta)}{H_R(w,\theta)}\right) \tag{8.4}$$

although $\alpha(\omega,\theta)$ is usually described in decibels (20 log α) and is called the interaural level difference (ILD). Both parameters α and ϕ are functions of frequency w for each direction θ. Their frequency characteristics are determined by the physical acoustics of the environment (and receiver) and are discussed below.

In the single-source case, if the direction θ can be determined from the ratio $I(w,\theta)$ and if the two received signals $r_L(t)$ and $r_R(t)$ are available, then after finding θ, the transmitted signal can be calculated using the received signal at either ear. For example, given θ, $H_L(w,\theta)$ is determined (i.e., assumed known) and $S(w)$ can be calculated using Eq.(8.1).

If there are two sources from two known directions, θ_1 and θ_2, and it is again assumed that the spatial transfer functions for each direction are known, then in general the sources can be determined from the two received signals. This fact can be appreciated by considering a single frequency w with the spatial transfer functions and the angles known. The following equations apply:

$$\begin{aligned} R_L(w) &= S_1(w)H_L(w,\theta_1) + S_2(w)H_L(w,\theta_2) \\ R_R(w) &= S_1(w)H_R(w,\theta_1) + S_2(w)H_R(w,\theta_2) \end{aligned} \tag{8.5}$$

Since the only unknowns are the source waveform values $S_1(w)$ and $S_2(w)$, these equations can be solved for the source waveforms as a function of frequency. Of course, for special circumstances, such as one in which the transfer functions are equal for both directions, the equations cannot be solved, but in general there is a unique solution with reasonable constraints on the source waveforms.

If there are more than two sources, there is not enough information from two receivers to determine directions or source waveforms. In this case more a priori information or assumptions about the nature of the source are required to make judgments about directions. This case has not been related to models of sound localization.

2.1.3 Multiple Receivers (e.g., Microphone Arrays)

The arguments made above for a single source and two receivers, which generally allow a unique solution to the location and waveform of a single source, can be generalized to multiple sources, as long as there is at least one more receiver than the number of sources (assuming that the receivers are distinctive). In general the directions and source waveforms can be determined for this case. Further, for the case when all source directions have been determined, one can generalize Eq. (8.5) to the case of *N* received signals and *N* sources so that one can determine the sources when their directions and the spatial transfer functions are known.

More general approaches are also possible with multiple receivers, which are called microphone arrays in the acoustic case. These general strategies are usually based on combinations (often linear) of the received signals (microphone outputs) to generate a resultant output signal, and either fixed or adaptive processing can be used to determine the combination weights (Greenberg et al. 2003). The spatial pattern of the output signal for this array, output amplitude as a function of the direction of a source, is known as the antenna pattern of the array processor. The antenna pattern generally has maxima and minima as a function of angle, and strategies are usually focused on a minimum, called a null, or on the pattern near the maximum, often called the beam. For example, a simple two-microphone array can be used to achieve the pattern in Figure 8.2.

The antenna pattern in Figure 8.2 is shown by the heavier curve, and the circles are reference coordinate lines in the circular coordinate system. For each angle from the horizontal line to the right (straight ahead), the plotted curve is the gain (or attenuation) of the output for a source in the far field at this angle. This pattern has a broad beam straight ahead (right side of figure) and two symmetric nulls at $\pm$ 125°. This pattern is generated by subtracting the output of two microphones separated by a distance p for a low frequency, with an internal delay $\tau_{int} = p / \sqrt{3}c$ imposed before subtraction.

Strategies based on finding maxima, either by physically turning the array or by changing the processing of a fixed array to change the direction of the beam, are called "steering the beam" strategies. To localize, for example, one can vary direction of the beam to find the direction for which a maximum output is found and assume that the beam is then pointing toward a source. In general, for beam-steering strategies, performance is improved with narrower beam patterns, and narrower patterns are obtained for larger arrays (relative to wavelength) and for more microphones in the array. For multiple sources, one would expect maxima at each source, although reflections in reverberant environments would also cause local maxima. In fact, the signal from reflections in some direction can be as strong as or stronger than the direct path signal, so that it is not surprising that reflected sound in reverberant environments can cause great difficulties for simple array processing schemes.

Another strategy is to focus on minima in the array output and exclude energy from unwanted sources by "steering the null." In addition to eliminating interfering sounds from an unwanted source, it is also possible to localize by search-

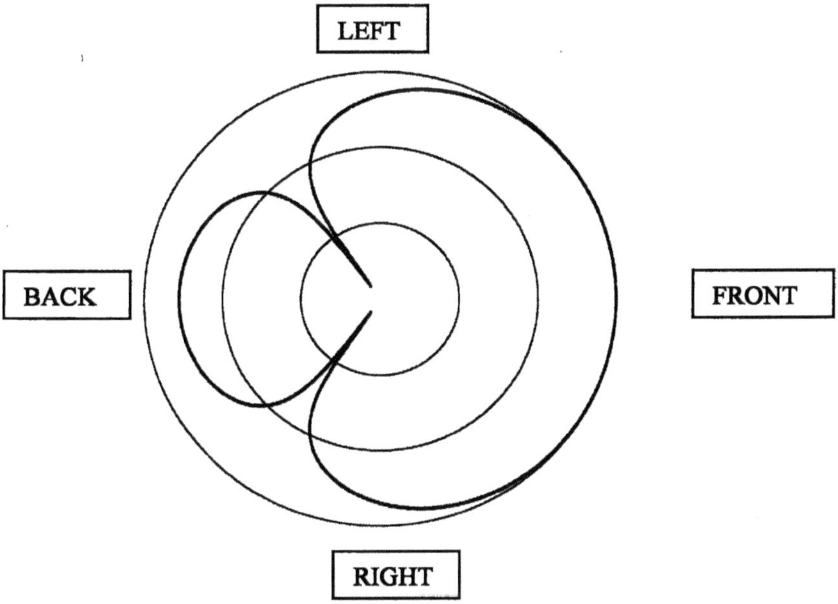

FIGURE 8.2. Antenna pattern for simple two-microphone array. The pattern should be viewed in circular coordinates. For each angle, the distance from the origin is proportional to the gain in decibels. The right side of the figure is straight ahead, the direction upward corresponds to the left of the array axis, etc. The gain is zero straight ahead and the concentric circles are approximately 10 dB apart.

ing for minima in the spatial response patterns. Minima in reverberant or multisource environments would be expected when the sources are in the direction of the null. One can also develop antenna patterns with multiple, independently steerable nulls, and eliminate multiple unwanted sources or unwanted reflections from a desired source.

Our primary interest in describing these simple examples of array-processing schemes is to provide analogies for the internal processing in models of human sound-localization behavior. This notion is considered further later. Another interest is the development of an external processor that could be developed as an artificial localizer. In fact, after locating the sources, a processor could be designed to estimate the signals from the estimated locations. If this were successful, the major problem would then be to design methods for the presentation of the signals to an observer such that he or she could sample the sources and choose the desired source at any time (the target source) for complete attention.

2.2 Physical Acoustics

Acoustic signals in the audio range have wavelengths that vary from about 15 mm (at 20 kHz) to approximately 15 m (at 20 Hz). This range of wavelengths make it clear that the dimensions of the body, head, and structures of the outer ear (pinna) are comparable to the wavelengths; thus, these structures have complex effects on the acoustic variables of pressure and volume velocity. The complexity of this pattern makes it difficult to compute explicit predictions of pressure distributions from the geometry and also imply that different individuals will have different STFs. Thus, most of our knowledge of physical acoustics relevant to sound localization comes from simple acoustic models (e.g., the spherical models of Kuhn 1977 and Shinn-Cunningham et al. 2000) or from empirical measurements (e.g., Shaw 1997) although there are recent computations that use boundary-element or finite-element methods to compute distributions of pressure given detailed geometry (Naka et al. 2004).

Since all acoustic information used for sound localization must come from the two received acoustic signals, it is natural to ask what attributes of these signals depend on the location of the source. First, the wave propagation delay that would be measured between any two separated microphones would be a function of the direction of arrival of the sound, and basically the same effect applies to the signals at the ears. In the frequency domain, this cue is represented by the interaural phase ϕ as a function of frequency as defined in Eq. (8.4). Second, considering the size of the head relative to the wavelength, one would expect that there would be shadowing of the sound wave by the head at high frequencies, so that sounds coming from the side would generate level differences between the ears. This cue is represented by the interaural level difference a defined as a function of frequency in Eq. (8.3). Third, the fact that the level difference between the ears varies with frequency makes it obvious that the level at a single ear also varies with frequency in a way that depends on the direction of the source. This directional dependence of the magnitude spectrum of the signal is predictable in general from the fact that the head, torso, and pinna are large or small relative to the wavelength as the frequency varies. Consistent with these considerations, there are three cues that are the basis of most discussions of sound localization mechanisms: the interaural time or phase difference (ITD or IPD), the ILD, and the "monaural spectral shape."

Although the shape of the head is generally ellipsoidal and irregular, it is instructive to consider the contours of interaural differences expected from simplified cases. Specifically, consider the contours of interaural differences expected for a single source in anechoic space when the head is modeled as a sphere with the ears at opposite ends of a line passing through the center. As shown in Figure 8.3, the line passing through the two ears is called the interaural axis. Because of the symmetry of this arrangement, it is easy to appreciate that a fixed (and achievable) value of ITD or ILD restricts possible source locations to circles perpendicular to the interaural axis with centers along this axis. (This follows from the thought experiment of rotating the sphere around the interaural axis and noting that there is no change in the relative path lengths or the relative shadowing of the head for this rotation.) When source locations are far enough away from the sphere to be out of the acoustic near field, the interaural differences lose their distance dependence, and locations of constant ITD or ILD

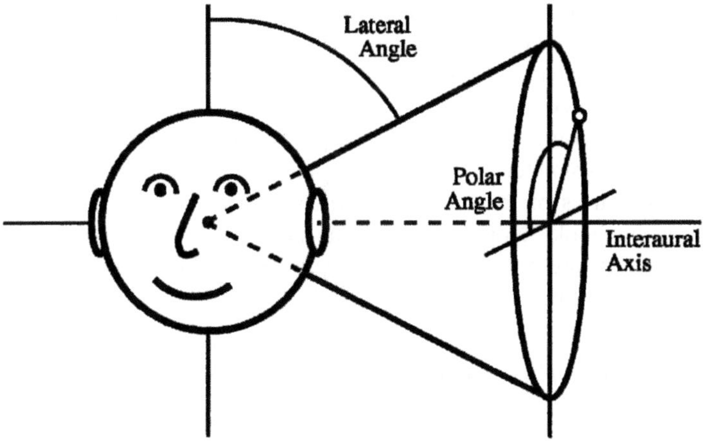

FIGURE 8.3. "Cone of confusion" for spherical head model.

become conic surfaces and are called "cones of confusion" (cf. Blauert 1997). If attention is restricted to a single plane, the ambiguity remains where the cone intersects the plane and would be referred to as front–back ambiguity (in the horizontal plane) or up–down ambiguity (in the vertical plane including the interaural axis). If the spherical model is relaxed, then the situation is more complicated: the spatial surfaces with constant ITD and ILD are more complex and not necessarily overlapping. However, the fundamental ambiguities still exist within the azimuthal and horizontal planes.

The structure of the cue variation in space leads to alternative coordinate systems and different ways of describing the influences of cue variations on perceived locations. The primary description used in this chapter (cf. Fig. 8.3) is defined with respect to the interaural axis and the median sagittal plane, the plane that is perpendicular to the interaural axis at the center of the head. The lateral angle is defined here as the angle from median sagittal plane to a line passing through the center of the head and the location to be characterized. (This angle is equal to $\pi/2$ minus the angle from the interaural axis to the location.) Within the horizontal plane, the lateral angle is the azimuthal angle from straight ahead to the source within the frontal hemisphere (which fact motivates the naming of this angle), and is equal to the angle from the back to the corresponding source location within the rear hemisphere. (Thus, the front–back ambiguity points have the same lateral angle.) If the lateral angle is combined with the angle around the cone of confusion within the plane that is perpendicular to the interaural axis and includes the source location, then any direction is defined. A complete three-dimensional characterization then requires additionally only the distance from the center of the head.

As described in this section, the most important cues for sound localization are the ITD, ILD, and the shape of the magnitude spectrum at each ear. The

ITD and ILD are functions of frequency, and only one magnitude spectrum along with the ILD as a function of frequency is sufficient to characterize the spectral information at both ears. These physical parameters vary with location and they have a clear influence on the perceived location of sounds, as is discussed below. Because of this observed sensitivity, physiological investigations have been focused on these parameters, which are also called physical cues for sound localization. Summaries of these cues for different directions are outlined in other chapters of this book. Note here only that the acoustic measurements of spatial transfer functions are not easy to make accurately and repeatedly for individual listeners (cf. Wightman and Kistler 1989a,b).

2.3 Peripheral Physiological Transformations

In the context of models of sound localization, there are two obvious motivations for studying the physiological transformations in the ascending auditory pathway. First, if the physiological mechanisms are compatible with the operations in black-box models, then they provide realistic mechanisms for achieving these operations. Second, understanding the physiological transformations may lead to improvements in our understanding of psychophysical behavior and improved models of sound localization. This may be particularly important for complex environments. In general, machine processing is better than human processing in simple environments, but human processing in complex environments is often better than machine processing. In this context, it has been speculated that initial stages of the auditory system may be optimized with respect to processing for complex systems since the most challenging situations for processing are those in complex environments. The relationship between the processing in the initial stages of the auditory system, processing in engineering solutions assuming various a priori assumptions about the stimuli and the environment, and processing consistent with psychophysical performance has not been worked out. This section briefly summarizes four types of physiological transformations: the peripheral encoding steps that are common to all auditory processing, the ITD-sensitive coincidence networks that are present in the medial superior olive (MSO), an ILD-sensitive population of neurons that may be related to the cells observed in the lateral superior olive (LSO) of most mammals, and spectral-notch sensitive neurons that are found in the dorsal cochlear nucleus (DCN) and that could be used to process spectral shape cues.

The lowest-level coding of sound to neural responses by mammals is relatively well characterized (e.g., the model of Heinz et al. 2001) and could be incorporated into models of sound localization, although most models of sound localization do not incorporate the details of this peripheral coding. Nevertheless, many attributes of the peripheral transformations are consistent with aspects of localization models. For example, bandpass filtering, nonlinearities, and envelope extraction that limits the processing of carrier timing information to frequencies below a few kiloHertz are consistent with many localization models.

Jeffress (1948, 1958) provided perhaps the best known and most referenced physiological model for sensitivity to interaural time differences. This model is based on neurons that have firing probabilities that are maximized by coincidence of inputs from the left and right. The Jeffress model assumes that different neurons have different "best-ITD" values so that different ITD values result in different distributions of activity over the population of neurons. Jeffress called this "a place model of sound localization," with the notion that different ITDs lead to different locations of maxima in the distribution. The basic structure suggested by Jeffress is consistent with much of the available anatomy and physiology from the MSO (Goldberg and Brown 1969; Yin and Chan 1990; Fitzpatrick et al. 2002). This basic neural behavior can be simulated by excitatory inputs from each side (Colburn et al. 1990) although better fits to the details of the available data are obtained from models that include inhibition as well as excitation (Colburn et al. 2004). This description of MSO activity is consistent with models based on cross-correlation functions of bandpass-filtered stimuli, and these models are discussed in Section 3.1 below and in the chapter by Trahiotis and colleagues (Trahiotis et al. 2005) Another view of MSO activity is that the best-ITD distribution is very narrow for neurons of similar best frequency. In any case, it is generally agreed that the population of MSO cells provides information about ITD across frequency although there are more MSO neurons tuned to low frequencies than to high frequencies. At high frequencies, this neural structure would respond to the temporal structures carried by the envelopes of the received signal. This structure naturally provides a mechanism for an ongoing temporal sequence of estimates of ITD or IPD for each frequency band of the stimulus.

The neurons in the LSO are fundamentally excited by the ipsilateral inputs and inhibited by contralateral inputs (IE cells) and are thus a natural structure for extracting information about interaural level differences. Although it is reported (Moore 2000) that the LSO is very small in human and that there is no associated medial nucleus of the trapezoid body, neurons at higher levels could provide mechanisms similar to those observed in the LSO in other mammals, so that the information provided by this type of mechanism is of interest in either case. A network diagram of a neural network of this type was suggested by van Bergeijk (1962) in the context of his model of binaural interaction, which is discussed below. As van Bergeijk pointed out, these neurons are sensitive to onset ITD as well as ILD, and a neural network of these cells provides sensitivity to combinations of ITD and IID for some stimuli. For transient (click) stimuli, both neural data and an associated psychophysical model was developed by Hall (1965). For low-frequency tones or high-frequency stimuli with envelopes, these IE cells would show minima of activity when the inputs arrive nearly simultaneously, and would increase their activity when the excitatory input led the inhibitory input at the cell. For periodic inputs the response would obviously be cyclic, with the points of maxima and minima depending on the timing of the synaptic activity for each type of input. When there is a level imbalance, and the rates of inputs of each type are imbalanced, we would expect that the side with cells with excitatory inputs (presumably the ipsilateral side at the LSO level and the contralateral side at the inferior colliculus [IC] level) would excite the population and the opposite side would inhibit. A count-comparison of these two sides would be sensitive to a combination of ITD and ILD, and would likely also depend on overall level. Localization models based on this idea are discussed in Section 3.3 below. Computational models of LSO neuron behavior, showing sensitivity to ILD and to ITD, have been developed from available data (Colburn and Moss 1981; Reed and Blum 1990; Joris and Yin 1998; Zacksenhouse et al. 1998; Fitzpatrick et al. 2002)

The spectral shape information is not generally contained within the interaural differences, and additional mechanisms are required to provide sensitivity to the level-frequency profile. Experiments have shown that auditory nerve fibers do not provide useful spectral information over a range of levels in their average rate responses (Sachs and Young 1979) and other suggested mechanisms including temporal response characteristics are being explored. For example, Carney and colleagues (2002) showed that level information across frequency can also be provided by monaural coincidence networks. In the DCN, wideband neurons show sensitivity to the locations of spectral notches, and this is consistent with other observations that suggest a role for the DCN in spectral processing (Young et al. 1997). More work remains to be done in these areas.

Overall, these physiological structures can be considered to provide a temporal sequence of values of ITD, ILD, and level for each frequency output, as well as indicators of broadband onset time and spectral notch locations. The computational models considered below are based on these inputs.

2.4 Overall Model Structure

A block diagram of a general model of human sound localization is shown in Figure 8.4. This diagram provides a conceptualization of factors involved in

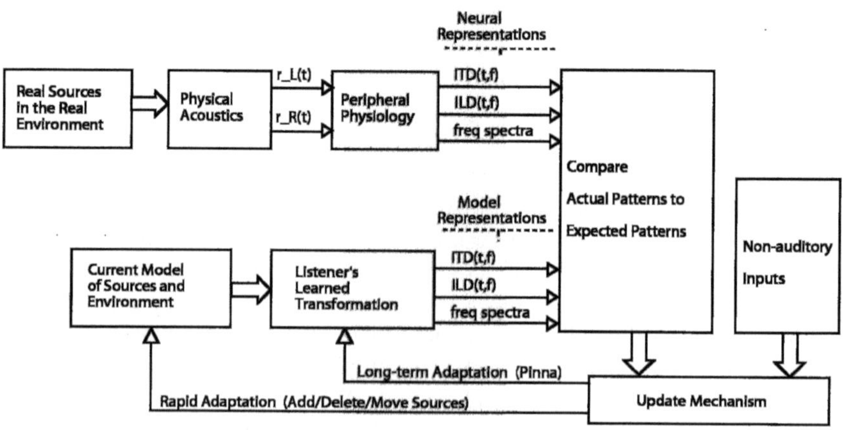

FIGURE 8.4. General model of localization.

localization. It is not developed to be able to make explicit predictions for behavior, but it illustrates many factors that have not yet been incorporated in most models. The upper sequence of blocks represents the physical and physiological processes that lead to neural representations of stimulus attributes that are known to be useful for localization judgments. More specifically, the sound source, environment, and physical acoustics determine the two pressure waveforms $r_L(t)$ and $r_R(t)$. These two time waveforms are processed by the auditory periphery and brainstem, frequency-by-frequency and time-interval-by-time-interval, to provide information adequate to estimate three parameters within each time-frequency tile: an ITD, an IID, and the magnitude spectrum at one ear (or, equivalently, the ITD and the levels at each ear). The ITD estimate provides an interaural phase for each low-frequency channel and an estimate of interaural envelope delay for each channel in the broader high-frequency channels.

In the model, the bottom-up representation within the neural patterns is interpreted relative to the patterns that would be expected for top-down generated hypotheses about the arrangement of sources. These hypothesized sources represent a combination of top-down influences, including locations of sources at previous times, locations expected from visual and other nonauditory inputs, expectations about changes in source positions, and so forth. The hypothetical source model is constantly being compared with the neural patterns and updated accordingly. Within this model, there is an interplay between expectation and input data, and judgments of source direction are influenced by both aspects. Note that this model includes changes in performance as the listener learns about the room or adapts to previous stimuli. Also note that the processing is based on time samples, and implicitly allows for movement of head and of sources. A similar conceptualization of integrated bottom-up and top-down processing is presented by Blauert (1997).

The factors emphasized in Figure 8.4 are complex and only vaguely specified. General models of this type are not sufficiently developed to allow predictions that can be compared to empirical data. In contrast, most of the rest of this chapter emphasizes models that are explicitly specified and that can be used to predict results of human psychophysical experiments. As one would expect, these explicit models have increasing difficulties as the complexity of the perceptual situation increases and more factors are included.

3. One-Dimensional Models—Azimuth

In this and the following section, attention is restricted to locations that can be described by variation is a single angle, azimuth here and elevation next. Models have been developed for these cases and addressed to experiments in which sound locations were restricted accordingly. The equations in Section 1 and 2 are applicable here with θ as a simple angle variable (as opposed to the more general vector definition conceptualized above). A general model for the analysis of localization performance in source-identification experiments is provided by Hartmann et al. (1998).

The primary physical cues for azimuthal judgments are the interaural time and level differences described above, ITD and ILD. There are situations, however, such as monaural listening, in which the spectral shape can play a major role. In this section, attention is focused on models of localization that are based on interaural difference cues. As noted earlier (cf. Section 2.2), interaural differences are ambiguous with respect to front and back in the azimuthal plane, and it is assumed that spectral information or head motion is used to resolve these positions. In the analysis here, the "horizontal polar coordinate system" (Middlebrooks 1992) is chosen, which specifies (cf. Fig. 8.3) the lateral angle (the angle of the location vector relative to the median sagittal plane) and the polar angle (the angle around the circle perpendicular to the interaural axis passing through the location). Predictions and data may be analyzed in these terms instead of the azimuthal angle in the horizontal plane, which includes both front–back and left–right aspects of position.

Lord Rayleigh (Strutt 1907) noted that ITD is a prominent cue for lateral angle and suggested that the perception of lateral position is likely based on interaural time and level differences. He also suggested the well-known "duplex theory of localization" in which the ITD is the relevant and useful cue for low frequencies and interaural amplitude ratio is the relevant and useful cue for localizing high frequencies. Another general rule (Wightman and Kistler 1992) that has had a major impact on thinking about localization states that, for broadband stimuli in relatively simple environments, ITD cues in the low-frequency components dominate azimuthal decisions.

3.1 Cross-Correlation Models

3.1.1 Jeffress Correlation Model

The basic structure of the Jeffress model discussed above, a network of coincidence detectors, each of which is tuned to a specific narrow band of frequencies and a specific ITD, provides a natural mechanism for the computation of a set of narrow-band cross-correlation functions. The cross-correlation function over the interval $[0,T]$ is defined here as

$$R_{LR}(\tau) = \int_0^T x_L(t)x_R(t - \tau)dt, \tag{8.6}$$

where $x_L(t)$ and $x_R(t)$ are the left and right peripherally filtered signals, respectively. Computational models of localization often include information extracted from cross-correlation functions, usually because cross-correlation is a natural mechanism for estimating time delay between two signals. The estimation of ITD using cross-correlation functions is an expected application of these ideas, which is even more popular because of Jeffress's hypothetical neural mechanism. One view of the interaural cross-correlation function considers the function as an antenna pattern, as described in Section 2.1.3, with angle replaced by the equivalent delay τ for a given angle. In this view, the argument τ of the function is an index of the direction to which the "beam is steered," and the argument that corresponds to the maximum value is the argument that compensates for the delay of the source. That is, when the cross-correlation function is maximized for an argument of τ_0, the right ear would be leading the left by τ_0 and we can thus think of steering the beam to the maximum as finding the ITD of a source.

Most modern models of binaural processing visualize the outputs of narrowband cross-correlation processors, as we are discussing here, as a two-dimensional display of cross-correlation (as a joint function of internal delay and frequency). This display changes with time and thereby presents dynamic information about the environment. The relation of this cross-correlation surface to localization judgments varies over different formulations. For example, models vary in the range of internal delays that are assumed to be available or in the assumed internal noise or statistical variability associated with different delays. Models differ in the attribute that is related to location at each frequency, such as the internal delay at the peak of the pattern or the centroid of the pattern across delay. Models also differ in the weighting functions that are applied to the pattern, and they may depend on internal delay and/or on interaural level differences. Finally, models differ in the way that information from the correlation display is combined over frequency and with other information such as level differences and monaural spectral shapes.

One of the earliest models of lateral position based on cross-correlation was the model of Licklider (1959), the triplex theory of hearing. Licklider presented a general formulation of how the stimulus waveforms might be processed to provide the information necessary for understanding hearing perceptions and abilities. The model includes a cross-correlation surface for localization judgments as well as an autocorrelation surface that is hypothesized to provide information for pitch judgments.

Another early localization model based on cross-correlation was presented by Sayers and Cherry (1957). Because this model integrates the left and right halves of the cross-correlation function in forming two statistics that are compared to derive laterality, this model is considered a count-comparison model and is discussed below in this context.

Extended discussions of cross-correlation models, in a more general context that includes applications to other binaural phenomena such as binaural detection and binaural pitch, can be found in Blauert (1997), Colburn (1996), and Stern and Trahiotis (1995, 1996). An introduction to models of localization based on the interaural cross-correlation function is provided in this volume by Trahiotis et al. (2005).

3.1.2 Combination of ITD and IID

The cross-correlation mechanism is a natural mechanism for representing or estimating ITD, as just described, but cross-correlation is not naturally sensitive to ILD. To address laterality in general, it is necessary to supplement a cross-correlation device with a mechanism for sensitivity to ILD. Cross-correlation models generally include one of several hypotheses about the influence of ILD. First, one may assume that the levels at each ear cause the temporal responses to shift earlier in time, so that the ILD would cause a shift in the effective ITD. Thus, an ILD would be translated to an ITD. This mechanism was suggested by Jeffress in his original presentation of the coincidence mechanism. Physiological evidence for this "latency hypothesis" at the level of the auditory nerve responses is present in the shifts of the envelope of post-stimulus-time histograms for impulsive stimuli, whereas there is no consistent shift in phase with level for the fine structure or for ongoing stimuli (Kiang 1965). For tones and narrow-band noises cases, the phase shift with level depends on the center frequency of the stimulus relative to the best frequency of the cell (Anderson et al. 1971). In addition to the lack of physiological evidence, psychophysical lateralization judgments suggest that ILD and ITD have separable effects, and plots of lateral position versus ITD shift differentially, with ILD translating the curves upward instead of along the ITD axis (Sayers and Cherry 1957). A second hypothesis is that ILDs have a multiplicative effect on the cross-correlation function. The multiplicative effect of ILD does not translate the cross-correlation function in the direction of ITD but would affect some of the measures that are hypothesized to be computed from the cross-correlation function, such as the position of the centroid (Stern and Colburn 1978). [Some physiological evidence for a multiplicative combination of ITD and ILD information is found in the work of Peña and Konishi (2001).] In this case, the peaks of the function are not affected significantly, but the centroid is shifted by ILD-weighting functions that are not symmetrical. Finally, a third hypothesis is that the cross-correlation function mechanism provides information about the ITD and another mechanism provides information about ILD. Then, the separately extracted information is combined to form estimates of laterality at more central levels of processing, the simplest form being a weighted linear combination (e.g., Hafter 1971).

Blauert and colleagues (cf. Blauert 1997) have evolved a general model of binaural sound processing that includes sound localization. After model stages corresponding to the peripheral filtering and rectification, stages that correspond to interaural cross-correlation function and an interaural level processor are included for each frequency. Blauert's model has evolved to include all of the elements of the model shown in Figure 8.4, including top-down processing to incorporate a priori expectations, cross-modality inputs, and other effects (cf., Blauert 1997).

More complex models related to cross-correlation provide integrated processing such that both ITD and ILD affect a cross-correlation-like function. For example, Lindemann (1986a,b) developed a model of lateralization that incorporated an inhibition mechanism so that stronger input from one side blocked the propagation along a delay line which was used to compute the correlation function and the resulting patterns depended on both the ITD and the ILD. This model also has predictions for the temporal order effects discussed below, as a consequence of the same inhibition mechanism that leads to the ILD effects. This model was further developed by Gaik (1993) to provide a mechanism that is preferentially sensitive to natural interaural combinations of ITD and ILD. Finally, this model was elaborated by Bodden (1993) to localize multiple sources as part of a "cocktail party processor." In this version of the model, the cross-correlation display is interpreted as containing the response to multiple sources. The resulting processor tracks peaks in the cross-correlation display through time and is able to separate and localize sources moving separately through space.

There have also been more abstract models of the combination of ITD and ILD. For example, Hafter and Carrier (1970) (see also Hafter 1971) formulated their model in order to describe lateral position (the left–right displacement of the image) and to explore this as a cue for the understanding of binaural detection phenomena. Their model is a simple linear combination of ITD and ILD with a weighting factor that depends on frequency. This simple model reflects the facts that either ITD or ILD alone lead to lateral displacement, and that combinations of ITD and ILD may be canceling (corresponding to time-intensity trading in laterality) or reinforcing.

A general issue that arises when unnatural combinations of ITD and ILD are presented artificially with headphones is that the cues considered separately are signaling two different directions. Thus, it is not surprising that human subjects often give different responses for these stimuli, and in some cases, listeners hear multiple source directions simultaneously. It is notable that the model of Gaik (1993) predicts multiple images when unnatural combinations are presented.

3.1.3 Bandwidth Effects

Another aspect of the localization of bands of noise is the dependence of the lateral position on the bandwidth of the noise for some external ITD values. This phenomenon is related to the phase ambiguity of very narrow bandwidths, most obviously with tonal stimuli for which the cross-correlation function is periodic. In these cases, the laterality of the stimulus is generally perceived near the center of the head (i.e., at the delay corresponding to the most central maxima), even though the actual ITD is not near zero. As the bandwidth increases, the stimulus cross-correlation function has a better defined maximum (smaller side peaks) at the actual ITD. In an internal delay versus frequency display, there is a pattern of maxima over frequency that distinguishes the cross-frequency pattern of maxima that is constant across frequency (straight) at the actual ITD and that varies over frequency for the other maxima. Stern et al. (1988) refer to the factors at work here as "straightness versus centrality." They modified Stern's "position model" to include a mechanism for boosting position estimates when maxima are part of a straight pattern. This mechanism can be realized as a second level of coincidence, applied to elements in the internal delay-frequency display, with coincidence across frequency for constant internal delays (Stern and Trahiotis 1992). Thus, internal delays with straight patterns have larger outputs than internal delays at other maxima. This model successfully describes the observed preference for localization at true ITD locations as the bandwidth increases. (These ideas are discussed further by Trahiotis et al. 2005.)

3.2 *Equalization-Cancellation (EC) Model*

Another mechanism for localization, one that is based on the idea of "steering the null" as discussed in Section 2.1.3, was described by Durlach (1972) in his equalization-cancellation (EC) model. Although the EC model was developed to describe human performance in binaural detection experiments, the basic mechanism can also be used to determine the direction of a sound source. The process of canceling the masker involves adjusting the phase shift and amplitudes of the received signals such that the masker components on the left and right inputs can be canceled by subtracting the equalized masker components. If one considers the operations involved in equalizing right and left components to give maximum cancellation, it is apparent that the internal amplitude and delay adjustments that give the minimum result after cancellation are those that match the external ILD and ITD. Thus, by finding the delay that gives the minimum output, the external delay is estimated. In other words, the EC mechanism, in searching for the delay that gives minimum output, is effectively "steering a null" in the output of the EC mechanism versus delay. One can also search for multiple sources by looking for multiple relative minima. This notion, which can be extended to the two-dimensional case, can also be applied to the binaural model of Breebaart et al. (2001a,b,c), a relative of the EC model.

The EC model can also be related to cross-correlation models. As noted previously (Colburn and Durlach 1978), the energy of the subtracted output in the EC model (Green 1966) is related to the interaural cross-correlation function as follows:

$$EC(\tau_i,\alpha_i) = \int_0^T [x_L(t) - \alpha_i x_R(t - \tau_i)]^2 dt = E_L + \alpha_i^2 E_R - 2\alpha_i R_{LR}(\tau_i) \quad (8.7)$$

where the peripherally filtered received signals, $x_L(t)$ and $x_R(t)$, have energies E_L and E_R and the correlation function $R_{LR}(\tau_i)$ [cf., Eq. (8.6)] arises from the cross-term in the expansion of the EC energy. Note that the interaural amplitude ratio α_i and the interaural time delay τ_i are internal processor parameters that are adjusted to compensate for the corresponding external parameters (ILD and ITD). The internal parameters appear explicitly in the equalized output expression, and the ITD and ILD are implicitly included in the received signals. The values of the internal parameters τ_i and α_i that lead to the minimum value of *EC* are the values that best compensate for the corresponding external parameters and that can be used to estimate the ITD and ILD as described above. In the complete formulation of the EC model, the waveforms are corrupted versions of the received acoustic signals, and the corrupting noise is an important part of the model for detection (but not discussed further here). The lack of ILD dependence of the cross-correlation function was noted above, and we note here that the asymmetry of the α_i weighting of the terms in Eq. (8.7) provides an appropriate sensitivity to the ILD in the value of *EC*. It can also be seen (after some consideration) that, for single external sources, the value of τ_i that maximizes the cross-correlation function also minimizes the EC-model output. Thus, for localization functions, one looks for maxima in the cross-correlation function but looks for minima in the EC-model output.

A multichannel version of the EC model was developed (Culling and Summerfield 1995) with the important constraint that the model selects equalization delays in each frequency channel independently. This model has been applied to various binaural pitch phenomena with good success. The application of the EC model to the localization of multiple sources and to temporal sequences of sources has not been completely developed.

3.3 Left–Right Count-Comparison Models

There is a long and continuing history of models of sound localization in the horizontal plane based on comparisons between activity levels in left and right hypothetical neural populations. In these models, the relative strengths of the responses of the two populations determine the relative sidedness of an acoustic image. These models do not attempt to describe changes in elevation, but they have been so influential in thinking about localization in the horizontal plane that a brief summary is given.

The earliest localization models that combine interaural cues to make a localization judgment were the models of von Békésy (1930) and van Bergeijk (1962), as discussed above in relation to neural mechanisms for extracting information about stimulus ITD and ILD. Both of these related models (the title of van Bergeijk's paper includes the phrase "Variation on a theme of Békésy") provide localization information through the relative level of activity in left–right symmetric neural populations, and both combine ITD and ILD in the relative activities of these populations. A more extended description of these models can be found in Colburn and Durlach (1978), but a partial description is given here because several more recent models can be thought of as further variations on this theme.

The models of von Békésy and van Bergeijk have not been quantitatively compared to localization data, except for the work of Hall (1965). He obtained physiological data from brainstem neurons using click stimuli that varied in ITD and ILD, then characterized the statistics of these responses. Using a population of these model cells, he successfully predicted psychophysical behavior in time and intensity discrimination for click stimuli.

Another model based on a count-comparison version of a cross-correlation model was formulated by Sayers and Cherry (1957). The left and right counts in this model were determined by combining separately over left lags (internal delays) and right lags from a running cross-correlation surface. More specifically, the "left count" is created by adding together correlations with left-lag values weighted by the left amplitude and the right count uses the right-lag values weighted by the right amplitude. This provides a left count that increases relative to the right count when either the left-ear stimulus leads or the left-ear stimulus is greater. Sayers and colleagues (Sayers and Cherry 1957; Sayers 1964; Sayers and Lynn 1968) compared this general mechanism to lateralization data with general success. Their analysis does not address aspects of localization beyond left–right judgments. They also made an important distinction about the relationship of model predictions to lateralization data. When the listener is requested to pay attention to the overall stimulus location, often in the context of an adjustment experiment, a centroid measure of the weighted cross-correlation function is appropriate. When the listener makes judgments for short bursts, locations may be better described by local maxima, with a significant role of context in peak selection.

Colburn and Latimer (1978) considered a model closely related to the Sayers and Cherry model as part of the sequence of binaural models based on auditory nerve patterns. One imagines an array of coincidence detectors (a la Jeffress, see above) that are separated according to their best-ITDs into two populations, and the total activity of each of these two populations is weighted by an ILD factor to form a decision variable (analogous to lateral position) that is used to discriminate changes in the stimulus parameters ITD and ILD. Colburn and Latimer used this formulation to predict asymmetries observed in the combined ITD and ILD dependence of ITD discrimination with tonal stimuli. They showed that a simple amplitude-weighting of the two populations, fixed for each ILD as described above, is not adequate to describe the discrimination data, even though Domnitz and Colburn (1977) showed consistency between the asymmetries observed in mean localization and in discrimination.

A recent argument for a left–right count-comparison model was put forth by McAlpine and colleagues (2001). Their motivations for this type of model were suggested by their data that indicate a coupling between the best-ITD of a coincidence cell and the best-frequency (or characteristic frequency CF) of that cell. Specifically, they assume that the best-ITD of each cell is approximately one eighth of the period of the cell's best frequency. Thus, the location information is represented by the relative strength of activity on right and left sides, and not by the distribution of activity over best-ITD as one would expect from the Jeffress description given above. An extension of this model has been suggested by Marquardt and McAlpine (2001). In this extended model, there are four count values per frequency: on each side the best-ITD count and its worst-ITD LSO-type-cell complement (i.e., an IE cell with the same ITD). These are pairs of coincidence and anti-coincidence cells on each side. Thus, for each frequency, there are four counts. These counts can be considered as four samples of the cross-correlation function and they could be used in a variety of ways for making decisions. Although this model is closely related to the localization models described earlier in this section, it is not yet published in detail and has not yet been specified as a model of sound localization.

4. One-Dimensional Models—Elevation

The judgment of elevation is much less reliable for human listeners than azimuthal judgments, and responses are generally more variable and less accurate. In these judgments, there are larger intersubject differences and greater dependence on the nature of the stimuli, particularly on the stimulus bandwidth and frequency content. Also, as described more fully below, models for elevation judgments are less developed than for azimuthal judgments. In the following subsections, available models are described as well as some conceptual notions that may form a basis for future models.

4.1 Spectral Cues

Attention to spectral information for vertical elevation (and for localization in general) was provided by Butler for many years (Roffler and Butler 1968a,b; Butler 1974, 1997). The importance of spectral shape on vertical localization judgments is seen clearly in judgments about narrow-band stimuli, for which there are preferred directions associated with the frequency of each band. That is, for narrow-band stimuli, the judgment of the elevation of the sound is primarily dependent on the frequency of the sound and not on the position of the source. One interpretation of this result is the notion of a "covert peak." For a given frequency, a covert peak corresponds to the spatial location at which the spatial transfer function has the largest value (for that frequency). It is suggested that localization judgments for narrow-band stimuli are based on the location of the covert peak for the presented frequency. Consistent with the covert-peak idea, it is observed that the "covert peak direction" for a given frequency roughly corresponds to the direction at which listeners perceive the location of a narrow band at that frequency. The covert-peak analysis leads to the idea of spectral regions (SRs) that are consistent with sets of frequencies and that localization could result from combinations of SRs corresponding to a selection of frequencies as might be associated with peaks in the spectrum (Butler 1997). Although these ideas have not been formulated into an explicit computational model, they continue to have strong influence on thinking about spectrally based sound localization, and are particularly important for elevation judgments and for monaural localization in general.

The notion of covert peaks is closely related to the "directional bands" that were discussed by Blauert (1968, reviewed in his 1997 book) and that refer to the tendency to perceive narrow-band stimuli at frequency-specific elevations. This notion can be extended to configurations of spectral shapes that are associated with specific elevations. It is also generally consistent with models described below based on comparing received spectra to those expected from different elevations (Middlebrooks 1992; Langendijk and Bronkhorst 2002).

Batteau (1967) noted that the pinna was likely to be a source of spectral coloration and suggested a simple model with a primary reflection from the pinna surface as a source of information leading to a vertical location cue. The basic idea is that the acoustic pathlength differences between the direct path to the ear canal and an indirect path that reflects off the back of the concha of the ear create a net impulse response that would result in interference at some frequencies and reinforcement at other frequencies. From the frequency response perspective, this leads to an HRTF that has peaks and troughs (or notches) whose relative spacing (and locations in frequency) indicate the pathlength difference, and therefore the elevation of the source. Batteau (1967) described his ideas in terms of the temporal characteristics of the impulse response and suggested that the internal processing to extract elevation from responses was based on time-domain processing, although spectral-domain processing can reach equivalent results. As von Békésy (1960) notes, the idea of pinna reflections and the interactions that they cause, suggested and sketched by Schelhammer (1684), is oversimplified for the frequencies involved in human hearing. However, Batteau's demonstrations and presentations had a major impact on thinking during the 1960s and 1970s (cf. Watkins 1978). In fact, Batteau's analysis of the effects of the pinna became more complex in his later work. The effects of the pinna shape on the spectrum of the proximal stimulus was explored extensively by Shaw and colleagues (e.g., Shaw and Teranishi 1968; Teranishi and Shaw 1968; Shaw 1997). An analysis of the pinna including both reflections and diffraction is presented in Lopez-Poveda and Meddis (1996). They also note that the resonance (and anti-resonance) structure of the pinna response depends on elevation in ways that are similar for different azimuths. This suggests that a system that treats azimuth and elevation separately could do a reasonable job of localization in two dimensions.

Of course, the pinna only partially determines the spectral response to a stimulus, and other physical attributes of the body (and the environment) also contribute. Most notably, the head has a substantial effect on the soundfield, and the effects of the pinna, head, and torso have been discussed extensively by Duda and colleagues (e.g., Algazi and Duda 2002). These physical influences on the signal are prominent and identifiable in anechoic spaces, and spherical models of the head are particularly useful for understanding the overall effects of source position on the received sound pressure waveforms, as we discuss further below.

It is well known that virtual environment simulations that are very effective for some listeners are not effective for others. It is also clear that the spatial transfer functions used in simulators must be selected to match individual listeners. This is presumed to be related to the variation in the physical shape of the head and pinnae over listeners. Middlebrooks (1999a) has shown that if HRTFs are appropriately scaled in frequency they are much more similar across listeners, and this transformation allows improved virtual acoustic simulations without using individualized HRTFs (Middlebrooks 1999b). Individual differences are presumably related to differences in physical sizes of the head and pinna but the picture is not yet clear.

The importance of inter-individual differences in spectral shape are related to to the relative insensitivity to the details of the spectral shape that is demonstrated by the work of Kulkarni and Colburn (1998). They demonstrated that the spatial attributes of a virtual stimulus were insensitive to the details of the spectral shape as represented by the presence of high-frequency components in the Fourier representation of the magnitude spectrum of the HRTF. This indicates that the overall shape of the HRTF is much more important than the local variations in the shape. This analysis suggests that only the primary features of the spectral shape are important and may allow a tuning of HRTFs to listeners in a way that maintains only the main features of the shape, such as the positions of the primary spectral notches. This insensitivity to detail is also consistent with the experiments and conclusions of Macpherson and Middlebrooks (2002b).

4.2 Models of Elevation Judgments

There is now widespread agreement about the importance of the spectral shape of the proximal stimulus (i.e., the magnitude spectrum of the received stimulus), although there is no consensus on what aspects of the spectral shape are critical cues for localization. The locations of peaks and notches are prominent and frequently noted (Blauert 1997; Rice et al. 1992; Kulkarni 1997); others have suggested regions of steep slopes (Hebrank and Wright 1974) and profiles of interaural spectral differences (Searle et al. 1975). Template matching paradigms (Middlebrooks 1992; Macpherson 1997; Langendijk and Bronkhorst 2002) are the most successfully developed concept so far. Although none of these cues have led to models that have been successfully applied to arbitrary stimuli and their perceived locations, they can predict performance successfully in limited contexts.

4.2.1 Template Matching Hypothesis

As a generalization of the narrow-band noise judgment data, which suggest direction judgments based on matching of the received (proximal) spectrum to locations for which that frequency is strong, the idea of template matching of spectra to location has been suggested (Middlebrooks 1992) and developed into explicit models that have been evaluated by Macpherson (1997) and by Langendijk and Bronkhorst (2002). The basic idea (or assumption) of these models is that the magnitude spectrum of the proximal signal is the result of a relatively flat stimulus spectrum modified by the spatial transfer function (STF). Strong variability in the stimulus spectrum challenges these models.

Langendijk and Bronkhorst (2002) evaluated a specific formulation of a template matching model with comparisons to performance with broadband stimuli. In addition to the baseline condition, cases were studied with the directionally related variations in the transfer function removed by replacing sections of the STF with constant values over a selection of frequency ranges. The empirical data were compared to predictions from several versions of template-matching models with generally successful results. The most successful model was based on comparing the proximal spectrum to the directional transfer function (roughly, the HRTF with the ear canal effects removed) and deriving a probability-like measure of match from either (1) the correlation of the proximal magnitude spectrum with transfer function magnitudes for candidate elevations or (2) a mean-square deviation measure of their agreement. In general, response elevations were near the peaks of the probability functions for most stimulus elevations. They also tested models based on the first or second derivative of the magnitude spectra and found them to be less successful. This finding was surprising given that the analysis of Zakarauskas and Cynader (1993b) suggested that derivatives of the spectra ideally provide a more useful characterization.

The template-matching idea for elevation judgments is generally based on the assumption that the stimulus spectrum is relatively smooth. Although this is generally true for long-term averages, it is not a safe assumption for short time intervals, particularly when there are multiple sources. Although template matching models have been among the most successful, they have not been fully investigated for these difficult circumstances.

4.2.2 The Primary-Notch Hypothesis

The hypothesis that the locations of the primary spectral notches in the HRTF control the perceived elevation of a sound source has been explored in several series of experiments by Bloom (1977), Watkins (1978), and Kulkarni (1997). There were three types of experiments, one in which the consequences of the removal of notches at different locations in the HRTF were evaluated with localization judgments, one in which subjects matched the position of a virtual stimulus to an external speaker by adjusting the frequency of a notch, and one in which subjects judged the location of a stimulus for various notch locations. In general, these experiments were consistent with the notion that the elevation of the perceived source was predominantly determined by the location of a broad notch in frequency. As the notch frequency was reduced from about 10 kHz to about 5 kHz, the elevation of the image decreased from roughly overhead to near the horizontal plane.

The notch-frequency hypothesis has also been discussed in the context of the responses of cells in the dorsal cochlear nucleus, as noted above. The projection cells in this nucleus have been shown to be particularly sensitive to notch frequencies (Young et al. 1997).

4.2.3 Interaural versus Monaural Spectral Shape

Even if the spectral shape at high frequencies is taken to be the cue for elevation, the manner in which the spectra at the two ears are combined and the relative importance of interaural differences are not specified. As noted in Section 2.1.2, if the received magnitude spectra at both ears are available, then the effects of the magnitude spectrum of the stimulus can be eliminated by taking the ratio of the magnitudes (i.e., the difference in decibels). This option is particularly relevant for situations in which the stimulus spectrum is unfamiliar and irregular. As noted in Section 2.1.1 above, the proximal stimulus at a single ear cannot in general be deconvolved to extract the transfer function unless information about the stimulus spectrum, such as the general smoothness or overall shape, is known. Although experiments have shown that this interaural spectral information [i.e., the ILD vs. frequency or $\alpha(\omega,\theta)$ in Eq.(8.3)] is useful for vertical localization in some cases (Searle et al. 1975), most models of localization focus on the spectral shape at a single ear. The relative importance of the frequency dependence of the ILD and its relation to spectral shape information in general has not been sorted out, and there is conflicting evidence about the usefulness of interaural differences in making vertical judgments. These issues are discussed further in the context of simultaneous judgments of azimuth and elevation.

5. Two-Dimensional Models—Direction

This section addresses models of two-dimensional localization, including azimuth and elevation together (or, equivalently, lateral angle and polar angle together). This is essentially a judgment of direction and is appropriate for anechoic situations in which the source is farther away than about a meter. This range is beyond "near-field" acoustic effects; in this range of anechoic space, there is no dependence on distance beyond overall level changes. Note that there are distance effects in reverberant spaces, and these are discussed in Section 6.

5.1 General Cue-Based Two-Dimensional Models

We first describe models that are based on weighted combinations of individual cues. Models of this type have their roots in physical acoustic analysis and measurements in combination with observations of the relationship between physical acoustics and the perception of image location, as carried out by Rayleigh (Strutt 1907) and others.

5.1.1 Overall Model Structure

A broad-based account of available localization data in terms of physical cues of the stimulus waveforms was carried out by Searle and colleagues (1976). These authors included six physically based cues in their analysis, the ITD and ILD already discussed and four others: monaural head shadow, interaural pinna amplitude response, monaural pinna response, and amplitude response due to shoulder reflections. The variances of these cues were estimated and used to describe a large number of experiments, with the assumption that the listener used all useful cues in each experiment. For conditions in which there were restrictions to the availability of cues, such as limited frequency ranges, weights to reflect each situation were determined. The authors specifically explored the dependence of the variance of the decision variable on the span or range of angles in identification experiments. They concluded that the variance of the contribution of each cue to the decision variable has a term that depends on the square of the range. This property, which was found in intensity perception experiments by Durlach and Braida (1969) and Braida and Durlach (1972), also applies to localization experiments. This observation was confirmed in a set of localization experiments specifically designed for this purpose (Koehnke and Durlach 1989). The authors of the general model for cue-weighted localization concluded that this general decision model can be successfully applied to describe available data. The model is incomplete in the sense that there are no rules for predicting the weight allocations in cases for which stimulus restrictions are imposed.

5.1.2 Artificial Neural Networks

Another approach to characterizing the information available in sets of physical cues for sound localization is to use artificial neural networks to perform localization using these cues. This approach has led to several interesting conclusions. Several studies have demonstrated (Palmieri et al. 1991; Neti et al. 1992; Janko et al. 1997) that ILD as a function of frequency is adequate to achieve localization comparable to or superior to that observed in human subjects and that smoothing of the frequency inputs with a cochlear model maintains this information. Of course, this information is available in the two monaural levels as a function of frequency as well.

Jin and colleagues (2000) analyzed neural network models of two-dimensional localization that operate on the outputs of a model of the peripheral auditory systems (the AIM model of Patterson et al. 1995). These neural networks are sophisticated and include time delays in the combination weights and several different cross-connection patterns. They find performance similar to that of humans when there is frequency-specific early processing and conclude that both frequency band processing and central combination of activity across frequency are important attributes of human sound localization performance.

In general, neural networks perform much better than human listeners, indicating that a key ingredient of models of sound localization is the set of assumptions about which cues are chosen for various situations. Neural network models have not yet been used to predict performance in complex environments.

5.2 Computational Two-Dimensional Models

This subsection addresses models that generate explicit responses from the pressure waveforms at the two ears. These models are computational in the sense that signal processing operations are combined to lead to a judged direction (including both azimuth and elevation).

Middlebrooks (1992), in a detailed analysis of localization judgments using high-frequency narrow-band stimuli (1/6-octave bands centered at 6, 8, 10, and 12 kHz), concluded that for these stimuli, the laterality of the perceived direction is determined by the ILD, and the elevation component (i.e., the angular position on the cone of confusion) is determined by finding the direction that gives the best match between the shape of the proximal stimulus spectrum (the magnitude spectrum of the signal received in the ear canal) and the shape of the direction-specific transfer function. Specifically, a predicted direction judgment was selected on the basis of a double correlation measure (called the "sensitivity index"). This index is based on the sum of two normalized measures: the measured ILD relative to the ILD at candidate directions, and the magnitude spectrum calculated from the received spectra at each ear relative to the directional transfer function at the candidate direction. The direction that has the highest sensitivity index is the model output for that stimulus. Analysis of the independence of the ILD and the monaural spectral-shape cues for this experiment are interpreted as consistent with the cues functioning independently. This notion is also consistent with the analysis of Lopez-Poveda and Meddis (1996) and with the observations and analysis of Kulkarni (1997).

The Middlebrooks model is successful for the narrow-band noise stimuli used in his experiment, and the author suggests that the model may be generalizable to broadband stimuli, maintaining the idea that the laterality component of the judgment would be based primarily on interaural differences (ITD and ILD) and the other (elevation and front–back) components of the judgment would be based primarily on the frequency dependence of the magnitude spectra of the monaural signals (with each spectrum weighted by the power reaching that ear).

A similar conclusion was reached by Morimoto et al. (2003), who further hypothesized that the elevation judgment for a given laterality could be based on the spectral shape for the median plane STF. This would be a relatively simple way to specify location in both azimuth and elevation and would have implications for what aspects of the received spectral shape is important. Their experiments were consistent with the adequacy of this representation.

The most complete model that has been applied to arbitrary combinations of elevation and azimuth is the template-matching model of Langendijk and Bronkhorst (2002) described in Section 4.2.1. Although they do not show results from the full two-dimensional application of their model, the formulation of the model is two-dimensional and they report that it is similarly successful when both azimuth and elevation are varied.

In a modeling study that explicitly incorporates responses of neurons in the IC of guinea pig to wideband stimuli, Hartung (2001) describes a model for localization based on a population of model IC neurons that are tuned to specific directions. The final localization responses in the model are based on the characteristic directions of the cells with the maximum responses. The model IC neurons are based on a weighted linear combination of frequency-specific ITD estimates and monaural spectral level estimates.

The general assumption is that the cues are combined optimally when full cues are available, although there is generally more information available than used. The fact that different cue combinations are used for different experiments leads to two conclusions: that sound localization is a complex task, and that listeners use aspects of the physical stimulus for judgment according to the circumstances. Thus, there is not a simple answer to the question of what cues are used for localization. Factors such as a priori knowledge of the stimulus spectrum make a difference in the way listeners approach the task.

5.3 Cue-Manipulation Analysis

There have been a number of headphone experiments in which stimuli are manipulated to present partial or contradictory cues to the listener. The basic idea is to interpret observed behavior as reflections of the relative strength or importance of available, often conflicting, cues. The recent development of virtual acoustic stimuli, which are presented by headphones or tubephones to generate pressures that are approximately the same as those in natural environments, provide opportunities for changing some aspects of these stimuli and measuring responses. For example, stimulus manipulations that have been used include (1) setting ITD or ILD to default (zero) values or altering the monaural spectral cues (Macpherson and Middlebrooks 2002a), or (2) flattening regions of the monaural spectra (Langendijk and Bronkhorst 2002) in order to present stimuli with or without spectral peaks or notches (Kulkarni 1997).

In an extensive study by Jin and colleagues (2004), localization judgments were made with stimuli that were modified from normal stimuli by constraining the left ear transfer functions to be constant. (The transfer functions in this study were directional transfer functions [DTFs], which are derived from anechoic HRTFs by removing components that are common to all directions.) Stimuli at the right ear were either modified to maintain the interaural transfer function or were maintained at their natural level. This study found that neither the right-ear-correct nor the interaural-level-correct stimuli gave normal localization behavior. In interpreting these results, it should be kept in mind that, when one cue is maintained here, others are corrupted. Thus, as in essentially all cue-manipulation studies, listeners are making judgments about unnatural stimuli with conflicting information about source position. When subjects respond to the stimuli with the correct interaural transfer function, they are also responding to stimuli with incorrect monaural spectra at both ears. Given that most stimuli with single sources have substantial redundancy (Carlile et al. 1999) in their information about source location, provided by a number of cues, the normal mode of human localization behavior is a response to numerous consistent cues. These studies illustrate the difficulties in designing a computational model that will accommodate conflicting cues.

6. Three-Dimensional Models (Direction and Distance)

For a full, three-dimensional model, distance as well as direction has to be computed. It is generally agreed that, similar to the elevation case, distance cues are ambiguous and not well understood. As one would expect, humans are relatively poor at judging distance with purely acoustical cues (Loomis et al. 1998). Another attribute of sound perception that is related to distance is the degree to which the perception is perceived as a source in space as opposed to a source "inside the head." Sources perceived outside the head are said to be "externalized" and the factors involved in externalization are obviously related to distance perception, but may also be related to the compatibility of different cues as well as to the likelihood of observed cues in natural environments.

There are few models that have been developed for localizing in both distance and direction; more typically, partial models are developed for specific dimensions and circumstances. For example, Bronkhorst and Houtgast (1999) proposed a model for distance perception in rooms based on a modified direct-to-reverberant energy ratio and show that this model is consistent with their empirical data. There are multiple cues for distance, however, and the appropriate model will depend on the circumstances.

Distance cues include most prominently the overall level of the received sound, interaural differences, and the level of the sound in combination with perceived sound quality that would imply the level of the sound at the source. (For example, whispered speech would be expected to be low in level at the source.) In environments with reverberation, the level of the reverberant sound relative to the direct sound also provides an important distance cue, as captured in the model mentioned above. Zahorik (2002) presents a useful analysis of distance cues and presents data for the combination and interpretation of these cues.

For sources close to the head (in the acoustic near field), interaural difference cues become prominent (Brungart et al. 1999). Brungart and colleagues conclude that ILDs in low-frequency signals are dominant cues for localization in the near field. An analysis of acoustic cues used in near-field localization based on a spherical model of the head was presented by Shinn-Cunningham and colleagues (2000).

7. Effects of Reverberation and Multiple Sources

As noted above, the level of reverberation is an important cue for judging distance in reverberant spaces. In addition, since reverberation changes the pattern of interaural differences as well as other qualities of the acoustics, an interpretation of the locations of sounds in the room (and other judgments) depend on knowledge of the room characteristics, like the reverberation time. There is evidence that listeners adapt to the reverberation level of the room when they have been in the room a relatively short time (Shinn-Cunningham 2000).

In a reverberant environment, the direct sound comes from the direction of the source and in addition each reflection has its own direction. The multiple reflections are distributed through time so that the combined reflected sound tends to be decorrelated at the two ears, causing the ITD and ILD to vary in time. This phenomenon was characterized in measurements in real rooms by Shinn-Cunningham and Kawakyu (2003). They showed that good localization performance can be achieved by averaging short-term estimates to give a long-term estimate. This can be seen from the distributions of IPD estimates in Figure 8.5. The dark x's in this figure illustrate the additional point that the earliest estimates (from the direct sound) are more reliable. The relative accuracy of early estimates is also consistent with the conclusions of the experiments of Hartmann and Rakerd (Hartmann 1983; Rakerd and Hartmann 1985, 1986). In their analysis of results, Hartmann and Rakerd also introduce the concept of

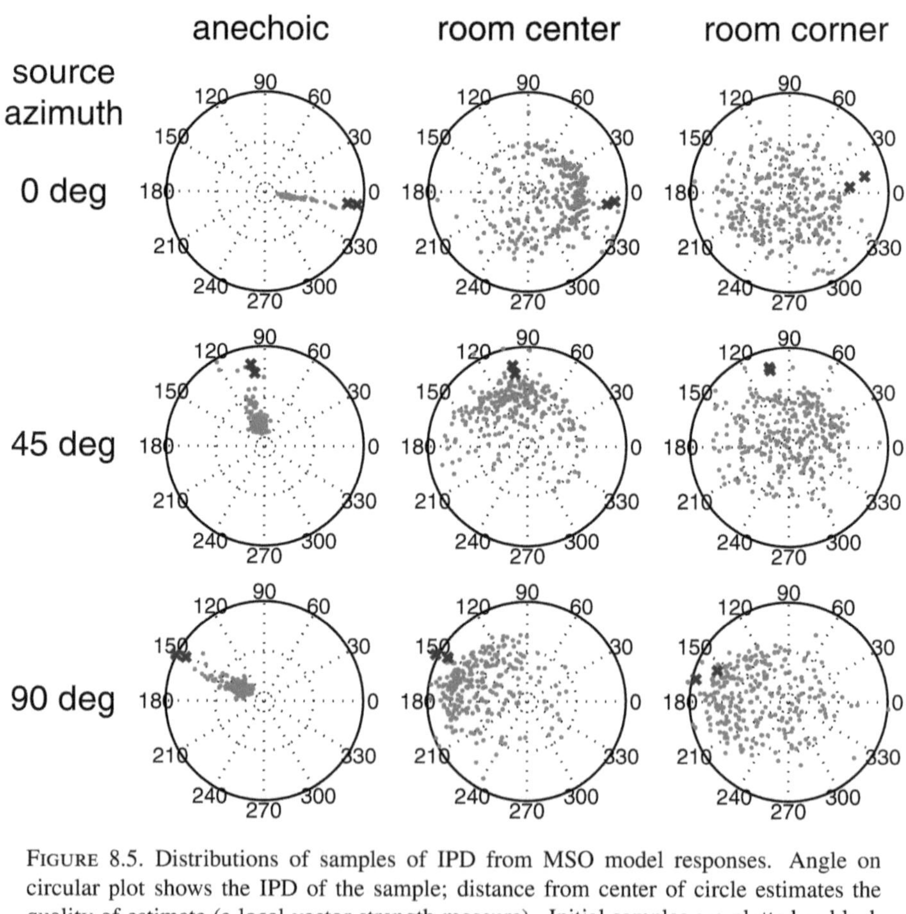

FIGURE 8.5. Distributions of samples of IPD from MSO model responses. Angle on circular plot shows the IPD of the sample; distance from center of circle estimates the quality of estimate (a local vector strength measure). Initial samples are plotted as black x's. (From Shinn-Cunningham and Kawakyu 2003; © 2003 IEEE.)

"plausibility," a top-down factor that seems to be needed to explain subject responses.

In a study of the importance of early reflections on the spatial aspects of sound perception in rooms, Bech (1998) argues that subjects distinguish between the timbral and spatial aspects of the sounds and that the only the first-order floor reflection contributes to the spatial aspects.

A combined series of experiments, measurements, and analyses of localization based on the statistics of interaural differences for targets in diffuse "cafeteria" noise was undertaken by Nix and Hohmann (1999). They developed a localization algorithm based on interaural difference distributions and applied it to localization in quiet and in this difficult noise condition with good results.

If there are two, fixed-position sources of independent noise stimuli with overlapping frequency bands, then the interaural differences in these bands are distributed over time in a pattern that depends on the interaural differences and the relative strength of the two sources within the bands. These varying interaural differences generally fall between the values corresponding to each source alone, and for Gaussian noise are roughly uniformly distributed when the intensities are equal. These stimuli have perceived spatial characteristics that depend on the context, which is not surprising given their distribution of interaural differences. For example, one wide image is typically perceived when the stimuli are wideband with identical spectra and they are turned on and off in synchrony. If, however, they are turned on and off separately, two images are typically perceived for a long as several seconds.

In a study of the ability of listeners to separate simultaneous sources with overlapping broadband spectra, Best et al. (2004) found that virtual sources were perceived separately when they had different horizontal locations but unseparated when they differed only in vertical dimensions. The separability was promoted best by ongoing ITDs. Langendijk et al. (2001) also studied localization of multiple broadband sources, and showed that simple template models operating on short temporal segments (consistent with the conclusions of Hofman and van Opstal 1998, described in Section 8.1) show many properties of the data.

Another set of algorithms for localization judgments with multiple sources have been developed by several groups (e.g., Bodden 1993; Albani et al. 1996; Liu et al. 2000) as part of processing models for the "cocktail party effect." In these algorithms, a binaural displayer extracts interaural differences (time, level, and correlation) and this information is analyzed frequency by frequency and over time to track the (temporally variable) estimated directions of sound sources. The associated "cocktail party problem" processors (cf. Bodden 1993; Wittkop et al. 1997; Liu et al. 2001; Lockwood et al. 2004) use the time-varying directional information to tune the directional sensitivity of the processor.

Another development in the modeling of multiple sources (Braasch 2002) is based on the idea that one can estimate the cross-correlation function of one of the sources if it is played alone for an interval of time. The location of the second source is then estimated from the estimated second-source cross-correlation function which is calculated as the difference between the cross-correlation function of the received signal (including both sources) and the cross-correlation function with the first source. This hypothesis is consistent with the outcomes of several experiments also reported by Braasch (2002). It is also consistent with the observation that the perception of independent noise stimuli from two different speakers is perceived as a single source when the stimuli are turned on together, but are perceived as two distinct sources in the appropriate locations if one or both sources are switched on and off asynchronously. This is closely related to the method for localization of multiple transients suggested by Zakarauskas and Cynader (1993b).

In general, the related problems of localization in reverberant environments and in the presence of other sources have not been worked out and much remains to be done.

8. Temporal and Motion Effects

8.1 Integration Time and Multiple Looks

Although localization studies are usually done with long-duration noise bursts or with very brief transients, a few studies have explored stimuli with complex temporal structures.

A study of spectrotemporal effects by Hofman and van Opstal (1998) led them to conclude that localization judgments are based on a series of estimates based on brief temporal intervals of the stimulus (say 5 or 10 ms). According to this concept, if the frequency content during the brief intervals is too restricted, as it might be with a slowly sweeping frequency modulation of a narrow-band stimulus, the short-term estimates would be unreliable and combinations of these judgments would not lead to good performance. On the other hand, if the stimulus is broadband over every brief interval, the estimates would be noisy but related to the actual location. If only a narrow band is present in each interval, the judgment of location from that interval would be related to the frequency of the band and overall judgments would be very poor. The data they present from a series of experiments of this type are consistent with these ideas.

This general model principle and the associated experiments raise the question of variability in the magnitude spectrum of brief samples of the wideband noise. Accordingly, a 1-s-duration, flat-spectrum random noise token was convolved with the left HRTF of a human subject (for the direction directly in front). The HRTF is from the CIPIC1[1] HRTF archive (Algazi et al. 2001) from measure-

1. CIPIC originally stood for the Center of Image Processing and Integrated Computing, in which the CIPIC Interface Laboratory was formed at UC Davis. Although administrative entities have changed, the name is preserved. The website can be found with the URL http://interface.cipic.ucdavis.educ/.

ments made on subject 003. The corresponding waveform was then analyzed in sequential, nonoverlapping, 5-ms chunks to generate a series of spectra. These spectra are displayed as the gray band in Figure 8.6. Also plotted on the same axes is the HRTF magnitude spectrum (thick black line). It may be noted that there is considerable variability in the spectrum computed from the 5-ms intervals of noise. To reduce the variability, a processor designed to estimate the HRTF would perform some form of averaging over individual time slices as described above. For localization processing, this problem is further complicated by the fact that real-world signals are not stationary and the time windows over which the ear would compute spectral estimates and the averaging process would not be fixed but rather depend upon the temporal characteristic of the signal waveform.

As described above (Section 7), the integration of interaural information over time is also important for azimuthal judgments in reverberant environments (Shinn-Cunningham and Kawakyu 2003; Best et al. 2004).

8.2 Precedence Effects

Temporal effects with leading and lagging stimulus components are often discussed in relation to localization performance in reverberant spaces because direct sounds generally reach the listener in advance of reflected sounds. Effects that might be related to this situation are generally referred to as "precedence effects," and some issues related to these cases were discussed in previous sections. The basic notion of the precedence effect is that localization information in earlier arriving (direct) sound components are emphasized relative to the information in later arriving (delayed or reflected) sound components. There have been many studies of the precedence effect, primarily with simple pairs of stimuli, as opposed to the multiple reflections in a natural environment. A summary of these experiments can be found in a recent review chapter (Litovsky et al. 1999).

A widely used conceptual model of the precedence effect, described by Zurek (1987), has a delayed suppressive component generated in response to onsets in the stimulus. This model is generally consistent with a great deal of data, although quantitative comparisons depend on specification of the "onset detector" (which could respond to any increase in the ongoing level of or significant change in the quality of the signal) and on the nature of the suppression mechanism. Recent studies of the precedence effect (Dizon 2001; Braasch and Blauert 2003; Dizon and Colburn 2005) have revealed that the emphasis of the leading sound occurs even when the only information available is provided by the ongoing parts of the waveform. In addition, these studies also make it clear that the emphasis on preceding components applies to components that differ in their interaural level differences as well as in their interaural time differences.

There have been only a few computational models of precedence effects, and it is not yet clear how much of this effect is understandable in terms of current

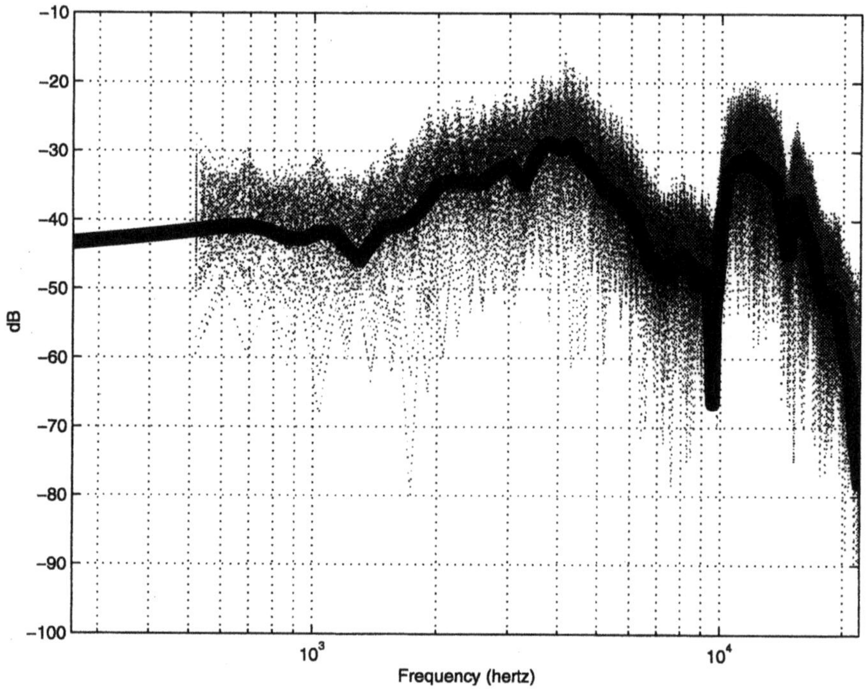

FIGURE 8.6. Variation of spectral shape over time samples.

physiological knowledge of the auditory system. It has been demonstrated that a substantial part of the short-time interactions in the azimuthal precedence effect are related to the interaction of lead and lag components in the responses of peripheral filters (Tollin and Henning 1998, 1999), and models of these effects (Tollin 1998; Hartung and Trahiotis 2001) fit much of the available data. Another apparent factor is the physiologically observed suppression of responses to later transients by earlier transients (Litovsky and Yin 1998). This is seen in the computational model of IC neurons of Cai and colleagues (Cai et al. 1998a,b).

Most studies in this area have been done with stimuli that differ only in azimuth (cf. Section 3.5), and the few studies with stimuli that differ in elevation indicates that more emphasis is given to lead relative to lag components in this dimension as well. The median-sagittal-plane elevation studies include those of Litovsky and colleagues (Litovsky et al. 1997), which show a similar pattern of lead-dominance in the elevation dimension as in the horizontal plane, but the experimental conditions are limited so far. Very few data have been reported for two- or three-dimensional sound localization performance. In addition, there are few studies that involve multiple reflections, which are present in real environments.

Overall, it is not yet clear to what degree the totality of observed "precedence effects" in azimuth are explainable in terms of the peripheral interactions or in terms of the inhibitory interactions observed in the brainstem and midbrain. It seems likely that these effects result from processing at several levels of the system, including interactions in the peripheral filtering of the cochlea, delayed inhibition pathways in the ascending auditory system, and cognitive effects at cortical levels of the nervous system. The phenomena of the buildup and breakdown (cf. Clifton and Freyman 1997) seem particularly top-down in their behavior. There have been models for isolated situations, primarily related to relatively peripheral processing, but an integrated model of these effects is lacking.

8.3 Head Motion and Moving Sources

When the head moves, the acoustic stimulus is modified because the positions of the sources are changed relative to the head and the STFs are thereby modified. Head motion thus modulates the acoustic stimulus, and can have a significant influence on the perceived location of sources. This effect was noted by Wallach (1939) in the context of perceived elevations that arise when the ITD change is chosen to match that for different elevations. (Note that a source overhead would lead to no ITD changes when the head was turned, whereas a source straight ahead would lead to the changes that we have already described for the horizontal plane.) Even though the acoustic effects of head movement may have a large impact on perception, this phenomenon has not been studied very extensively relative to other localization cues.

There are several advantages of head motion coordinated with acoustic inputs. In addition to the elevation information, the externalization of virtual auditory images can be achieved with relatively crude auditory reproductions (Loomis, personal communication), front–back and up–down ambiguities are resolved (Begault et al. 2001), and the level of realism improves substantially.

Judgments of moving sources have been studied in limited contexts (Miele and Hafter 2002), and much remains to be done in this area. Several experiments are reviewed by Saberi and Hafter (1997).

9. Plasticity and Cross-Modality Interactions

Another area of great importance for understanding human behavior is the general area of changes in performance due to training and experience. Although this area has potential for elaborate and interesting models, only limited studies have addressed this aspect of sound localization. The most extensive studies, which are both physiological and behavioral in nature, have been with the barn owl localization system. This system is discussed in depth by Kubke and Carr in this volume, so that it is not treated here (cf., Kubke and Carr, Chapter 6).

For human subjects, recent studies that are having significant influence on thinking include those of Hofman and colleagues (Hofman et al. 1998) and those of Shinn-Cunningham and colleagues (1998a,b).

Finally, it is obvious that the senses work together to create our picture of the world. There have been numerous, interesting studies of auditory–visual interactions. See, for example, the study of Alais and Burr (2004) who explore the relative strength of audio and visual inputs for sound localization when the quality of each modal input is varied. Most models of sound localization do not include cross-modal interactions, so these topics are not considered further here.

10. Concluding Comments

A crude summary of the state of sound-localization modeling is that the simplest cases can be successfully modeled when interaural differences are used for lateral judgments and spectral shape is used for judgments about elevation and front versus back. In this context, simple cases correspond to situations in which there is a single source with a broadband, relatively flat stimulus spectrum, the room is relatively anechoic, and stimuli at the ears contain the normal full spectrum of cues. Generally, models are inadequate or only partially tested for cases in which there are multiple sources, significant reverberation, manipulated (conflicting) cues, or (for elevation judgments) unusual or misleading stimulus spectra.

If attention is restricted to lateral position judgments (i.e., azimuthal judgments with front–back confusions resolved) of single sources in anechoic space, existing models of sound localization are quite successful. The models are computationally explicit and are consistent with psychophysical experimental results, much physiological data, and our understanding of signal processing. Fundamentally, laterality judgment in these simple cases is predominantly dependent on the ITD in the low-frequency components of the received stimuli, and coincidence or correlation mechanisms are used to extract this information. However, the detailed mechanisms of the sensitivity to the complementary ILD cue are not yet fully understood, and the interaction of ITD and ILD is complicated, particularly in cases in which the relationship is not natural.

Localization judgments of elevation, which are generally understood to be based on the magnitude of the received spectra, are not as accurate as azimuthal (or laterality) judgments. In these cases, the received spectrum may be used along with interaural differences, and the adequacy of alternative cues sets for observed performance is easy to demonstrate in many circumstances. In these cases, knowledge of the magnitude spectrum of the source is very beneficial, and interesting cases include those with randomly varying spectra. It is not surprising that performance in these cases is also variable and often relatively poor compared to the simple cases.

When the situation is more complex, such as reverberant environments and multiple sources, performance is not understood and is relatively poor. In these cases, the localization problem is not well defined and predicted performance depends on assumptions about the environment and source characteristics. Complex situations when the stimulus and environment are uncertain is an important but relatively new area of study. The study of these situations could lead to understanding why human performance can be better than machine performance in these cases. It is believed that top-down processing is required to give good performance in the cases, although research in this area is still relatively new.

Acknowledgments. Thanks to the graduate students in the Hearing Research Center at Boston University for their suggestions, including Dan Shub and Gerald Ng, and particularly Yi Zhou and Jacob Scarpaci, who also helped with the figures. This work was supported by NIH (Grant R01 DC00100.

References

Albani S, Peissig J, Kollmeier B (1996) Model of binaural localization resolving multiple sources and spatial ambiguities. In: Kollmeier B (ed), Psychoacoustics, Speech and Hearing Aids. Singapore: World Scientific, pp. 233–242.

Algazi VR, Duda RO (2002) Approximating the head-related transfer function using simple geometric models of the head and torso. J Acoust Soc Am 112:2053–2064.

Algazi VR, Duda RO, Thompson DM, Avendano C (2001) The cipic hrtf database. In: Proceedings of the 2001 Workshop on Applications of Signal Processing to Audio and Electroacoustics New Paltz, NY: IEEE, pp. 99–102.

Anderson DJ, Rose JE, Hind JE, Brugge JF (1971) Temporal position of discharges in single auditory nerve fibers within the cycle of a sine-wave stimulus: frequency and intensity effects. J Acoust Soc Am 49:1131–1139.

Batteau D (1967) The role of the pinna in human localization. Proc R Soc Lond B Biol Sci 168:158–180.

Bech S (1998) Spatial aspects of reproduced sound in small rooms. J Acoust Soc Am 103:434–445.

Begault D, Wenzel E, Anderson M (2001) Direct comparison of the impact of head tracking, reverberation, and individualized head-related transfer functions on the spatial perception of a virtual speech source. J Audio Eng Soc 49:904–916.

Best V, Schaik A van, Carlile S (2004) Separation of concurrent broadband sound sources by human listeners. J Acoust Soc Am 115:324–3367.

Blauert J (1968) Localization in the median plane. Acustica 39:96–103.

Blauert J (1997) Spatial Hearing (revised ed.). Cambridge, MA: MIT Press.

Bloom PJ (1977) Determination of monaural sensitivity changes due to the pinna by use of minimum-audible-field measurements in the lateral vertical plane. J Acoust Soc Am 61:820–828.

Bodden M (1993) Modeling human sound-source localization and the cocktail party effect. Acta Acustica 71:283–290.

Braasch J (2002) Localization in presence of a distracter and reverberation in the frontal horizontal plane. II. Model algorithms. Acta Acustica 88:956–969.

Braasch J, Blauert J (2003) The precedence effect for noise bursts of different bandwidths. II. Comparison of model algorithms. Acoust Sci Tech 24:293–303.

Braida LD, Durlach NI (1972) Intensity perception. II. Resolution in one-interval paradigms. J Acoust Soc Am 51:483–502.

Breebaart J, van de Par S, Kohlrausch A (2001a) Binaural processing model based on contralateral inhibition. I. Model structure. J Acoust Soc Am 110:1074–1088.

Breebaart J, van de Par S, Kohlrausch A (2001b) Binaural processing model based on contralateral inhibition. II. Dependence on spectral parameters. J Acoust Soc Am 110:1089–1104.

Breebaart J, van de Par S, Kohlrausch A (2001c) Binaural processing model based on contralateral inhibition. III. Dependence on temporal parameters. J Acoust Soc Am 110:1105–1117.

Bronkhorst AW, Houtgast T (1999) Auditory distance perception in rooms. Nature 397: 517–520.

Brungart DS, Durlach NI, Rabinowitz WM (1999) Auditory localization of nearby sources. II. Localization of a broadband source. J Acoust Soc Am 106:1956–1968.

Butler RA (1974) Does tonotopicity subserve the perceived elevation of a sound? Fed Proc 33:1920–1923.

Butler RA (1997) Spatial referents of stimulus frequencies: their role in sound localization. In: Gilkey RH, Anderson T (eds), Binaural and Spatial Hearing in Real and Virtual Environments. Mahwah, NJ: Lawrence Erlbaum.

Cai H, Carney LH, Colburn HS (1998a) A model for binaural response properties of inferior colliculus neurons. I. A model with interaural time difference-sensitive excitatory and inhibitory inputs. J Acoust Soc Am 103:475–493.

Cai H, Carney LH, Colburn HS (1998b) A model for binaural response properties of inferior colliculus neurons. II. A model with interaural time difference-sensitive excitatory and inhibitory inputs and an adaptation mechanism. J Acoust Soc Am 103: 494–506.

Carlile S, Delaney S, Corderoy A (1999) The localisation of spectrally restricted sounds by human listeners. Hear Res 128:175–189.

Carney LH, Heinz M, Evilsizer ME, Gilkey RH, Colburn HS (2002) Auditory phase opponency: a temporal model for masked detection at low frequencies. Acta acustica 88:334–347.

Clifton RK, Freyman RL (1997) The precedence effect: Beyond echo suppression. In: Gilkey R, Anderson T (eds), Binaural and Spatial Hearing. Hillsdale, NJ: Lawrence Erlbaum.

Colburn HS (1996) Computational models of binaural processing. In: Hawkins H, McMullen T (eds), Auditory Computation. New York: Springer-Verlag.

Colburn HS, Durlach NI (1978) Models of binaural interaction. In: Carterette E, Friedman M (eds), Handbook of Perception: Hearing, Vol. 4. New York: Academic Press.

Colburn HS, Latimer JS (1978) Theory of binaural interaction based on auditory-nerve data. III. Joint dependence on interaural time and amplitude differences in discrimination and detection. J Acoust Soc Am 61:525–533.

Colburn HS, Moss PJ (1981) Binaural interaction models and mechanisms. In: Syka J (ed), Neuronal Mechanisms in Hearing. New York: Plenum Press.

Colburn HS, Han Y, Culotta C (1990) Coincidence model of MSO responses. Hear Res 49:335–346.

Colburn HS, Zhou Y, Dasika V (2004) Inhibition in models of coincidence detection. In: Pressnitzer, deCheveigne A, McAdams S (eds), Auditory Signal Detection. Hanover, PA: Sheridan.

Culling JF, Summerfield Q (1995) Perceptual separation of concurrent speech sounds: absence of across-frequency grouping by common interaural delay. J Acoust Soc Am 98:785–797.

Dizon RM (2001) Behavioral and computational modeling studies of the precedence effect in humans. Unpublished doctoral dissertation, Biomedical Engineering Department, Boston University.

Dizon RM, Colburn HS (2005) Left-right localization with ongoing lead-lag stimuli. J Acoust Soc Am (submitted).

Domnitz RH, Colburn HS (1977) Lateral position and interaural discrimination. J Acoust Soc Am 61:1586–1598.

Durlach NI (1972) Binaural signal detection: equalization and cancellation theory. In: Tobias JV (ed), Foundations of Modern Auditory Theory, Vol. 2. New York: Academic Press.

Durlach NI, Braida LD (1969) Intensity perception. I. Preliminary theory of intensity resolution. J Acoust Soc Am 46:372–383.

Fitzpatrick D, Kuwada S, Batra R (2002) Transformations in processing interaural time differences between the superior olivary complex and inferior colliculus: beyond the Jeffress model. Hear Res 68:79–89.

Gaik W (1993) Combined evaluation of interaural time and intensity differences: psychoacoustic results and computer modeling. J Acoust Soc Am 94:98–110.

Goldberg JM, Brown PB (1969) Response of binaural neurons of dog superior olivary complex to dichotic tonal stimuli: some physiological mechanisms of sound localization. J Neurophysiol 32:613–636.

Green DM (1966) Signal detection analysis of ec model. J Acoust Soc Am 44:833–838.

Greenberg JE, Desloge JG, Zurek PM (2003) Evaluation of array-processing algorithms for a headband hearing aid. J Acoust Soc Am 113:1646–1657.

Hafter ER (1971) Quantitative evaluation of a lateralization model of masking-level differences. J Acoust Soc Am 50:1116–1122.

Hafter ER, Carrier SC (1970) Masking-level differences obtained with a pulsed-tonal masker. J Acoust Soc Am 47:1041–1047.

Hall JL II (1965) Binaural interaction in the accessory superior-olivary nucleus of the cat. J Acoust Soc Am 37:814–823.

Hartmann WM (1983) Localization of sound in rooms. J Acoust Soc Am 774:1380–1391.

Hartman WM, Rakerd B, Gaalaas JB (1998) On the source-identification method. J Acoust Soc Am 104:3546–3557.

Hartung K (2001) A computational model of sound localization based on neurophysiological data. In: Greenberg S, Slaney M (eds), Computational Models of Auditory Function. Amsterdam: Ios Press.

Hartung K, Trahiotis C (2001) Peripheral auditory processing and investigations of the 'precedence effect' which utilize successive transient stimuli. J Acoust Soc Am 110: 1505–1513.

Hebrank J, Wright D (1974) Spectral cues used in the localization of sound sources on the median plane. J Acoust Soc Am 56:1829–1834.

Heinz MG, Zhang X, Bruce IC, Carney LH (2001) Auditory nerve model for predicting performance limits of normal and impaired listeners. ARLO 2:91.

Hofman P, van Opstal A (1998) Spectro-temporal factors in two-dimensional human sound localization. J Acoust Soc Am 103:2634–2648.

Hofman P, Riswick JV, van Opstal A (1998) Relearning sound localization wiht new ears. Nat Neurosci 1:417–421.

Janko J, Anderson T, Gilkey R (1997) Using neural networks to evaluate the viability of monaural and interaural cues for sound localization. In: Gilkey RH, Anderson T (eds), Binaural and Spatial Hearing in Real and Virtual Enviornments. Mahwah, NJ: Lawrence Erlbaum.

Jeffress LA (1948) A place theory of sound localization. J Comp Physiol Psychol 41: 35–39.

Jeffress LA (1958) Medial geniculate body—a disavowal. J Acoust Soc Am 30:802–803.

Jin C, Schenkel M, Carlile S (2000) Neural system identification model of human sound localization. J Acoust Soc Am 108:1215–1235.

Jin C, Corderoy A, Carlile S, van Schaik A (2004) Contrasting monaural and interaural spectral cues for human sound localization. J Acoust Soc Am 115:3124–3141.

Joris PX, Yin TC (1998) Envelope coding in the lateral superior olive. iii. Comparison with afferent pathways. J Neurophysiol 79:253–269.

Kiang NY-S (1965) Discharge patterns of single fibers in the cat's auditory nerve. Cambridge, MA: MIT Press.

Koehnke J, Durlach NI (1989) Range effects in the identification of lateral position. J Acoust Soc Am 86:1176–1178.

Kuhn GF (1977) Model for the interaural time difference in the azimuthal plane. J Acoust Soc Am 82:157–167.

Kulkarni A (1997) Sound localization in real and virtual acoustical environments. Ph.D. Thesis, Department of Biomedical Engineering, College of Engineering, Boston University.

Kulkarni A, Colburn HS (1998) Role of spectral detail in sound-source localization. Nature 396:747–749.

Langendijk EH, Bronkhorst AW (2002) Contribution of spectral cues to human sound localization. J Acoust Soc Am 112:1583–1596.

Langendijk EH, Kistler DJ, Wightman FL (2001) Sound localization in the presence of one or two distracters. J Acoust Soc Am 109:2123–2134.

Licklider JCR (1959) Three auditory theories. In: Koch ES (ed), Psychology: A Study of a Science, Study 1, Vol 1. New York: McGraw-Hill.

Lindemann W (1986a) Extension of a binaural cross-correlation model by contralateral inhibition. I. Simulation of lateralization for stationary signals. J Acoust Soc Am 80: 1608–1622.

Lindemann W (1986b) Extension of a binaural cross-correlation model by contralateral inhibition. II. The law of the first wavefront. J Acoust Soc Am 80:1623–1630.

Litovsky R, Yin T (1998) Physiological studies of the precedence effect in the inferior colliculus of the cat. II. Neural mechanisms. J Neurophysiol 80:1302–1316.

Litovsky RY, Rakerd B, Yin TC, Hartmann WM (1997) Psychophysical and physiological evidence for a precedence effect in the median sagittal plane. J Neurophysiol 77:2223–2226.

Litovsky R, Colburn HS, Yost WA, Guzman S (1999) The precedence effect. J Acoust Soc Am 106:1633–1654.

Liu C, Wheeler BC, O'Brien WD Jr, Bilger RC, Lansing CR, Feng AS (2000) Localization of multiple sound sources with two microphones. J Acoust Soc Am 108:1888–1905.

Liu C, Wheeler BC, O'Brien WD Jr, Lansing CR, Bilger RC, Jones DL, Feng AS (2001)

A two-microphone dual delay-line approach for extraction of a speech sound in the presence of multiple interferers. J Acoust Soc Am 110:3218–3231.

Lockwood, ME, Jones DL, Bilger RC, Lansing CR, O'Brien WD, Wheeler BC, Feng AS (2004) Performance of time- and frequency-domain binaural beamformers based on recorded signals from real rooms. J Acoust Soc Am 115:379–391.

Loomis JM, Klatzky R, Philbeck J, Golledge R (1998) Assessing auditory distance perception using perceptually directed action. Percept Psychophys 60:966–980.

Lopez-Poveda EA, Meddis R (1996) A physical model of sound diffraction and reflections in the human concha. J Acoust Soc Am 100:3248–3259.

Macpherson EA (1997) A comparison of spectral correlation and local feature-matching models of pinna cue processing. J Acoust Soc Am 101:3104.

Macpherson EA, Middlebrooks JC (2002a) Listener weighting of cues for lateral angle: the duplex theory of sound localization revisited. J Acoust Soc Am 111:2219–2236.

Macpherson EA, Middlebrooks JC (2002b) Vertical-plane sound localization probed with ripple-spectrum noise. J Acoust Soc Am 114:430–445.

Marquardt T, McAlpine DA (2001) Simulation of binaural unmasking using just four binaural channels. In: Abstracts of 24th meeting of ARO, Mt. Royal, NJ.

McAlpine D, Jiang D, Palmer AR (2001) A neural code for low-frequency sound localization in mammals. Nat Neurosci 4:396–401.

Middlebrooks J (1992) Narrow-band sound localization related to external ear acoustics. J Acoust Soc Am 92:2607–2624.

Middlebrooks J (1999a) Individual differences in external-ear transfer function reduced by scaling in frequency. J Acoust Soc Am 106:1480–1492.

Middlebrooks J (1999b) Virtual localization improved by scaling non-individualized external-ear tranfer functions in frequency. J Acoust Soc Am 106:1493–1509.

Middlebrooks JC, Green DM (1991) Sound localization by human listeners. Annu Rev Psychol 42:135–159.

Miele JA, Hafter ER (2002) Trajectory perception in the free field. J Acoust Soc Am 111:2536.

Moore J (2000) Organization of the human superior olivary complex. Microsc Res Tech 51:403–412.

Morimoto M, Iida K, Itoh M (2003) Upper hemisphere sound localization using head-related transfer functions in the median plane and interaural differences. Acoust Sci Tech 24:267–275.

Naka Y, Oberai AA, Shinn-Cunningham BG (2004) The finite element method with the dirichlet-to-neumann map for sound-hard rectangular rooms. In: Proceedings of the International Congress on Acoustics, Kyoto, Japan.

Neti C, Young ED, Schneider MH (1992) Neural network models of sound localization based on directional filtering by the pinna. J Acoust Soc Am 92:3140–3156.

Nix J, Hohmann V (1999) Statistics of binaural parameters and localization in noise. In: Dau T, Hohmann V, Kollmeier B (eds), Psychophysics, Physiology and Models of Hearing. Singapore: World Scientific, pp. 263–266.

Palmieri F, Datum M, Shah A, Moiseff A (1991) Sound localization with a network trained with the multiple extended Kalman algorithm. In: Proceedings of the International Joint Conference on Neural Networks, Seattle, WA, pp. 1125–1131.

Patterson RD, Allerhand MH, Giguere C (1995) Time-domain modeling of peripheral auditory processing: a modular architecture and a software platform. J Acoust Soc Am 98:1980–1984.

Peña J, Konishi M (2001) Auditory spatial receptive fields created by multiplication. Science 292:249–252.

Rakerd B, Hartmann WM (1985) Localization of sound in rooms. II: The effects of a single reflecting surface. J Acoust Soc Am 78:524–533.

Rakerd B, Hartmann WM (1986) Localization of sound in rooms. III: Onset and duration effects. J Acoust Soc Am 80:1695–1706.

Reed MC, Blum JJ (1990) A model for the computation and encoding of azimuthal information by the lateral superior olive. J Acoust Soc Am 88:1442–1453.

Rice JJ, May BJ, Apirou GA, Young ED (1992) Pinna-based spectral cues for sound localization in cat. Hear Res 58:132–152.

Roffler S, Butler RA (1968a) Factors that influence the localization of sound in the vertical plane. J Acoust Soc Am 43:1255–1259.

Roffler S, Butler RA (1968b) Localization of tonal stimuli in the vertical plane. J Acoust Soc Am 43:1260–1266.

Saberi K, Hafter ER (1997) Experiments on auditory motion discrimination. In: Gilkey RH, Anderson T (eds), Binaural and Spatial Hearing in Real and Virtual Enviornments. Mahwah, NJ: Lawrence Erlbaum.

Sachs MB, Young ED (1979) Encoding of steady-state vowels in the auditory nerve: representation in terms of discharge rate. J Acoust Soc Am 66:470–479.

Sayers BM (1964) Acoustic-image lateralization judgments with binaural tones. J Acoust Soc Am 36:923–926.

Sayers BM, Cherry EC (1957) Mechanisms of binaural fusion in the hearing of speech. J Acoust Soc Am 29:973–987.

Sayers BM, Lynn P (1968) Interaural amplitude effects in binaural hearing. J Acoust Soc Am 44:973–978.

Searle CL, Braida LD, Cuddy DR, Davis MF (1975) Binaural pinna disparity: another auditory localization cue. J Acoust Soc Am 57:448–455.

Searle CL, Braida LD, Davis MF, Colburn HS (1976) Model for auditory localization. J Acoust Soc Am 60:1164–1175.

Shaw EA (1997) Acoustical features of the human external ear. In: Gilkey RH, Anderson T (eds), Binaural and Spatial Hearing in Real and Virtual Enviornments. Mahwah, NJ: Lawrence Erlbaum.

Shaw EA, Teranishi R (1968) Sound pressure generated in an external-ear replica and real human ears by a nearby point source. J Acoust Soc Am 44:240–249.

Shinn-Cunningham BG (2000) Learning reverberation: considerations for spatial auditory displays. In: Proceedings of the International Conference on Auditory Displays, Atlanta, GA, pp. 126–134.

Shinn-Cunningham BG, Kawakyu K (2003) Neural representation of source direction in reverberant space. In: Proceedings of the 2003 IEEE Workshop on Applications of Signal Processing to Audio and Acoustics. New York: IEEE Press.

Shinn-Cunningham BG, Durlach NI, Held RM (1998a) Adapting to supernormal auditory localization cues I: Bias and resolution. J Acoust Soc Am 103:3656–3666.

Shinn-Cunningham BG, Durlach NI, Held RM (1998b) Adapting to supernormal auditory localization cues II: Changes in mean response. J Acoust Soc Am 103:3667–3676.

Shinn-Cunningham BG, Santarelli SG, Kopco N (2000) Tori of confusion: binaural cues for sources within reach of a listener. J Acoust Soc Am 107:1627–1636.

Stern RM Jr, Colburn HS (1978) Theory of binaural interaction based on auditory-nerve data. iv. A model for subjective lateral position. J Acoust Soc Am 64:127–140.

Stern RM Jr, Trahiotis C (1992) Role of consistency of interaural timing over frequency in binaural lateralization. In: Cazals Y (ed), Auditory Physiology and Perception. Oxford: Pergamon Press.
Stern RM Jr, Trahiotis C (1995) Models of binaural interaction. In: Moore BC (ed), Handbook of Perception and Cognition, Vol. 6. Hearing. New York: Academic Press, pp. 347–386.
Stern RM Jr, Trahiotis C (1996) Models of binaural perception. In: Gilkey RH, Anderson T (eds), Binaural and spatial hearing in real and virtual environments. Mahwah, NJ: Lawrence Erlbaum, pp. 499–531.
Stern RM Jr, Zeiberg A, Trahiotis C (1988) Lateralization of complex binaural stimuli: a weighted-image model. J Acoust Soc Am 84:156–165.
Strutt JW (3rd Baron of Rayleigh) (1907) On our perception of sound direction. Philos Mag 13:214–232.
Teranishi R, Shaw EA (1968) External-ear acoustic models with simple geometry. J Acoust Soc Am 44:257–263.
Tollin DJ (1998) Computational model of the lateralisation of clicks and their echoes. In: Proceedings of the NATO Advanced Study Institute on Computational Hearing. NATO, pp. 77–82.
Tollin DJ, Henning GB (1998) Some aspects of the lateralization of echoed sound in man. I. The classical interaural delay-based precedence effect. J Acoust Soc Am 104: 3030–3038.
Tollin DJ, Henning GB (1999) Some aspects of the lateralization of echoed sound in man. II. The role of the stimulus spectrum. J Acoust Soc Am 105:838–849.
Trahiotis C, Bernstein LR, Stern RM, Buell TK (2005) Interaural correlation as the basis of a working model of binaural processing: an introduction. In: Popper A, Fay R (eds), Sound Source Localization. New York: Springer.
van Bergeijk W (1962) Variation on a theme of von Békésy: a model of binaural interaction. J Acoust Soc Am 34:1431–1437.
von Békésy G (1930) Zur theorie des horens. Physikalische Zeitschrift 31:857–868.
von Békésy G (1960) Experiments in Hearing. New York: McGraw-Hill.
Wallach H (1939) On sound localization. J Acoust Soc Am 10:270–274.
Watkins AJ (1978) Psychoacoustical aspects of synthesized vertical locale cues. J Acoust Soc Am 63:1152–1165.
Wightman FL, Kistler DJ (1989a) Headphone simulation of free-field listening. I: Stimulus synthesis. J Acoust Soc Am 85:858–867.
Wightman FL, Kistler DJ (1989b) Headphone simulation of free-field listening. II: Psychophysical validation. J Acoust Soc Am 87:868–878.
Wightman FL, Kistler DJ (1992) The dominant role of low-frequency interaural time differences in sound localization. J Acoust Soc Am 91:1648–1661.
Wittkop T, Albani S, Hohmann V, Peissig J, Woods WS, Kollmeier B (1997) Speech processing for hearing aids: noise reduction motivated by models of binaural interaction. Acta Acustica 83:684–699.
Yin TCT, Chan JCK (1990) Interaural time sensitivity in the medial superior olive of the cat. J Neurophysiol 64:465–488.
Young ED, Rice JJ, Apirou GA, Nelken I, Conley RA (1997) Head-related transfer functions in cat: neural representation and the effects of pinna movement. In: Gilkey RH, Anderson T (eds), Binaural and Spatial Hearing in Real and Virtual Environments. Mahwah, NJ: Lawrence Erlbaum.

Zacksenhouse M, Johnson D, Williams J, Tsuchitani C (1998) Single-neuron modeling of LSO unit responses. J Neurophysiol 79:3098–3110.
Zahorik P (2002) Assessing auditory distance perception using virtual acoustics. J Acoust Soc Am 111:1832–1846.
Zakarauskas P, Cynader MS (1993b) A computational theory of spectral cue localization. J Acoust Soc Am 94:1323–1331.
Zurek PM (1987) The precedence effect. In: Yost WA, Gourevitch G (eds), Directional Hearing. New York: Springer-Verlag.

Index

SPRINGER HANDBOOK OF AUDITORY RESEARCH *(continued from page ii)*

Volume 22: Evolution of the Vertebrate Auditory System
Edited by Geoffrey A. Manley, Arthur N. Popper and Richard R. Fay

Volume 23: Plasticity of the Auditory System
Edited by Thomas N. Parks, Edwin W Rubel, Arthur N. Popper, and Richard R. Fay

Volume 24: Pitch: Neural Coding and Perception
Edited by Christopher J. Plack, Andrew J. Oxenham, Richard R. Fay, and Arthur N. Popper

Volume 25: Sound Source Localization
Edited by Arthur N. Popper, and Richard R. Fay

For more information about the series, please visit www.springer-ny.com/shar.

MIX
Papier aus verantwortungsvollen Quellen
Paper from responsible sources
FSC® C105338

If you have any concerns about our products,
you can contact us on
ProductSafety@springernature.com

In case Publisher is established outside the EU,
the EU authorized representative is:
Springer Nature Customer Service Center GmbH
Europaplatz 3, 69115 Heidelberg, Germany

Printed by Libri Plureos GmbH
in Hamburg, Germany